Big Data Analytics and Information Science for Business and Biomedical Applications

Big Data Analytics and Information Science for Business and Biomedical Applications

Editors

Farouk Nathoo
S. Ejaz Ahmed

MDPI • Basel • Beijing • Wuhan • Barcelona • Belgrade • Manchester • Tokyo • Cluj • Tianjin

Editors
Farouk Nathoo
Mathematics and Statistics
University of Victoria
Victoria
Canada

S. Ejaz Ahmed
Faculty of Mathematics Science
Brock
St. Catharines, ON
Canada

Editorial Office
MDPI
St. Alban-Anlage 66
4052 Basel, Switzerland

This is a reprint of articles from the Special Issue published online in the open access journal *Entropy* (ISSN 1099-4300) (available at: www.mdpi.com/journal/entropy/special_issues/Big_Data_Biomedical).

For citation purposes, cite each article independently as indicated on the article page online and as indicated below:

LastName, A.A.; LastName, B.B.; LastName, C.C. Article Title. *Journal Name* **Year**, *Volume Number*, Page Range.

ISBN 978-3-0365-3193-9 (Hbk)
ISBN 978-3-0365-3192-2 (PDF)

© 2022 by the authors. Articles in this book are Open Access and distributed under the Creative Commons Attribution (CC BY) license, which allows users to download, copy and build upon published articles, as long as the author and publisher are properly credited, which ensures maximum dissemination and a wider impact of our publications.

The book as a whole is distributed by MDPI under the terms and conditions of the Creative Commons license CC BY-NC-ND.

Contents

About the Editors vii

Preface to "Big Data Analytics and Information Science for Business and Biomedical Applications" ix

Jingqiao Wu, Xiaoyue Feng, Renchu Guan and Yanchun Liang
Cancer Research Trend Analysis Based on Fusion Feature Representation
Reprinted from: *Entropy* 2021, 23, 338, doi:10.3390/e23030338 1

Tong Su, Yafei Wang, Yi Liu, William G. Branton, Eugene Asahchop and Christopher Power et al.
Sparse Multicategory Generalized Distance Weighted Discrimination in Ultra-High Dimensions
Reprinted from: *Entropy* 2020, 22, 1257, doi:10.3390/e22111257 15

S. Ejaz Ahmed, Saeid Amiri and Kjell Doksum
Ensemble Linear Subspace Analysis of High-Dimensional Data
Reprinted from: *Entropy* 2021, 23, 324, doi:10.3390/e23030324 49

Jiwei Zhao and Chi Chen
A Nuisance-Free Inference Procedure Accounting for the Unknown Missingness with Application to Electronic Health Records
Reprinted from: *Entropy* 2020, 22, 1154, doi:10.3390/e22101154 65

Alex Pijyan, Qi Zheng, Hyokyoung G. Hong and Yi Li
Consistent Estimation of Generalized Linear Models with High Dimensional Predictors via Stepwise Regression
Reprinted from: *Entropy* 2020, 22, 965, doi:10.3390/e22090965 87

Eugene A. Opoku, Syed Ejaz Ahmed, Yin Song and Farouk S. Nathoo
Ant Colony System Optimization for Spatiotemporal Modelling of Combined EEG and MEG Data
Reprinted from: *Entropy* 2021, 23, 329, doi:10.3390/e23030329 115

Hao Mei, Yaqing Xu, Jiping Wang and Shuangge Ma
Evaluation of Survival Outcomes of Endovascular Versus Open Aortic Repair for Abdominal Aortic Aneurysms with a Big Data Approach
Reprinted from: *Entropy* 2020, 22, 1349, doi:10.3390/e22121349 151

Xuan Cao and Kyoungjae Lee
Variable Selection Using Nonlocal Priors in High-Dimensional Generalized Linear Models With Application to fMRI Data Analysis
Reprinted from: *Entropy* 2020, 22, 807, doi:10.3390/e22080807 167

Yaojin Sun and Hamparsum Bozdogan
Segmentation of High Dimensional Time-Series Data Using Mixture of Sparse Principal Component Regression Model with Information Complexity
Reprinted from: *Entropy* 2020, 22, 1170, doi:10.3390/e22101170 189

Lukas Börjesson and Martin Singull
Forecasting Financial Time Series through Causal and Dilated Convolutional Neural Networks
Reprinted from: *Entropy* 2020, 22, 1094, doi:10.3390/e22101094 215

About the Editors

Farouk Nathoo

Farouk Nathoo received his B.Sc. in Mathematics and Statistics (combined honours) from the University of British Columbia in 1998, his M.Math. from the University of Waterloo in 2000 and his Ph.D. in statistics at Simon Fraser University in 2006. He joined the Department of Mathematics and Statistics at the University of Victoria in 2006, and became a Full Professor in 2021. He currently holds the Tier 2 Canada Research Chair in Biostatistics for Spatial and High-Dimensional data. His research interests focus on Bayesian Methods, High-dimensional Data, Statistical Computation, Neuroimaging Statistics, and Machine Learning.

S. Ejaz Ahmed

Ejaz Ahmed is professor and Dean of the Faculty of Mathematics and Science. Before joining Brock, he was a professor and head of Mathematics at the University of Windsor and University of Regina. Prior to that, he had a faculty position at the University of Western Ontario. Further, he is a Senior Advisor to Sigma Analytics (Data Mining and Research), Regina.

His research achievements have been recognized with honours and awards, including the prestigious status of Fellow of the American Statistical Association, editor/associate editorship to influential scientific journals, adjunct/visiting professorships, and invited scholarly talks around the globe. He has supervised numerous Ph.D./Master students and Post-doctoral Fellows. He is an elected member of the International Statistical Institute and a Fellow of the Royal Statistical Society. He served as a Board of Director and Chairman of the Education Committee of the Statistical Society of Canada.

Preface to "Big Data Analytics and Information Science for Business and Biomedical Applications"

In today's data-centric world, there is a host of buzzwords appearing everywhere in digital and print media related to 'Big Data'. We encounter data in every walk of life, and the information it contains can be used to improve society, business, health, and medicine. This presents a substantial opportunity for humanity in general and a great challenge for statisticians and data scientists. Making sense of modern data structures and extracting meaningful information from them is not an easy task. The rapid growth in the size, scope, and complexity of data in a host of disciplines has created the need for innovative statistical strategies for analyzing and visualizing such data.

An enormous trove of digital data has been produced by biomedical research worldwide, including genetic variants genotyped or sequenced at genome-wide scales, gene expression measured under different experimental conditions, biomedical imaging data, including neuroimaging data, electronic medical records (EMR) of patients, and much more.

Analysis of such data will not only deepen our understanding of complex human traits and diseases, but will also shed light on disease prevention, diagnosis, and treatment. Undoubtedly, comprehensive analysis of Big Data in genomics and neuroimaging calls for statistically rigorous methods that can be applied in complex settings with data arising from multiple platforms. Various statistical methods have been developed to accommodate the features of genomic studies, as well as studies examining the function and structure of the brain.

Meanwhile, statistical theories have also correspondingly been developed. Alongside biomedical applications, there has been a tremendous increase and interest in the use of Big Data in business and financial applications. Financial time series analysis and prediction problems present many challenges for the development of statistical methodology and computational strategies for streaming data. The analysis of Big Data in biomedical, as well as business and financial research, has drawn much attention from researchers worldwide.

This book provides a platform for an in-depth discussion of powerful statistical methods developed for the analysis of Big Data in these areas. Both applied and theoretical contributions to these areas are showcased.

With a focus on statistical and machine learning, Wu et al. develop a novel approach for text feature learning motivated by trend analysis in cancer research examining 260,000 cancer studies and the corresponding low-dimensional text representations. Su et al. extend the lasso-penalized distance-weighted discrimination approach for binary classification to multicategory classification problems with an approach that considers all classes simultaneously. They establish uniqueness of the solution and obtain a non-asymptotic error bound in the case of group lasso penalization for ultra-high dimensional data.

With a focus on high-dimensional regression, Ahmed, Amiri, and Doksum focus on developing efficient prediction methods within the setting of high-dimensional regression where the number of predictor variables is larger than the sample size. They investigate the performance of ensemble linear subspace methods that combine the results of linear models applied to smaller subsets of predictor variables selected by random selection and find settings where ensemble methods perform relatively well. Zhao and Chen consider regression analysis within the context of missing data and develop a conditional likelihood approach incorporating an instrumental variable that avoids the specification of the process generating the missing data. A data perturbation technique is also developed for inference in the high-dimensional case. Pet al. consider estimation and variable selection in ultra high-dimensional settings and develop an approach combining forward selection and backward elimination with stopping criteria that control false positives and false negatives. The authors obtain probability bounds on variable selection and show consistent estimation.

With a focus on biomedical data, Opoku et al. develop an algorithm based on ant colony system optimization for improving solutions to the ill-posed neuroelectromagnetic inverse problem from combined EEG and MEG data. Mei et al. use emulated clinical trial data and deep learning to compare endovascular aortic repair and emergent open aortic repair as procedures for abdominal aortic aneurysm and find that the former leads to improved expected survival. Cao and Lee consider the variable selection problem for fMRI data and develop a Bayesian approach based on non-local priors with posterior computation based on a combination of the Laplace approximation and stochastic search.

Finally, with a focus on financial modeling, Sun and Bozdogan consider the segmentation of high-dimensional time series data and develop an approach based on sparse principal component regression and mixture model cluster analysis. This approach is applied to find change-points in the adjusted closing price of the S&P 500 index over a span covering the years 1999 to 2019.

Börjesson and Singull consider deep learning for forecasting financial time series in a study comparing multi-channel convolutional neural networks represented as nonlinear autoregressive models and standard autoregressive time series modeling. A number of financial indices are considered including the S&P 500 covering the years 2010 to 2019.

In summary, this collection comprises a variety of contributions to the state-of-the-art on statistical methodology for Big Data and high-dimensional problems for biomedical and financial applications and beyond. We hope that it will in turn inspire new methods and applications.

Farouk Nathoo, S. Ejaz Ahmed
Editors

Article

Cancer Research Trend Analysis Based on Fusion Feature Representation

Jingqiao Wu [1], Xiaoyue Feng [2], Renchu Guan [1,2] and Yanchun Liang [1,2,*]

1. Zhuhai Sub Laboratory of Key Laboratory of Symbolic Computation and Knowledge Engineering of the Ministry of Education, Zhuhai College of Jilin University, Zhuhai 519041, China; wujingqiao17@gmail.com (J.W.); guanrenchu@jlu.edu.cn (R.G.)
2. Key Laboratory of Symbolic Computation and Knowledge Engineering of the Ministry of Education, College of Computer Science and Technology, Jilin University, Changchun 130012, China; fengxy@jlu.edu.cn
* Correspondence: ycliang@jlu.edu.cn; Tel.: +86-18686604031

Abstract: Machine learning models can automatically discover biomedical research trends and promote the dissemination of information and knowledge. Text feature representation is a critical and challenging task in natural language processing. Most methods of text feature representation are based on word representation. A good representation can capture semantic and structural information. In this paper, two fusion algorithms are proposed, namely, the Tr-W2v and Ti-W2v algorithms. They are based on the classical text feature representation model and consider the importance of words. The results show that the effectiveness of the two fusion text representation models is better than the classical text representation model, and the results based on the Tr-W2v algorithm are the best. Furthermore, based on the Tr-W2v algorithm, trend analyses of cancer research are conducted, including correlation analysis, keyword trend analysis, and improved keyword trend analysis. The discovery of the research trends and the evolution of hotspots for cancers can help doctors and biological researchers collect information and provide guidance for further research.

Keywords: feature representation; feature fusion; trend analysis; text mining

1. Introduction

Since the completion of the Human Genome Project and with the rapid development of high-throughput biotechnology, the amount of data in the fields of biology, medicine, genetics, and chemistry has exponentially grown. As of January 2021, the number of entries in PubMed (Biomedical Literature Retrieval System) has exceeded 30 million [1]. However, given the large-scale, rapid growth and massive amounts of data in various formats, people can do little with the data. It is a major challenge for clinicians or biological researchers to obtain cutting-edge information about research from tens of thousands of publications. Traditional methods, the knowledge of which was manually acquired from literature and images, can no longer meet researchers' needs for understanding the current hotspots and trends of biomedical research [2,3]. It has become urgent to use intelligent algorithms to quickly and effectively acquire and discover biomedical knowledge.

Cancer research has attracted much attention in the medical field. Among all cancers, lung cancer poses the greatest threat to human health; it is characterized by its rapid spread and high probability of death. In recent years, according to statistical data around the world, the possibility of people suffering from lung cancer has greatly increased. Besides, gastric cancer, colorectal cancer, breast cancer, and liver cancer are also high-risk cancers that have been studied in the medical field in recent years. Traditional trend analysis can only be completed after reading and sorting out many documents published in the field in recent years by experts. This approach may hinder the dissemination of information and knowledge and cause omissions in the retrieval of papers by experts, which may affect the results of the extraction of research hotspots or trend analysis. The usage of machine learning models to automatically discover biomedical research trends can make up for this deficiency [4,5].

Text feature learning is an important task in the field of natural language processing, and it is the basis of many downstream applications, such as text clustering and classification [6]. Most existing text feature representation learning is based on words, that is, word vector representation. It obtains word representation by mapping words from a one-dimensional space to a continuous vector space. The word representation methods include neural networks, word co-occurrence methods, methods that rely on probability, and interpretable knowledge base methods. A good low-dimensional mapping representation often improves the performance of downstream tasks [7]. Feature fusion is the integration of multiple different feature information to obtain more prominent feature information [8–11]. Multimodal features from text, audio and vision can be fused with fusion technique [12]. There are two types of fusion technique, early fusion, and late fusion. Early fusion concatenates the features together at first and late fusion combines results [13]. We adopt early fusion for text clustering.

Based on text representation methods, we propose a multi-view feature fusion strategy. The hotspots and trend analysis were conducted on 260,000 cancer studies using the proposed method. Our contribution mainly includes the following points. (1) The fusion of the improved vector representation model Ti-W2v algorithm and Tr-W2v algorithm were proposed. (2) A correlation analysis algorithm based on similarity is proposed to analyze the relationship among five cancer types. (3) A keyword trend analysis model and its improved model are proposed. Taking lung cancer as an example, the keyword analysis model analyzes the overall research hotspots. (4) Taking lung cancer as an example, the trend of lung cancer research is further analyzed from three perspectives, including gene proteins, therapeutic drugs and methods. The results can help guide the literature summary and further work of relevant researchers.

The remainder of this paper is organized as follows: Section 2 lists the materials and methods. Section 3 describes the experimental details, presents the experimental results, and gives the error analysis. Section 4 discusses the results. Section 5 concludes our work.

2. Materials and Methods

2.1. Background

Traditional text representation models commonly include models based on word frequency, TF-IDF, TextRank, and word embedding. The text feature representation model based on word frequency is the simplest. It calculates the number of occurrences of each word in the text and obtains the text vector with the word frequency of each word [14]. The expression based on the word frequency algorithm is shown in Equation (1):

$$wordcount(i,j) = n_{i,j} \qquad (1)$$

where $n_{i,j}$ is the occurrence number of word t_i in document d_j.

The text feature representation model based on TF-IDF considers the frequency of occurrence of each word in the training texts and the number of other training text containing the word, that is, the frequency of the reverse text [15,16]. The expression of the TF-IDF algorithm is shown as Equation (2):

$$tfidf(i,j) = tf(i,j) * idf(i) = \frac{n_{i,j}}{\sum_k n_{k,j}} * \log\frac{|D|}{1+|\{j:t_i \epsilon d_j\}|} \qquad (2)$$

where $|D|$ represents the total number of files in the corpus. $1+|\{j:t_i \in d_j\}|$ represents numbers of documents containing the term t_i, we add 1 here to prevent the denominator from being 0. The TF-IDF text representation model is an algorithm based on word frequency, which pays more attention to the number of times the words appeared in the document and does not consider the relative position between them.

The TextRank-based text feature representation model is a graph-based sorting algorithm for text [17]. Its core idea is that a word is more important if it appears after many words. Besides, if a word is followed by another word with a high TextRank value, the TextRank value of this word is accordingly higher. The TextRank model is an algorithm based on graphs. Let G = (V,E) be a directed graph with the set of vertices V and set of edges E, where E is a subset of V*V. For a given vertex V_i, let $In(V_i)$ be the set of vertices that point to it (predecessors), and let $Out(V_i)$ be the set of vertices that vertex V_i points to (successors). The score of a vertex V_i is defined as followed Equation (3):

$$WS(V_i) = (1-d) + d * \sum_{V_j \in In(V_i)} \frac{w_{ji}}{\sum_{V_k \in Out(V_j)} w_{jk}} WS(V_j) \qquad (3)$$

where d is a damping factor that can be set between 0 and 1; w_{ji} is the weight between V_j and V_i. The TextRank model focuses more on the degree of co-occurrence between words in a fixed-length window. This considers the relative position of words to a certain extent, so when the number of documents is small, the TextRank algorithm can express text information more accurately, while the TF-IDF algorithm cannot do this.

The text feature representation model based on word embedding maps words to another space through a certain mapping rule and generates expressions in a new space [18]. The word embedding text representation model is an algorithm based on a neural network. The hidden attributes between words in the text, such as the similarity and part of speech between words, are emphasized. As the characteristics of neural networks, the word embedding text representation model is difficult to be explained, but its final effect is better than TF-IDF and the TextRank algorithm. The obtained word vectors can measure the semantic and other relevant features between words. Therefore, word embedding methods to represent text features has been a hotspot in recent years. The most commonly used word embedding tool is Word2Vec [19–21], which contains two training modes: the CBOW training mode and the Skip-gram training mode.

2.2. Method

As shown in Figure 1, our proposed framework consists of two modules, a feature fusion module, and a research trend analysis module. The feature fusion module contains two fusion strategies, and the research trend analysis includes three trend analysis methods.

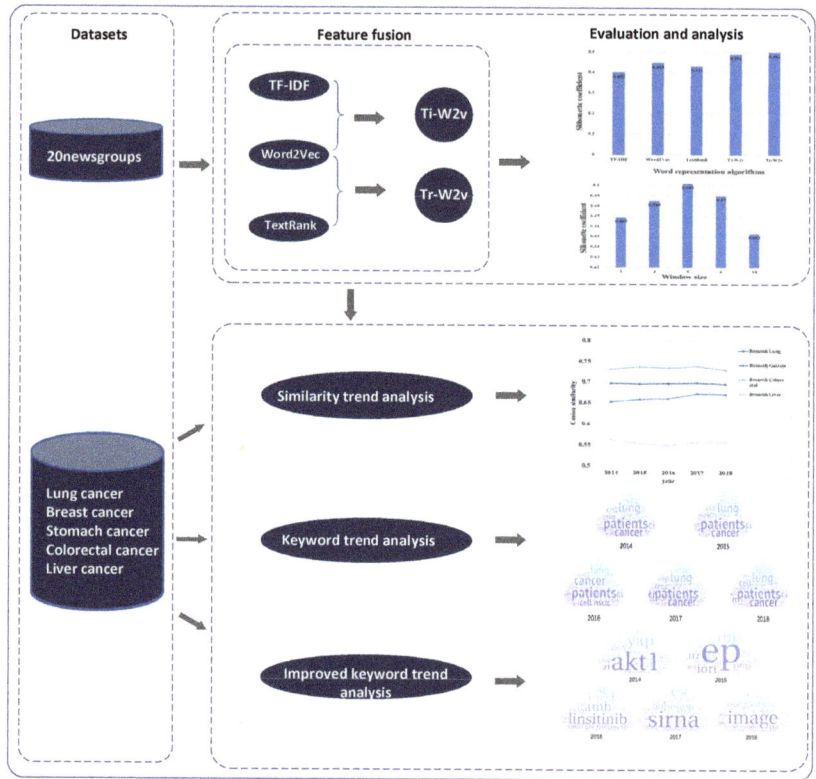

Figure 1. Framework of our work.

2.2.1. Feature Fusion Representation Model

The word-based text representation method needs to obtain the representation of each word first. Then, word vectors can be used to obtain a text representation. The classic method is used to add all word vectors and the average of all vectors as the text vector. In this method, all words in the text are considered equally important. This is obviously far-fetched because the importance of words in the text is different. The representation algorithms of TF-IDF and TextRank represent a text by calculating the weight of words in the text, but the analysis point and calculation method of the two are quite different. The Word2Vec algorithm can determine the semantic information of words and does not consider the importance of words. To retain the advantages of the above methods, we propose a multi-view fusion strategy, which combines Word2Vec with TF-IDF and TextRank. In this fusion strategy, Word2Vec is chosen as the representation method of words. The weights of words in text are given by TF-IDF and TextRank. We named the fusion method Ti-W2v and Tr-W2v, and the details are given in the following sections.

Ti-W2v is an improved algorithm that combined TF-IDF and Word2Vec. TF-IDF is adopted to calculate the weight coefficient of each word in the text, and the embedding vector of the text can be generated with the product of the weights and embedding vectors of Word2Vec for all words. The advantage is that different words in the text can be given different degrees of importance, closer to the actual situation than average embedding. For a corpus, D is the corpus, and D = $\{D_1, D_2, \ldots, D_k\}$, D_i is the ith document. V_i is the representation of D_i. w_{ij} is the jth word in D_i, v_{ij} is the vector of w_{ij} obtained by word2vec. TI_{ij} is the weight of w_{ij}, obtained with TF-IDF as Equation (4):

$$TI_{ij} = \frac{n_{i,j}}{\sum_k n_{k,j}} * \log \frac{|D|}{1 + |\{j : t_i \in d_j\}|} \quad (4)$$

V_i is defined using Equation (5):

$$V_i = TI_{ij} \times v_{ij} \quad (5)$$

The TF-IDF algorithm is based on word frequency. It measures the importance of words based on text word frequency and global reverse text frequency. It is suitable for cases in which the number of documents is relatively large. While, in TextRank, the importance of words is decided by their relative position. It does not depend on other documents and considers the co-occurrence of each word. Based on the fusion strategy, we propose the Tr-W2v algorithm, which combines TextRank and Word2Vec. First, TextRank is used to calculate the weight coefficients of different words in the text, and then the Word2Vec embedding vectors of the words by weight are added to obtain the text vector. TR_{ij} is the weight of w_{ij}, obtained with TextRank as Equation (6):

$$TR_{ij} = (1-d) + d * \sum_{V_m \in In(V_j)} \frac{w_{mi}}{\sum_{V_k \in Out(V_m)} w_{mk}} WS(V_m) \quad (6)$$

as Ti-W2v, V_i is defined with Equation (7):

$$V_i = TI_{ij} \times v_{ij} \quad (7)$$

2.2.2. Cancer Research Trend Analysis Model

Based on the fusion-improved feature representation model proposed in the previous section, we propose three trend analysis models. We first propose a similarity trend analysis model based on the five high-incidence cancer datasets. A keyword trend analysis model and an improved keyword analysis model are proposed based on the lung cancer dataset. Then, lung cancer-related gene proteas, treatment methods, and drugs, and other hotspots related to lung cancer were analyzed.

Correlation Analysis Based on Similarity

We use the Tr-W2v algorithm to obtain the corresponding text vectors of abstracts on the five major cancer in the last five years. Then, the text vectors of various cancers are integrated into a vector for a certain year of this type of cancer in units of years (addition and average). The cosine similarities of different cancers are calculated in different years, and the correlation of different cancers are analyzed for the past five years through cosine similarity. Figure 2 shows the flowchart of the algorithm.

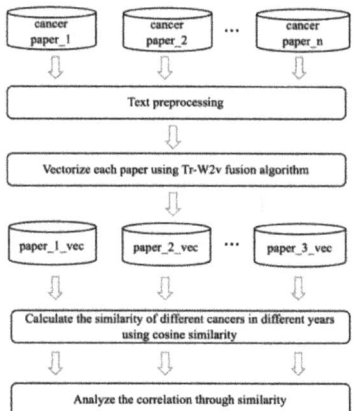

Figure 2. Flowchart of similarity trend analysis.

Keyword Trend Analysis Model

Taking the lung cancer dataset as an example, we use the TextRank algorithm to obtain the top 10% of keywords and corresponding weights in each document. Then, all the keywords and corresponding weights of the year are integrated into units of years. The method of integration is as follows: for the keywords that have not appeared, we add them and the corresponding weights directly to the keywords of the year. For the keywords that have appeared, we add and merge their weights as their new weights. Finally, the top 50 keywords were obtained as hotspots of the year through keyword reordering. Figure 3 shows the flowchart of the algorithm.

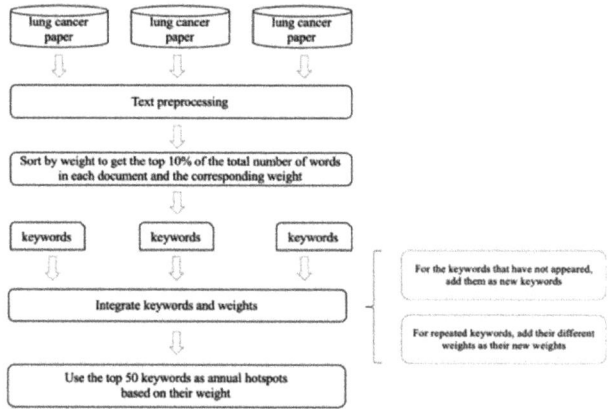

Figure 3. Flowchart of keyword trend analysis.

Improved Keyword Trend Analysis Model

The keyword analysis model proposed in the previous section can coarsely analyze the annual research hotspots of single types of cancer (taking lung cancer as an example). For more detailed trend analysis, we propose an improved keyword analysis on this basic model. As in the correlation analysis, we first use the Tr-W2v algorithm to obtain the text vector corresponding to lung cancer of each year. Further, the k-means clustering algorithm is adopted, and k categories are generated. The keyword integration operation in the previous section is utilized to obtain hotspots of different clusters. Then, the top keyword of each category is extracted and integrated into the distribution of hotspots of that year. Figure 4 shows flowchart of the algorithm.

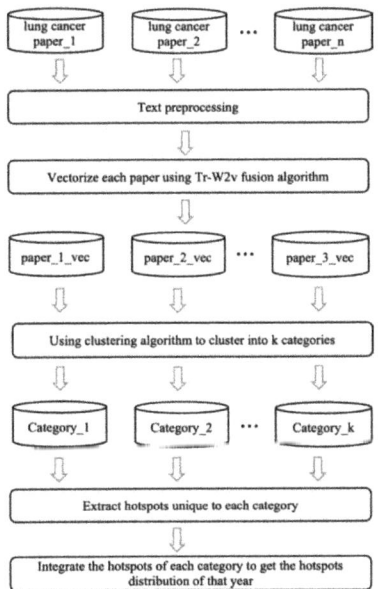

Figure 4. Flowchart of improved keyword trend analysis.

3. Results

3.1. Datasets

For comparing the effect of representation methods, we use the second edition of the well-known public classification dataset 20 newsgroups [22]. In the analysis of cancer research trends, we retrieve PubMed articles using MeSH terms and obtain experimental datasets that include data from the past five years on the five most common cancers in China (lung cancer, breast cancer, gastric cancer, colorectal cancer, and liver cancer) [23,24]. Table 1 shows the distribution of the number of research papers for the five major cancers in the most recent five years.

Table 1. Number of research papers for the five cancers.

Cancer	2014	2015	2016	2017	2018
Lung	9322	9966	9446	9508	10,149
Breast	12,328	12,825	12,600	12,286	12,743
Gastric	3747	3572	3637	3414	3561
Colorectal	8950	9174	8778	8617	8868
Liver	6651	6871	6517	6431	6555

3.2. Results

In the cancer dataset, the most papers on lung cancer and breast cancer were published in 2018. The number of papers published in 2014 is the largest for gastric cancer. For colorectal and liver cancer, the number of papers published in 2015 is the largest. The number of cancer papers has not increased over the years. It shows a stable trend, and in some years, the trend is slightly lower than in previous years; however, the total number of cancer research papers is still rising slightly.

3.2.1. Comparison Results of Feature Fusion Methods

To compare the results of feature fusion methods, we conduct clustering experiments on five text representation algorithms, including TF-IDF, Word2Vec, TextRank, Ti-W2v, and Tr-W2v. First, the five algorithms are used to vectorize the text of the data set. Then, we use the classical k-means clustering algorithm to evaluate the effects of the five word-representation algorithms. We select ten categories from 20 newsgroups dataset as the experimental dataset. The number of clusters in k-means is set to 10, the initialization method defaults to k-means++, and the maximum iteration number is set to 300. Using the silhouette coefficient of clustering as a measurement [25]. The result of TF-IDF, Word2Vec, TextRank, Ti-W2v, and Tr-W2v are 0.402, 0.449, 0.433, 0.491, and 0.502, respectively. Figure 5 shows the two-dimensional clustering visualization effect of the data.

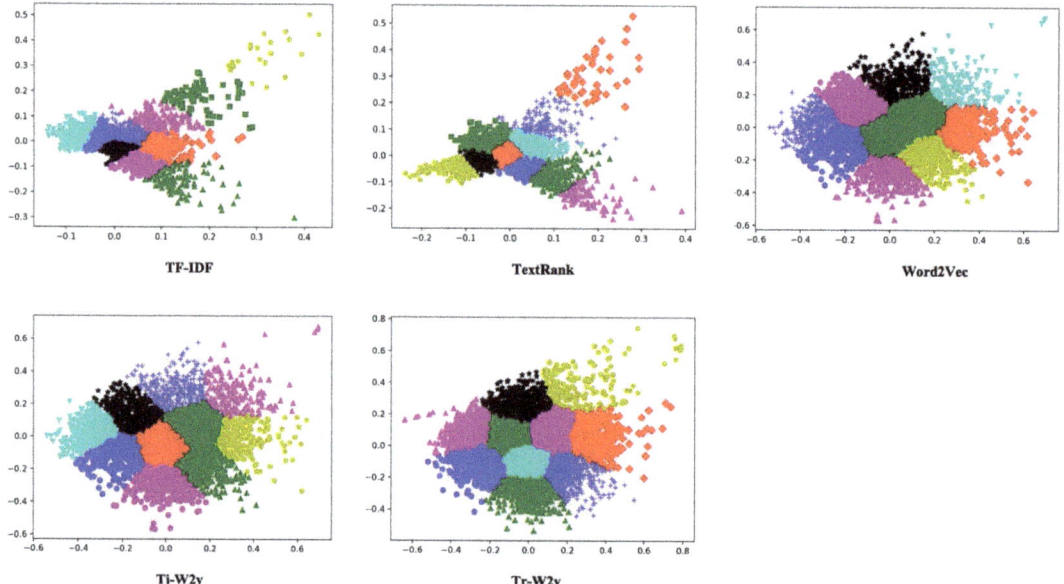

Figure 5. Two-dimensional clustering visualization results based on five word-representation algorithms.

In the clustering experiment, the clustering silhouette coefficients based on TF-IDF, Word2Vec, TextRank, Ti-W2v, and Tr-W2v are 0.402, 0.449, 0.433, 0.491, and 0.502, respectively. Among them, the Tr-W2v algorithm has the best result. The effect of Word2Vec vector is 11.7% higher than that of TF-IDF. The effect of the TextRank vector is 7.7% higher than that of TF-IDF. Ti-W2v has a 9.4% improvement over Word2Vec. Tr-W2v has a 2.2% improvement over Ti-W2v. The choice of word vectors plays a vital role in text representation. It is best to use the Word2Vec method to improve the results. Additionally, the word vector fusion method also has a certain effect. The effect based on the TextRank fusion text vector is better than that of the TF-IDF fusion text vector. They are both better than Word2Vec. Although the improvement effect is not as obvious as the replacement of word vectors, there is also a certain degree of improvement. In general, Tr-W2v fusion text vectors have the best clustering effect, which also reflects that it can better represent the text information. We evaluated the effect of the Tr-W2v algorithm's TextRank window size. When the window size is 2, 4, 5, 6, 10, the results are 0.469, 0.485, 0.502, 0.490, and 0.453, respectively. We choose 5 as the window size.

The results of the fusion feature experiment show that the fusion vector obtained by the Tr-W2v algorithm has the best result in clustering experiments, and the Ti-W2v algorithm is slightly inferior to the Tr-W2v algorithm; however, both are better than the traditional text representation model. It may be caused by the difference between the TF-IDF algorithm and the TextRank algorithm. The TF-IDF algorithm only considers word frequency information and does not consider the relationship between words. Compared with the TF-IDF algorithm, TextRank can obtain important information, such as the relative position of words within a single text, so the integration of TextRank and Word2Vec will perform better. In addition, we can see that the TF-IDF and TextRank vector clustering effects are slightly different from the other three vector clustering effects. The TF-IDF and TextRank algorithms represent vectors by word frequency and word co-occurrence position, respectively. Meanwhile, the other three algorithms are based on Word2Vec's low-dimensional dense vectors. Therefore, the clustering shape based on TF-IDF and the TextRank algorithm are more decentralized, while the other three algorithms are more uniform and regular.

3.2.2. Cancer Trend Analysis Results

Next, the experimental results of the cancer research trend analysis model based on the fusion-improved feature representation model are listed below.

Correlation Analysis Results Based on Similarity

A correlation analysis is conducted based on similarity to determine the relationships among the five cancer types. Figure 6 shows the results for the most recent five years.

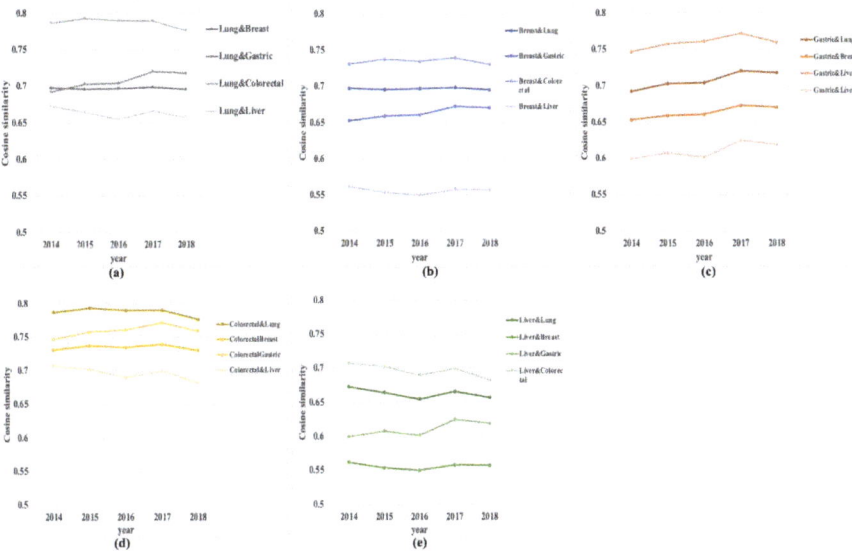

Figure 6. Correlation of the top five high-risk cancers. (**a**) Correlation between lung cancer and the other four cancers, (**b**) correlation between breast cancer and the other four cancers, (**c**) correlation between gastric cancer and the other four cancers, (**d**) correlation between colorectal cancer and the other four cancers, and (**e**) correlation between liver cancer and the other four cancers.

From Figure 6, we can conclude that colorectal cancer is most closely related to the other four cancers. The following reasons indicate that smoking may cause lung cancer; long-term smokers are more likely to die from colorectal cancer than nonsmokers [26]. There are many repeated research directions for the treatment of breast cancer and colorectal cancer [27,28]. The stomach and colorectal are organs of the digestive tract system, and many studies are conducted simultaneously [29,30]. The above studies can confirm the close relationship between colorectal cancer and four other cancers from the side. Lung cancer, breast cancer, gastric cancer, and colorectal cancer have the lowest similarity with liver cancer. In addition, lung cancer has the highest similarity to colorectal cancer among all relations, and breast cancer has the lowest similarity to liver cancer. This also shows that among the top five high-incidence cancers, lung cancer is most closely linked to colorectal cancer, while breast cancer is relatively less linked to liver cancer.

Results of the Keyword Trend Analysis Model

Taking lung cancer as an example, Figure 7 shows the visualization results of the annual hotspots word cloud of lung cancer obtained by the keyword trend analysis model and the improved keyword trend analysis model.

Figure 7. Hotspots of lung cancer were obtained by keyword trend analysis model (**a**) and improved keyword trend analysis model (**b**).

It can be found that the hotspots in the past five years have focused on the *patient, cancer, cell, lung, study, tumor, CI, nsclc,* and so on. According to the principles of the TextRank algorithm and the characteristics of the lung cancer research literature, the above results are normal because the central idea of the TextRank algorithm is that the more times a word and other important words co-occurrence within a certain length, the more important the word is. The above vocabulary in the literature of lung cancer research uses TextRank's weight ranking mechanism, and the above vocabulary may exist in the important vocabulary and hotspot area of the study in the literature on lung cancer. Public hotspots are basically the same each year using the keyword trend analysis model. The results are too rough to study the trend of lung cancer in the past five years. The improved keyword trend analysis model is optimized with more details. The results are significantly different from the keyword analysis methods before improvement (see the right-hand side of Figure 7). Based on improved methods, hotspots in different categories are combined to generate hotspots of one year. Then, the differential hotspots between years are chosen to represent the public hotspot of each year. Therefore, the hotspots of each year are clearly distinguished, which makes it easier and clearer to analyze the research trends of lung cancer in the recent five years.

Results of Analysis on Research Trend

Based on improved keyword trend analysis results, to present the research trend on lung cancer, research trends in different areas are listed in Figures 8–10. Figure 8 shows the research trends of related gene proteins and invertase factors in lung cancer research in the past five years. Figure 9 shows the hot research trends of lung cancer-related treatment drugs and methods in the past five years, and Figure 10 shows the other hot topics of lung cancer in the past five years' research trends.

The number of genes and proteins in the human body is very large, and many studies invest in gene protein-related research on lung cancer each year. It can be seen from Figure 8 that the research hotspots for genes and proteins are various in different years. The unique hot research terms in 2014 included the ATK1 gene, YAP gene, PKM2 gene, LSCC gene, and PDCD5 gene. The unique hot research terms in 2015 included PMS separation enhancer protein, BBP gene, Bsm gene, and THOR long noncoding RNA. The unique hot research terms in 2016 included SFTPD gene, p110α protein, DDX17 gene, Globo H glycoprotein, and LHX6 gene. The unique hot research terms in 2017 included SiRNA, TGF-β transforming growth factor, luciferase, RDM1 protein, and SNHG15 long-chain noncoding RNA. The unique hot research terms in 2018 included miRNA-223 and LKB1 gene.

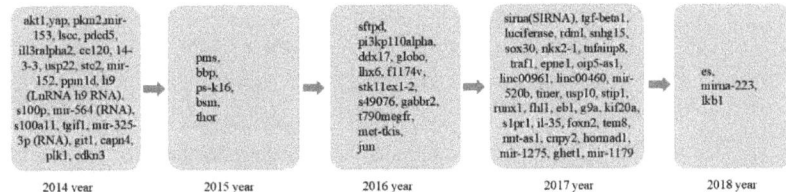

Figure 8. Research trends of lung cancer research related gene protein and invertase factor in the last five years.

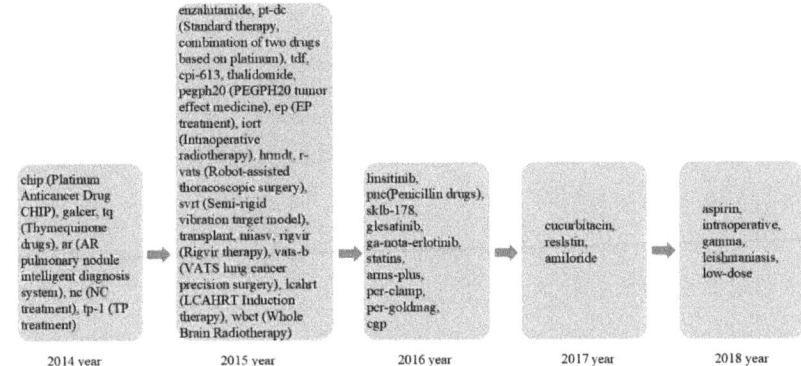

Figure 9. Research trends of lung cancer research related therapeutic drugs and methods in the last five years.

With the development of science and technology, different lung cancer related treatment drugs and treatment methods have emerged. It can be seen from Figure 9 that the research hotspots for different drugs and treatments vary by year. The unique research words in 2014 included metal platinum anticancer drugs CHIP, galactosylceramide, thymoquinone drugs, AR lung nodule intelligent diagnosis system treatment, NC treatment, and TP treatment. The unique research hotspot words in 2015 included enzalutamide drugs, tenofovir dipivoxil drugs, dibenzylthiocaprylic acid drugs, thalidomide drugs, PEGPH2O tumor effect drugs, EP regimen treatment, intraoperative radiotherapy, and robot-assisted thoracoscopic surgery. The unique research hotspot words in 2016 included linsitinib drugs, penicillin drugs, SKLB drugs, sitagliptin drugs, erlotinib drugs, statins, ARMS quantitative treatment, and PCR-clamp method detection. The unique research hotspot words in 2017 included cucurbitacin, human resistin, and dimethyl amiloride. The unique research hotspot words in 2018 included aspirin drugs, intraoperative radiotherapy for lung cancer, gamma knife treatment, leishmaniasis, and low-dose lung CT technology.

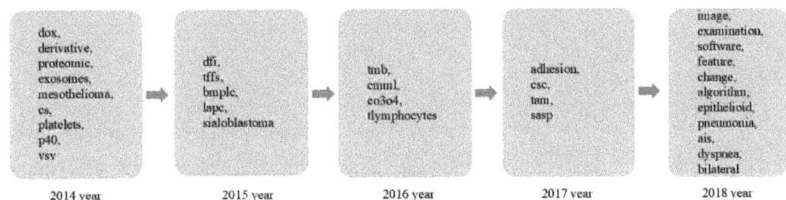

Figure 10. Research trends of other related hotspots in the last five years.

It can be seen from Figure 10 that the research hotspots for other factors related to lung cancer in different years are also different. The unique hot research terms in 2014 included DOX, derivative, exosomes, and mesothelioma. The unique hot research terms in

2015 included DFI and bmplc. The unique hot research terms in 2016 included TMB tumor mutation load, monocyte leukemia CMML, and Co3O4 nanoparticles. The hot research terms in 2017 included adhesion, csc, and tam. The unique hot research terms in 2018 included image, examination, software, feature, algorithm, epithelioid, pneumonia, and bilateral. Regarding the hot words in 2018, it is noteworthy that with the advancement of science and technology, computer software and artificial intelligence algorithms play an increasingly important role in lung cancer research, such as artificial intelligence for image diagnosis of the lung.

4. Limitation

There are some limitations in our work. We only took five cancers as examples and discussed their relevance. More cancer data should be added. Further, the word2vec is chosen as the embedding method. Advanced text representation methods such as BERT (Bidirectional Encoder Representations from Transformers) [31] and BioBERT [32] might be a better choice. For the design of the experiment, we apply three methods to analyze the trend of cancer, and more diversified test methods could be used in future work.

5. Conclusions

Text feature representation models play an essential role in natural language processing. Improving these models helps machines better understand relevant text information and promote downstream tasks. Considering the words' degree of importance, we combined the TF-IDF and TextRank with word2vec. Results demonstrate the effectiveness of the fusion models. Meanwhile, the combined model is adopted to present research trend analysis of cancers. The proposed models can help researchers find research hotspots in biology, medicine, information retrieval, and natural language processing.

Author Contributions: Conceptualization, X.F.; methodology, R.G.; software, J.W.; validation, J.W.; formal analysis, X.F.; resources, J.W.; data curation, J.W.; writing—original draft preparation, J.W. and X.F.; writing—review and editing, Y.L.; visualization, J.W.; supervision, Y.L.; project administration, Y.L.; funding acquisition, Y.L., R.G., and X.F. All authors have read and agreed to the published version of the manuscript.

Funding: This research was funded by the National Natural Science Foundation of China (grant number 61972174), the Science and Technology Planning Project of Guangdong Province (grant number 2020A0505100018), the Guangdong Key Project for Applied Fundamental Research (grant number 2018KZDXM076), the Guangdong Premier Key-Discipline Enhancement Scheme (grant number 2016GDYSZDXK036), and the Educational Commission of Jilin Province (grant number JJKH20200992KJ).

Institutional Review Board Statement: Not applicable.

Informed Consent Statement: Not applicable.

Data Availability Statement: Publicly available datasets were analyzed in this study. This data can be found here: https://www.ncbi.nlm.nih.gov/pmc/ (accessed on 20 November 2020).

Conflicts of Interest: The authors declare no conflict of interest.

References

1. PubMed. Available online: https://pubmed.ncbi.nlm.nih.gov/ (accessed on 7 January 2021).
2. Jensen, L.J.; Saric, J.; Bork, P. Literature Mining for the Biologist: From Information Retrieval to Biological Discovery. *Nat. Rev. Genet.* **2006**, *7*, 119–129. [CrossRef]
3. Gonzalez, G.H.; Tahsin, T.; Goodale, B.C.; Greene, A.C.; Greene, C.S. Recent Advances and Emerging Applications in Text and Data Mining for Biomedical Discovery. *Brief. Bioinform.* **2016**, *17*, 33–42. [CrossRef] [PubMed]
4. He, G.; Liang, Y.; Chen, Y.; Yang, W.; Liu, J.S.; Yang, M.Q.; Guan, R. A Hotspots Analysis-Relation Discovery Representation Model for Revealing Diabetes Mellitus and Obesity. *BMC Syst. Biol.* **2018**, *12*, 116. [CrossRef]
5. Guan, R.; Wen, X.; Liang, Y.; Xu, D.; He, B.; Feng, X. Trends in Alzheimer's Disease Research Based upon Machine Learning Analysis of PubMed Abstracts. *Int. J. Biol. Sci.* **2019**, *15*, 2065–2074. [CrossRef] [PubMed]

6. Guan, R.; Zhang, H.; Liang, Y.; Giunchiglia, F.; Huang, L.; Feng, X. Deep Feature-Based Text Clustering and Its Explanation. *IEEE Trans. Knowl. Data Eng.* **2020**, 1–13. [CrossRef]
7. Collobert, R.; Weston, J.; Bottou, L.; Karlen, M.; Kavukcuoglu, K.; Kuksa, P. Natural Language Processing (Almost) from Scratch. *J. Mach. Learn. Res.* **2011**, *12*, 2493–2537.
8. Hatzivassiloglou, V.; Gravano, L.; Maganti, A. An Investigation of Linguistic Features and Clustering Algorithms for Topical Document Clustering. In Proceedings of the 23rd annual international ACM SIGIR Conference on Research and Development in Information Retrieval, Athens, Greece, 24–28 July 2000; Association for Computing Machinery: New York, NY, USA, 2000; pp. 224–231.
9. Nam, S.; Jeong, S.; Kim, S.-K.; Kim, H.-G.; Ngo, V.; Zong, N. Structuralizing Biomedical Abstracts with Discriminative Linguistic Features. *Comput. Biol. Med.* **2016**, *79*, 276–285. [CrossRef]
10. Sarkar, K. Sentence Clustering-Based Summarization of Multiple Text Documents. *TECHNIA Int. J. Comput. Sci. Commun. Technol.* **2009**, *2*, 325–335.
11. Tang, B.; Cao, H.; Wang, X.; Chen, Q.; Xu, H. Evaluating Word Representation Features in Biomedical Named Entity Recognition Tasks. Available online: https://www.hindawi.com/journals/bmri/2014/240403/ (accessed on 24 February 2021).
12. Gogate, M.; Dashtipour, K.; Adeel, A.; Hussain, A. CochleaNet: A Robust Language-Independent Audio-Visual Model for Real-Time Speech Enhancement. *Inf. Fusion* **2020**, *63*, 273–285. [CrossRef]
13. Gogate, M.; Dashtipour, K.; Bell, P.; Hussain, A. Deep Neural Network Driven Binaural Audio Visual Speech Separation. In Proceedings of the 2020 International Joint Conference on Neural Networks (IJCNN), Glasgow, UK, 19–24 July 2020; pp. 1–7.
14. Salton, G. Developments in Automatic Text Retrieval. *Science* **1991**, *253*, 974–980. [CrossRef] [PubMed]
15. Qin, P.; Xu, W.; Guo, J. A Novel Negative Sampling Based on TFIDF for Learning Word Representation. *Neurocomputing* **2016**, *177*, 257–265. [CrossRef]
16. Wang, D.; Liang, Y.; Xu, D.; Feng, X.; Guan, R. A Content-Based Recommender System for Computer Science Publications. *Knowl. Based Syst.* **2018**, *157*, 1–9. [CrossRef]
17. Mihalcea, R.; Tarau, P. TextRank: Bringing Order into Text. In Proceedings of the 2004 Conference on Empirical Methods in Natural Language Processing, Barcelona, Spain, 25–26 July 2004; Association for Computational Linguistics: Barcelona, Spain, 2004; pp. 404–411.
18. Mikolov, T.; Sutskever, I.; Chen, K.; Corrado, G.; Dean, J. Distributed Representations of Words and Phrases and Their Compositionality. In Proceedings of the 26th International Conference on Neural Information Processing Systems, Lake Tahoe, NV, USA, 5–10 December 2013; Curran Associates Inc.: Red Hook, NY, USA, 2013; Volume 2, pp. 3111–3119.
19. Goldberg, Y.; Levy, O. Word2vec Explained: Deriving Mikolov et al.'s Negative-Sampling Word-Embedding Method. *arXiv* **2014**, arXiv:1402.3722.
20. Rong, X. Word2vec Parameter Learning Explained. *arXiv* **2016**, arXiv:1411.2738.
21. Mikolov, T.; Chen, K.; Corrado, G.; Dean, J. Efficient Estimation of Word Representations in Vector Space. *arXiv* **2013**, arXiv:1301.3781.
22. Aliguliyev, R.M. Performance Evaluation of Density-Based Clustering Methods. *Inf. Sci.* **2009**, *179*, 3583–3602. [CrossRef]
23. Bray, F.; Ferlay, J.; Soerjomataram, I.; Siegel, R.L.; Torre, L.A.; Jemal, A. Global Cancer Statistics 2018: GLOBOCAN Estimates of Incidence and Mortality Worldwide for 36 Cancers in 185 Countries. *CA Cancer J. Clin.* **2018**, *68*, 394–424. [CrossRef]
24. Lu, Z.; Kim, W.; Wilbur, W.J. Evaluation of Query Expansion Using MeSH in PubMed. *Inf. Retr.* **2009**, *12*, 69–80. [CrossRef]
25. Aranganayagi, S.; Thangavel, K. Clustering Categorical Data Using Silhouette Coefficient as a Relocating Measure. In Proceedings of the International Conference on Computational Intelligence and Multimedia Applications (ICCIMA 2007), Sivakasi, India, 13–15 December 2007; Volume 2, pp. 13–17.
26. Peppone, L.J.; Mahoney, M.C.; Cummings, K.M.; Michalek, A.M.; Reid, M.E.; Moysich, K.B.; Hyland, A. Colorectal Cancer Occurs Earlier in Those Exposed to Tobacco Smoke: Implications for Screening. *J. Cancer Res. Clin. Oncol.* **2008**, *134*, 743–751. [CrossRef]
27. Papadimitriou, N.; Dimou, N.; Tsilidis, K.K.; Banbury, B.; Martin, R.M.; Lewis, S.J.; Kazmi, N.; Robinson, T.M.; Albanes, D.; Aleksandrova, K.; et al. Physical Activity and Risks of Breast and Colorectal Cancer: A Mendelian Randomisation Analysis. *Nat. Commun.* **2020**, *11*, 597. [CrossRef]
28. Wang, M.; Chen, H. Chaotic Multi-Swarm Whale Optimizer Boosted Support Vector Machine for Medical Diagnosis. *Appl. Soft Comput.* **2020**, *88*, 105946. [CrossRef]
29. Moniuszko, T.; Wincewicz, A.; Koda, M.; Domysławska, I.; Sulkowski, S. Role of Periostin in Esophageal, Gastric and Colon Cancer (Review). *Oncol. Lett.* **2016**, *12*, 783–787. [CrossRef]
30. Sumer, F.; Karakas, S.; Gundogan, E.; Sahin, T.; Kayaalp, C. Totally Laparoscopic Resection and Extraction of Specimens via Transanal Route in Synchronous Colon and Gastric Cancer. *Il Giornale di Chirurgia* **2018**, *39*, 82–86. [PubMed]
31. Devlin, J.; Chang, M.-W.; Lee, K.; Toutanova, K. BERT: Pre-Training of Deep Bidirectional Transformers for Language Understanding. *arXiv* **2019**, arXiv:1810.04805.
32. Lee, J.; Yoon, W.; Kim, S.; Kim, D.; Kim, S.; So, C.H.; Kang, J. BioBERT: A Pre-Trained Biomedical Language Representation Model for Biomedical Text Mining. *Bioinformatics* **2020**, *36*, 1234–1240. [CrossRef] [PubMed]

Article

Sparse Multicategory Generalized Distance Weighted Discrimination in Ultra-High Dimensions

Tong Su [1], Yafei Wang [2], Yi Liu [2], William G. Branton [3], Eugene Asahchop [3], Christopher Power [3], Bei Jiang [2], Linglong Kong [2,*] and Niansheng Tang [1,*]

[1] Key Lab of Statistical Modeling and Data Analysis of Yunnan Province, Yunnan University, Kunming 650091, China; sutong_366@sina.com
[2] Department of Mathematical and Statistical Sciences, University of Alberta, Edmonton, AB T6G 2G1, Canada; yafei2@ualberta.ca (Y.W.); yliu16@ualberta.ca (Y.L.); bei1@ualberta.ca (B.J.)
[3] Department of Medicine (Neurology), University of Alberta, Edmonton, AB T6G 2G1, Canada; wbranton@ualberta.ca (W.G.B.); asahchop@ualberta.ca (E.A.); cp9@ualberta.ca (C.P.)
* Correspondence: lkong@ualberta.ca (L.K.); nstang@ynu.edu.cn (N.T.)

Received: 30 September 2020; Accepted: 2 November 2020; Published: 5 November 2020

Abstract: Distance weighted discrimination (DWD) is an appealing classification method that is capable of overcoming data piling problems in high-dimensional settings. Especially when various sparsity structures are assumed in these settings, variable selection in multicategory classification poses great challenges. In this paper, we propose a multicategory generalized DWD (MgDWD) method that maintains intrinsic variable group structures during selection using a sparse group lasso penalty. Theoretically, we derive minimizer uniqueness for the penalized MgDWD loss function and consistency properties for the proposed classifier. We further develop an efficient algorithm based on the proximal operator to solve the optimization problem. The performance of MgDWD is evaluated using finite sample simulations and miRNA data from an HIV study.

Keywords: high dimension; multicategory classification; DWD; sparse group lasso; L_2-consistency; proximal algorithm

1. Introduction

Classification problems appear in diverse practical applications, such as spam e-mail classification, disease diagnosis and drug discovery, among many others (e.g., [1–3]). In these classification problems, the goal is to predict class labels based on a given set of variables. Recent research has focused extensively on linear classification: see [4,5] for comprehensive introductions. Among many linear classification methods, support vector machines (SVMs) (see [6,7]) and distance-weighted discrimination (DWD) (see [8–10]) are two commonly used large-margin based classification methods.

Owing to the recent advent of new technologies for data acquisition and storage, classification with high dimensional features, i.e., a large number of variables, has become a ubiquitous problem in both theoretical and applied scientific studies. Typically, only a small number of instances are available, a setting we refer to as high-dimensional, low-sample size (HDLSS), as in [11]. In the HDLSS setting, a so-called "data-piling" phenomenon is observed in [8] for SVMs, occurring when projections of many training instances onto a vector normal to the separating hyperplane are nearly identical, suggesting severe overfitting. DWD was originally proposed to overcome data-piling in the HDLSS setting. In binary classification problems, linear SVMs seek a hyperplane maximizing the smallest margin for all data points, while DWD seeks a hyperplane minimizing the sum of inverse margins over all data points. Reference [8] suggests replacing the inverse margins by the q-th power of the inverse margins in a generalized DWD method; see [12] for a detailed description. Formally, for a training data set $\{(y_i, X_i)\}_{i=1}^N$ of N observations, where $X_i \in \mathbb{R}^p$ and $y_i \in \{-1, 1\}$, binary generalized linear DWD seeks a proper separating hyperplane $\{X : a + X^\top b = 0\}$ through the optimization problem

$$\underset{a,b}{\arg\max} \sum_{i=1}^{N} \frac{1}{d_i^q}$$
$$\text{s.t. } d_i = y_i \left(a + X_i^T b\right) + \eta_i \geq 0, \forall i,$$
$$\eta_i \geq 0, \forall i, \sum_i \eta_i \leq c,$$
$$\|b\|_2^2 = 1,$$
(1)

where a and b are the intercept and slope parameters, respectively. The slack variable η_i is introduced to ensure that the corresponding margin d_i is non-negative and the constant $c > 0$ is a tuning parameter to control the overlap between classes. Problem (1) can also be written in a loss-plus-penalty form (e.g., [12]) as

$$(\hat{a}, \hat{b}) = \underset{a,b}{\arg\min} \left[\frac{1}{N} \sum_{i=1}^{N} \phi_q \left\{ y_i \left(a + X_i^\top b\right) \right\} + \lambda \|b\|_2^2 \right],$$
(2)

where

$$\phi_q(u) = \begin{cases} 1 - u, & \text{if } u \leq Q \\ \varphi_q(u), & \text{if } u > Q, \end{cases}$$
(3)

with $Q = \frac{q}{q+1}$, $q > 0$ and $\varphi_q(u) = (1 - Q)(Qu^{-1})^q$. When $q = 1$, (1) becomes the standard DWD problem in [8] while problem (2) appears in [9,13].

The binary classification problem (1) is well studied. However, in many applications such as image classification [1], cancer diagnosis [2] and speech recognition [3], to name a few, problems with more than two categories are commonplace. To solve these multicategory problems with the DWD classifier, approaches based on either formulation (1) or (2) are common. One common strategy is to extend problem (1) to multiple classes by solving a series of binary problems in a one-versus-one (OVO) or one-versus-rest (OVR) method (e.g., [14]). Instead of reducing the multicategory problem to a binary one, another strategy based on problem (1) considers all classes at once. As shown in [14], this approach generally works better than the OVO and OVR methods. Based on an extension of problem (2), [15] proposes multicategory DWD, written in a loss-plus-penalty form as

$$\underset{a_k,b_k}{\min} \frac{1}{N} \sum_{i=1}^{N} \phi_q \left(a_{y_i} + X_i^\top b_{y_i}\right) + \lambda \sum_{k=1}^{K} \|b_k\|_2^2$$
$$\text{s.t. } \sum_{k=1}^{K} a_k = 0; \sum_{k=1}^{K} b_{jk} = 0, \quad \forall j = 1, \ldots, p,$$
(4)

with $y_i, k \in \{1, \ldots, K\}$ and where a_k and $b_k = (b_{1k}, \cdots, b_{pk})$ are the intercept and slope parameters for each category k, respectively. Although these methods can be applied to multicategory classification in the HDLSS setting, both problems (2) and (4) use the L_2 penalty and do not perform feature selection. As discussed in [16], for high dimensional classification, taking all features into consideration does not work well for two reasons. First, based on prior knowledge, only a small number of variables are relevant to the classification problem: a good classifier in high dimensions should have the ability to sparsely select important variables and discard redundant ones. Second, classifiers using all available variables in high-dimensional settings may have poor classification performance.

Much of the SVM literature has considered variable selection in high-dimensional classification problems to improve performance (e.g., [17–19]). Among the DWD literature, to the best of our knowledge, only [16] considered variables selection and classification simultaneously. Wang and Zou [16] considered an L_1 rather than an L_2 penalty in problem (2) to improve interpretability through sparsity in the binary classification. Moreover, [16] made selections based on the strengths of input variables within individual classes but ignored the strengths of input variable groupings, thereby selecting more factors than necessary for each class. To overcome this weakness in this paper, we developed a multicategory generalized DWD method that is capable of performing variable selection and classification simultaneously. Our approach incorporates sparsity and group structure information via the sparse group lasso penalty (see [20–24]).

Although DWD is well studied, it is less popular than the SVM for binary classification, arguably for computational and theoretical reasons. For an up-to-date list of works on DWD mostly focused on the $q = 1$ case, see [14,15]. Theoretical asymptotic properties of large-margin classifiers in high dimensional settings were studied in [25], and [26] derived an expression for asymptotic generalization error. In terms of computation, [8] solved the standard DWD problem in (1) as a second-order cone programming (SOCP) problem using a primal-dual interior-point method that is computationally expensive when N or p is large. To overcome computational bottlenecks, [12] proposed an approach based on a novel formulation of the primal DWD model in (1): this method, proposed in [12], does not scale to large data sets and requires further work. Lam et al. [27] designed a new algorithm for large DWD problems with $q \geq 2$ and $K = 2$ based on convergent multi-block ADMM-type methods (see [28]). Wang and Zou [16] solved the lasso-penalized binary DWD problem by combining majorization–minimization and coordinate descent methods: the lasso penalty does not directly permit a SOCP solution. In fact, solution identifiability in the generalized DWD problem with $q > 1$ requires more constraints and remains an open research problem (see [8]). To the best of our knowledge, no work focusing on computational aspects of lasso penalized multicategory generalized DWD (MgDWD) exists. The same holds for sparse group lasso-penalized MgDWD.

The theoretical and computational contributions of this paper are as follows. First, we establish the uniqueness of the minimizer in the population form of the MgDWD problem. Second, we prove a non-asymptotic L_2 estimation error bound for the sparse group lasso-regularized MgDWD loss function in the ultra-high dimensional setting under mild regularity conditions. Third, we develop a fast, efficient algorithm able to solve the sparse group lasso-penalized MgDWD problem using proximal methods.

The rest of this paper is organized as follows. In Section 2.1, we introduce the MgDWD problem with sparse group lasso penalty. In Sections 2.2 and 2.3, we establish theoretical properties of the population classifier and regularized empirical loss. We propose a computational algorithm in Section 2.4. Section 3 illustrates the finite sample performance of our method through simulation studies and a real data analysis. Proofs for major theorems are given in the Appendix A.

2. Methodology

2.1. Model Setup

We begin with some basic set-up and notation. Consider the multicategory classification problem for a random sample $\{(y_i, X_i)\}_{i=1}^N$ of N independent and identically distributed (i.i.d.) observations from some distribution $\mathbb{P}(y, X)$. Here, y is the categorical response taking values in $\mathcal{Y} = \{1, \ldots, K\}$, and $X = (x_1, \ldots, x_p)^\top \in \mathcal{X} \subset \mathbb{R}^p$ is the covariate vector. We wish to obtain a proper separating hyperplane $\{X \in \mathcal{X} | a_k + X^\top b_k = 0\}$ for each category $k \in \mathcal{Y}$, where a_k and $b_k = (b_{1k}, \ldots, b_{pk})^\top$ are intercept and slope parameters, respectively.

In this paper, we consider MgDWD with sparse group lasso regularization. That is, we estimate a classification boundary by solving the constrained optimization problem

$$\min_{a_k, b_k} \frac{1}{N} \sum_{i=1}^{N} \phi_q\left(a_{y_i} + X_i^\top b_{y_i}\right) + \lambda_1 \sum_{k=1}^{K} \sum_{j=1}^{p} |b_{jk}| + \lambda_2 \sum_{j=1}^{p} \sqrt{\sum_{k=1}^{K} b_{jk}^2} \tag{5}$$

$$\text{s.t.} \sum_{k=1}^{K} a_k = 0; \sum_{k=1}^{K} b_{jk} = 0, \quad \forall j = 1, \ldots, p,$$

where ϕ_q is as defined in (3).

To approach this problem, we apply the concept of a "margin vector" to extend the definition of a (binary) margin to the multicategory case. Denote the margin vector of an observation X_i as $F_i = (f_{i1}, \ldots, f_{iK})^\top$, with $f_{ik} = a_k + X_i^\top b_k$ satisfying $\sum_{k=1}^{K} f_{ik} = 0$. Let $E_i = (e_{i1}, \ldots, e_{iK})^\top$ be the class indicator vector with $e_{ik} = \mathbb{1}\{y_i = k\}$. The multicategory margin of the data point (y_i, X_i) is then given as $f_{iy_i} = a_{y_i} + X_i^\top b_{y_i} = E_i^\top F_i$. Therefore, the MgDWD loss can be rewritten as

$$\phi_q(a_{y_i} + X_i^\top b_{y_i}) = \phi_q(E_i^\top F_i) = E_i^\top \phi_q(F_i) = \sum_{k=1}^{K} \mathbb{1}\{y_i = k\} \phi_q(a_k + X_i^\top b_k). \tag{6}$$

Based on (6), Lemma 1 describes the Fisher consistency of the MgDWD loss.

Lemma 1. *Given* $X = u$*, the minimizer of the conditional expectation of (6) is* $\tilde{F}(u) = \left(\tilde{f}_1(u), \ldots, \tilde{f}_K(u)\right)^\top$*, satisfying*

$$\underset{k \in \mathcal{Y}}{\operatorname{argmax}} \tilde{f}_k(u) = \underset{k \in \mathcal{Y}}{\operatorname{argmax}} \Pr\{y = k | X = u\},$$

where

$$\tilde{f}_k(u) = \begin{cases} Q \sqrt[q]{\dfrac{\Pr\{y = k | X = u\}}{\Pr\{y = k_* | X = u\}}}, & k \neq k_* \\ -Q \sum_{l \neq k_*} \sqrt[q]{\dfrac{\Pr\{y = l | X = u\}}{\Pr\{y = k_* | X = u\}}}, & k = k_*. \end{cases}$$

and $k_* = \underset{k \in \mathcal{Y}}{\operatorname{argmin}} \Pr\{y = k | X = u\}$.

Consequently, $\tilde{f}_k(u)$ can be treated as an effective proxy of $\Pr\{y = k | X = u\}$ and, for any new observation X_*, a reasonable prediction of its label y_* is

$$\hat{y}_* = \underset{k \in \mathcal{Y}}{\operatorname{argmax}} \{a_k + X_*^\top b_k\}.$$

Speaking to the sparse group lasso (SGL) regularization in (5), the L_1 penalty encourages an element-wise sparse estimator that selects important variables for each category, indicated by $\hat{b}_{jk} \neq 0$. Assuming that parameters in different categories share the same information, we use an L_2 penalty to encourage a group-wise sparsity structure that removes covariates that are irrelevant across all categories, that is, where $\hat{\beta}_j = (b_{1j}, \ldots, b_{Kj})^\top = 0$. Specifically, let $x_j = (x_{1j}, \cdots, x_{Nj})^\top$ and $B = (b_{jk}) \in \mathbb{R}^{p \times K}_{jk}$, where the k-th column b_k is the slope vector for the category label k and the j-th row $\beta_j^\top$ is the group coefficient for the variable x_j. If x_j is noise in the classification problem or is not relevant to category label k, then the entry b_{jk} of $\mathbf{B}$ should be shrunk to exactly zero. The SGL penalty of (5) can be written as a convex combination of the lasso and group lasso penalties in terms of β_j as

$$\lambda_1 \sum_{k=1}^{K} \sum_{j=1}^{p} |b_{jk}| + \lambda_2 \sum_{j=1}^{p} \sqrt{\sum_{k=1}^{K} b_{jk}^2} = \lambda \sum_{j=1}^{p} \{\tau \|\beta_j\|_1 + (1-\tau)\|\beta_j\|_2\}, \tag{7}$$

where $\lambda > 0$ is the scale of the penalty and $\tau \in [0,1]$ tunes the propensity between the element-wise and group-wise sparsity structure.

2.2. Population MgDWD

In this subsection, some basic results pertaining to unpenalized population MgDWD are given. These results are necessary for further theoretical analysis.

Denote the marginal probability mass of y as $\Pr(y = k) = \pi_k$ with $\pi_k > 0$ and $\sum_{k=1}^{K} \pi_k = 1$, and the conditional probability density functions of $\mathbf{X}$ given $y = k$ by $g(\mathbf{X} \mid y = k) = g_k(\mathbf{X})$. Let $\Theta = (\theta_1, \ldots, \theta_K)$ be the collection of coefficient vectors $\theta_k = (a_k, b_k^\top)^\top$ for all labels and $\mathbf{Z} = (1, \mathbf{X}^\top)^\top$. The population version of the MgDWD problem in (6) is

$$\mathcal{L}(\vartheta) = \mathbb{E}\{\mathbb{I}(\mathcal{Y})^\top \phi_q(\Theta^\top \mathbf{Z})\} = \sum_{k=1}^{K} \pi_k \int_{\mathcal{X}} \phi_q(\mathbf{Z}^\top \theta_k) g_k(x) dx, \tag{8}$$

where $\vartheta = \text{vec}\{\Theta\}$ is the vectorization of the matrix Θ and $\mathbb{I}(\mathcal{Y}) = (\mathbb{1}\{y = 1\}, \ldots, \mathbb{1}\{y = K\})^\top$ is a random vector. Denote the true parameter value ϑ^* as a minimizer of the population MgDWD problem, namely,

$$\vartheta^* \in \underset{\vartheta \in \mathscr{C}}{\arg\min}\, \mathcal{L}(\vartheta),$$

where $\mathscr{C} = \{\vartheta \in \mathbb{R}^{K(p+1)} \mid \mathbf{C}\vartheta = \mathbf{0}_K\}$ is the set of sum-constrained ϑ with $\mathbf{C} = \mathbf{1}_K^\top \otimes \mathbf{I}_{p+1}$, where $\otimes$ denotes the Kronecker product.

To facilitate our theoretical analysis, we first define the gradient vector and Hessian matrix of the population MgDWD loss function. We then introduce some regularity conditions necessary to derive theoretical properties of this problem. Let $\text{diag}\{v\}$ be a diagonal matrix constructed from the vector v, and let $\circ$ and $\oplus$ be the Hadamard product and the direct matrix sum, respectively. Denote the gradient vector of the population MgDWD loss function (8) as

$$\mathcal{S}(\vartheta) = \mathbb{E}\left(\{\mathbb{I}(\mathcal{Y}) \circ \phi_q'(\Theta^\top \mathbf{Z})\} \otimes \mathbf{Z}\right) = \text{vec}(S_1, \ldots, S_K),$$

with

$$S_k = \mathbb{E}\{\mathbb{1}\{y = k\}\phi_q'(\mathbf{Z}^\top \theta_k)\mathbf{Z}\} = \pi_k \int_{\mathcal{X}} \phi_q'(\mathbf{Z}^\top \theta_k) \mathbf{Z} g_k(\mathbf{X}) d\mathbf{X},$$

and its Hessian matrix as

$$\mathcal{H}(\vartheta) = \mathbb{E}\left\{\text{diag}\{\mathbb{I}(\mathcal{Y}, \mathcal{X})\} \circ \phi_q''(\Theta^\top \mathbf{Z})\} \otimes (\mathbf{Z}\mathbf{Z}^\top)\right\} = \bigoplus_{k=1}^{K} H_k,$$

where ϕ_q'' denotes the second derivative of the function ϕ_q; $\mathbb{I}(\mathcal{Y}, \mathcal{X}) = \mathbb{I}(\mathcal{Y}) \circ \mathbb{I}(\mathcal{X})$ is a random vector with $\mathbb{I}(\mathcal{X}) = (\mathbb{1}\{\mathbf{X} \in \mathcal{X}_1\}, \ldots, \mathbb{1}\{\mathbf{X} \in \mathcal{X}_k\})^\top$ and $\mathcal{X}_k = \{\mathbf{X} \in \mathcal{X} \mid \mathbf{Z}^\top \theta_k > Q\}$; and

$$H_k = \mathbb{E}\{\mathbb{1}\{y = k, \mathbf{X} \in \mathcal{X}_k\}\phi_q''(\mathbf{Z}^\top \theta_k)\mathbf{Z}\mathbf{Z}^\top\} = \pi_k \int_{\mathcal{X}_k} \phi_q''(\mathbf{Z}^\top \theta_k)\mathbf{Z}\mathbf{Z}^\top g_k(\mathbf{X}) d\mathbf{X}.$$

The block structure of $\mathcal{H}(\vartheta)$ implies a parallel relationship between each category. The relationship between the θ_k is reflected by the sum-to-zero constraint in the definition of $\mathscr{C}$.

We assume the following regularity conditions.

(C1) The densities of X given $y = k \in \mathcal{Y}$, i.e., the $g_k(X)$, are continuous and have finite second moments.

(C2) $0 < \Pr\{X \in \mathscr{X}_k^* | y = k\} < 1$ for all $k \in \mathcal{Y}$, where $\mathscr{X}_k^* = \{X \in \mathscr{X} | Z^\top \theta_k^* > Q\}$.

(C3) $\mathrm{Var}\{X \mid X \in \mathscr{X}_k^*, y = k\} \succ \mathbf{O}_p$ for all $k \in \mathcal{Y}$.

Remark 1. *Condition (C1) ensures that $\mathcal{L}$, $\mathcal{S}$ and $\mathcal{H}$ are well defined and continuous in ϑ. For the theoretically optimal hyperplane $\{X \in \mathscr{X} | Z^\top \theta_k^* = 0\}$, the case with $\theta_k^* = \mathbf{0}_{p+1}$ leaves $\mathscr{X}$ useless for classification. On the other hand, when $a_k^* \neq 0$ and $b_k^* = \mathbf{0}_p$, the hyperplane is the empty set and is similarly meaningless. Condition (C2) is proposed to avoid the case where $b_k^* = \mathbf{0}_p$ so that ϑ^* always contains information relevant to the classification problem. For bounded random variables, condition (C2) should be assumed with caution. Condition (C3) implies the positive definiteness of $\mathcal{H}(\vartheta^*)$.*

By convexity and the second-order Lagrange condition, the following theorem shows that the local minimizer of the population MgDWD problem exists and is unique.

Theorem 1. *Under the regularity conditions (C1)-(C3), the true parameter $\vartheta^* \in \mathscr{C}$ is the unique minimizer of $\mathcal{L}(\vartheta)$ with $b_k^* \neq \mathbf{0}_p$, and*

$$\mathcal{L}(\vartheta^*) = \sum_{k=1}^{K} A(k,q) \pi_k,$$

with $0 \leq u(k,q) \leq A(k,q) \leq v(k,q) \leq 1$, where

$$A(k,q) = 1 - \mathbb{E}\left\{ \mathbb{1}\{X \in \mathscr{X}_k^*\}\left\{1 - \left(\frac{Q}{Z^\top \theta_k^*}\right)^q\right\} \,\Big|\, y = k \right\},$$

$$u(k,q) = \Pr\{X \notin \mathscr{X}_k^* | y = k\} + Q^{2q} \Pr\{Q < Z^\top \theta_k^* \leq Q^{-1} | y = k\},$$

$$v(k,q) = \Pr\{Z^\top \theta_k^* \leq 1 \mid y = k\} + \inf_{\epsilon > 0} \left(\frac{Q}{1+\epsilon}\right)^q \Pr\{Z^\top \theta_k^* > 1 + \epsilon \mid y = k\}.$$

The bounds in Theorem 1 show how q affects the loss function $\mathcal{L}(\vartheta^*)$. The upper bound $v(k,q)$ is a decreasing function of q with

$$\lim_{q \to 0} v(k,q) = 1 \text{ and } \lim_{q \to \infty} v(k,q) = \Pr\{Z^\top \theta_k^* \leq 1 \mid y = k\}.$$

In the lower bound $u(k,q)$, the first term $\Pr\{X \notin \mathscr{X}_k^* | y = k\}$ is an increasing function of q and the last term $Q^{2q} \Pr\{Q < Z^\top \theta_k^* \leq Q^{-1} | y = k\}$ is a decreasing function of q, with

$$\lim_{q \to 0} u(k,q) = 1 \text{ and } \lim_{q \to \infty} u(k,q) = \Pr\{Z^\top \theta_k^* \leq 1 \mid y = k\}.$$

Consequently, for the given population $\mathbb{P}(y, X)$, a larger q encourages the population MgDWD estimator to focus more on the regions $\{X \notin \mathscr{X}_k, y = k\}$ that correspond to misclassifications. As a result, the estimator's performance will be similar to the hinge loss as $q \to \infty$. Setting q too small will lead to an ineffective classifier due to the unreasonable penalty placed on the well classified region $\{X \in \mathscr{X}_k, y = k\}$. This variation in the lower bound with respect to q provides a necessary condition for the existence of an optimal q.

Remark 2. *The explicit relationship between q and ϑ^* is complicated. While it may be more desirable to prove that a greater value of q results in a smaller value of the loss function $\mathcal{L}(\vartheta)$, there is no explicit formula for the optimal value ϑ^* in terms of q.*

2.3. Estimator Consistency

Under the unpenalized framework presented in the previous subsection, all covariates will contribute to the classification task for each category: this scenario may lead to a classifier that overfits to the training data set. In this subsection, we study the consistency of the estimator for (5) in ultra-high dimensional settings.

To achieve structural sparsity in the estimator, the regularization parameter λ in (7) must be large enough to dominate the gradient of the empirical MgDWD loss evaluated at the theoretical minimizer $\vartheta^* = \text{vec}\{\Theta^*\}$ with high probability. On the other hand, λ should also be as small as possible to reduce the bias incurred by the SGL regularization term

$$P(\beta) = \sum_{j=1}^{p} \tau \|\beta_j\|_1 + (1-\tau)\|\beta_j\|_2.$$

Lemma 2 provides a suitable choice of λ under the following assumption.

(A1) The predictors $X = (x_1, \ldots, x_p) \in \mathbb{R}^p$ are independent sub-Gaussian random vectors satisfying $\mathbb{E}X = \mathbf{0}_p$, and where $\text{Var}(X) = \Sigma$, there exists a constant $\kappa > 0$ such that for any $\gamma \in \mathbb{R}^p$, $\mathbb{E}\exp(\gamma^\top \Sigma^{-1/2} X) \leq \exp(\|\gamma\|_2^2 \kappa^2 / 2)$. From here on, we define ς_1^2 as the largest eigenvalue of Σ.

Lemma 2. *Denote $S(\vartheta^*) = (\mathbf{I}_K \otimes \mathbf{Z}^\top)\text{diag}(\text{vec}\{\mathbf{E}\})\text{vec}\{\phi_q'(\mathbf{Z}\Theta^*)\}$, where $\mathbf{E} = (E_1, \ldots, E_N)^\top$, $\mathbf{Z} = (\mathbf{Z}_1, \ldots, \mathbf{Z}_N)^\top$ with $\mathbf{Z}_i = (1, \mathbf{X}_i^\top)^\top$, and $\mathbf{I}_K$ is the identity matrix of size K. Under condition (A1),*

$$\tilde{P}\{\mathbf{P}S(\vartheta^*)\} \leq \tau \Lambda_1 + (1-\tau)\Lambda_2$$

with probability at least $1 - 2(Kp)^{1-c_1^2} - p^{1-c_2^2}$, where

$$\mathbf{P} = (\mathbf{I}_K - K^{-1}\mathbf{1}_K\mathbf{1}_K^\top) \otimes \mathbf{I}_{p+1},$$

$$\Lambda_1 = \max\{\varsigma_1\kappa, 1\}\left(1 - \frac{1}{K}\right)\sqrt{\frac{2\log(pK)}{N}},$$

$$\Lambda_2 = \max\{2\sqrt{2}\varsigma_1\kappa, 1\}\left\{c_2\sqrt{\left(1 - \frac{1}{K}\right)\frac{2\log(p)}{N}} + \sqrt{\frac{K-1}{N}}\right\},$$

for constants $c_1, c_2 > 1$.

It is difficult to obtain a closed form for the conjugate of the SGL penalty, say, $\bar{P}(v) = \sup_{u \in \mathscr{C} \setminus \{0\}} \frac{\langle u, v \rangle}{P(u)}$. Instead, we use a regularized upper bound $\tilde{P}(v) \geq \bar{P}(v)$. Based on Lemma 2, we propose a theoretical tuning parameter value

$$\lambda = c_0 \sqrt{\frac{\log(pK)}{N}}, \qquad (9)$$

where c_0 is some given constant satisfying $\lambda > \tau \Lambda_1 + (1-\tau)\Lambda_2$.

Before we can derive an error bound for the estimator in (5), we impose two additional assumptions.

(A2) For the true parameter value ϑ^*, there is a (s_e, s_g)-sparse structure in the coefficients $\mathbf{B}^*$ with element-wise and group-wise support sets

$$\mathscr{E} = \{(j,k) \in \{1,\ldots,p\} \times \{1,\ldots,K\} | b_{jk}^* \neq 0\} \text{ and } \mathscr{G} = \{j \in \{1,\ldots,p\} | \boldsymbol{\beta}_j^* \neq \mathbf{0}_K\}$$

with cardinality $|\mathscr{E}| = s_e$ and $|\mathscr{G}| = s_g$, respectively.

(A3) There exist some positive constants ς_2 and ς_3 such that

$$\varsigma_2^2 = \max_{\gamma \in \mathscr{V}} \frac{\|\text{diag}\{\text{vec}(\mathbf{E}^\top)\}(\mathbf{Z} \otimes \mathbf{I}_K)\gamma\|_2^2}{N\|\gamma\|_2^2} \text{ and } \varsigma_3^2 = \min_{\gamma \in \mathscr{U}} \frac{\gamma^\top \mathcal{H}(\vartheta^*)\gamma}{\gamma^\top \gamma}$$

with $\mathscr{V} = \{v \in \mathbb{R}^{K(p+1)} | 0 < \|v\|_0 \leq s_e + K\}$ and

$$\mathscr{U} = \left\{\delta \in \mathbb{R}^{K(p+1)} \, \Big| \, \frac{\tau}{1-\tau}\|\delta_{\mathscr{E}_+}\|_1 + \sum_{j \in \mathscr{G}_+} \|\delta_j\|_2 \geq C_0\left(\frac{\tau}{1-\tau}\|\delta_{\mathscr{E}^c}\|_1 + \sum_{j \notin \mathscr{G}} \|\delta_j\|_2\right)\right\},$$

where $C_0 = \frac{(c_0-1)}{(c_0+1)}$, $\mathscr{E}^c$ is the complement of $\mathscr{E}$, $\mathscr{E}_+ = \mathscr{E} \cup \{l = 1 + (k-1)(p+1) | k = 1,\ldots,K\}$, and $\mathscr{G}_+ = \mathscr{G} \cup \{0\}$.

Under the choice of λ given in (9), we show the L_2-consistency of the estimator in (5).

Theorem 2. *Suppose that conditions (A1)-(A3) hold. Then with $\lambda = c_0 \sqrt{\frac{\log(pK)}{N}}$ in (5), we have that*

$$\|\hat{\vartheta} - \vartheta^*\|_2 \leq \left\{C_1 \sqrt{s_e + K} + C_2 \sqrt{s_g + 1}\right\} \sqrt{\frac{\log(pK)}{N}}$$

with probability at least $1 - 2(Kp)^{2(s_e+K)(1-c_3^2)}$, where $C_1 = 2\varsigma_3^{-2}\{c_0\tau + (\sqrt{2} + 2c_3)\varsigma_2\}$ and $C_2 = 2\varsigma_3^{-2}c_0(1-\tau)$.

Remark 3. *The sub-Gaussian distribution assumption (A1) is common in high-dimensional scenarios. This assumption characterizes the tail behavior of a collection of random variables including Gaussian, Bernoulli, and bounded variables as special cases. Assumption (A2) describes structural sparsity at two levels. The element-wise size $s_e < p$ is the size of the underlying generative model, and the group-wise size $s_g < pK$ is the size of the signal covariate set. Both s_e and s_g are allowed to depend on the sample size N. As a result, the dimension p is allowed to increase with the sample size N. Assumption (A3) guarantees that eigenvalues are positive in this sparse scenario.*

Remark 4. *In practice, the tuning parameters λ and τ in (7) are commonly chosen by M-fold cross validation. That is, we choose the pair (τ, λ) with the highest prediction accuracy among the sub-data sets $\mathcal{D}_m$, specifically,*

$$CV(\tau, \lambda) = \sum_{m=1}^{M} \sum_{i \in \mathcal{D}_m} \mathbb{1}\{y_i = \hat{y}_i(\tau, \lambda)\}$$

where $\hat{y}_i(\tau, \lambda) = \underset{k \in \mathscr{Y}}{\arg\max} \, \mathbf{Z}_i^\top \hat{\theta}_k(\tau, \lambda)$.

2.4. Computational Algorithm

In this section, we propose an efficient algorithm to solve problem (5). Our approach uses the proximal algorithm (see [29]) for solving high dimensional regularization problems. In two main steps, this approach obtains a solution to the constrained optimization problem by applying the proximal operator to the solution to the unconstrained problem.

Since regularization is not needed for the intercept terms $\boldsymbol{\alpha} = (a_1, \ldots, a_K)^\top$, it can be separated from the coefficients in $\mathbf{B}$. The empirical MgDWD loss of (8) is given by

$$L(\boldsymbol{\vartheta}) = \frac{1}{N}\sum_{i=1}^{N} \mathbf{E}_i^\top \phi_q(F_i) = \frac{1}{N}\mathrm{tr}\left\{\mathbf{E}\phi_q(\mathbf{F}^\top)\right\} = \frac{1}{N}\mathrm{vec}\{\mathbf{E}^\top\}^\top \mathrm{vec}\{\phi_q(\mathbf{F}^\top)\}$$

where $\mathbf{F} = (f_{ik})_{N \times K} = \mathbf{Z}\boldsymbol{\Theta} = \mathbf{1}_N \boldsymbol{\alpha}^\top + \mathbf{X}\mathbf{B}$. Various properties of the loss function $L(\boldsymbol{\vartheta})$ follow below.

Lemma 3. *The loss function $L(\boldsymbol{\vartheta})$ has Lipschitz continuous partial derivatives. In particular, for $S(\boldsymbol{\alpha}) = \frac{\partial L(\boldsymbol{\theta})}{\partial \boldsymbol{\alpha}} = \frac{1}{N}\left\{\mathbf{E} \circ \phi_q'(\mathbf{F})\right\}^\top \mathbf{1}_N$ and any $\boldsymbol{u}, \boldsymbol{v} \in \mathbb{R}^K$, we have that*

$$\|S(\boldsymbol{u}) - S(\boldsymbol{v})\|_2 \le \sqrt{\frac{n_{\max}}{N}} \frac{(q+1)^2}{q} \|\boldsymbol{u} - \boldsymbol{v}\|_2,$$

where $n_{\max}$ is the largest group sample size. For $S(\mathbf{B}) = \frac{\partial L(\boldsymbol{\theta})}{\partial \mathbf{B}} = \frac{1}{N}\left\{\mathbf{E} \circ \phi_q'(\mathbf{F})\right\}^\top \mathbf{X}$ and any $\mathbf{U}, \mathbf{V} \in \mathbb{R}^{p \times K}$, we have that

$$\|\mathrm{vec}\{S(\mathbf{U}) - S(\mathbf{V})\}\|_2 \le \frac{\max_k \|\mathrm{diag}(\boldsymbol{e}_k)\mathbf{X}\|_2^2}{N} \frac{(q+1)^2}{q} \|\mathrm{vec}\{\mathbf{U} - \mathbf{V}\}\|_2,$$

where $\boldsymbol{e}_k$ is the k-th column of $\mathbf{E}$ and indicates the observations belonging to the k-th group.

Hence, following the majorization–minimization scheme, we can majorize the empirical MgDWD loss $L(\boldsymbol{\vartheta})$ by a quadratic function, that is,

$$L(\boldsymbol{\vartheta}) \le L(\boldsymbol{\vartheta}_*) + S(\boldsymbol{\alpha}_*)^\top (\boldsymbol{\alpha} - \boldsymbol{\alpha}_*) + \frac{L_{\boldsymbol{\alpha}}}{2}\|\boldsymbol{\alpha} - \boldsymbol{\alpha}_*\|_2^2$$
$$+ \mathrm{vec}\{S(\mathbf{B})\}^\top \mathrm{vec}\{\mathbf{B} - \mathbf{B}_*\} + \frac{L_{\mathbf{B}_*}}{2}\|\mathrm{vec}\{\mathbf{B} - \mathbf{B}_*\}\|_2^2,$$

for some $\boldsymbol{\vartheta}_* = \mathrm{vec}\{(\boldsymbol{\alpha}_*, \mathbf{B}_*^\top)^\top\}$, where $L_{\boldsymbol{\alpha}}$ and $L_{\mathbf{B}}$ denote the Lipschitz constants in Lemma 3. Instead of minimizing $L(\boldsymbol{\vartheta})$ directly, we apply gradient descent to minimize its surrogate upper bound function. The gradient descent updates are given by

$$\boldsymbol{\alpha}_* = \boldsymbol{\alpha} - \frac{q(q+1)^{-2}}{\sqrt{n_{\max}N}}\left\{\mathbf{E} \circ \phi_q'(\mathbf{F})\right\}^\top \mathbf{1}_N, \tag{10}$$

$$\mathbf{B}_* = \mathbf{B} - \frac{q(q+1)^{-2}}{\max_k \|\mathrm{diag}(\boldsymbol{e}_k)\mathbf{X}\|_2^2}\left\{\mathbf{E} \circ \phi_q'(\mathbf{F})\right\}^\top \mathbf{X}. \tag{11}$$

Next, we address the problem's constraints and regularization simultaneously by applying the proximal operator. For $\boldsymbol{\alpha}_*$, it is clear that

$$\boldsymbol{\alpha}_{\mathrm{new}} = \underset{\boldsymbol{\alpha}^\top \mathbf{1}_K = 0}{\mathrm{argmin}} \|\boldsymbol{\alpha} - \boldsymbol{\alpha}_*\|_2^2 = \mathbf{P}_K \boldsymbol{\alpha}_*, \tag{12}$$

where $\mathbf{P}_K = \mathbf{I}_K - k^{-1}\mathbf{1}_K \mathbf{1}_K^\top$. For $\mathbf{B}_* = (\boldsymbol{\beta}_{1*}, \ldots, \boldsymbol{\beta}_{p*})^\top$, the minimization problem can be expressed as

$$\begin{aligned}\mathbf{B}_{\mathrm{new}} &= \underset{\mathbf{B}\mathbf{1}_K = \mathbf{0}_p}{\mathrm{argmin}} \frac{1}{2}\|\mathrm{vec}\{\mathbf{B} - \mathbf{B}_*\}\|_2^2 + \frac{\lambda_1}{L_{\mathbf{B}}}\|\mathrm{vec}\{\mathbf{B}\}\|_1 + \frac{\lambda_2}{L_{\mathbf{B}}}\|\mathrm{vec}\{\mathbf{B}\}\|_{1,2} \\ &= \underset{\mathbf{B}\mathbf{1}_K = \mathbf{0}_p}{\mathrm{argmin}} \sum_{j=1}^p \frac{1}{2}\|\boldsymbol{\beta}_j - \boldsymbol{\beta}_{j*}\|_2^2 + \frac{\lambda_1}{L_{\mathbf{B}}}\|\boldsymbol{\beta}_j\|_1 + \frac{\lambda_2}{L_{\mathbf{B}}}\|\boldsymbol{\beta}_j\|_2,\end{aligned} \tag{13}$$

which implies that we can implement minimization for p groups in parallel. The following theorem provides the solution to (13).

Theorem 3. *Let $\rho_1, \rho_2 \geq 0$ and $\beta_* \in \mathbb{R}^K$. Then the constrained regularization problem*

$$\min_{\beta \in \mathbb{R}^K} \frac{1}{2}\|\beta - \beta_*\|_2^2 + \rho_1\|\beta\|_1 + \rho_2\|\beta\|_2$$

$$\text{s.t. } \beta^\top \mathbf{1}_K = 0$$

has a solution of the form

$$\beta^* = \left\{1 - \frac{\rho_2}{\|\mathbf{P}_K(\beta_* - \rho_1 u)\|_2}\right\}_+ \mathbf{P}_K(\beta_* - \rho_1 u) \tag{14}$$

for some $u \in \partial\|\beta\|_1$.

In the special case with $\rho_2 = 0$, the constrained regularization problem in Theorem 3 reduces to the constrained lasso problem with solution $\tilde{\beta}^* = \mathbf{P}_K(\beta_* - \rho_1 u)$. Combined with (14), the proximal operator $\mathcal{U}$, given by

$$\beta^* = \mathcal{U}(\tilde{\beta}^*, \rho_2) = \left\{1 - \frac{\rho_2}{\|\tilde{\beta}^*\|_2}\right\}_+ \tilde{\beta}^*, \tag{15}$$

can be introduced to realize the group sparsity of $\tilde{\beta}^*$.

For the standard lasso problem, the subgradient u has a closed form given by $\tilde{\beta}^* = \beta_* - \rho_1 u = \mathcal{S}(\beta_*, \rho_1)$, with $\mathcal{S}(u, v) = \text{sign}(u)(|u| - v)_+$. However, under the constraint on $\tilde{\beta}^*$, the naive solution $\mathbf{P}_K\mathcal{S}(\beta_*, \rho_1)$ is misleading in that it satisfies the constraint but does not achieve shrinkage, let alone loss function minimization. The term $\mathbf{P}_K u$ is suggestive of the intersection between the subdifferential set $\partial\|\beta\|_1$ and the constraint set $\{\beta \in \mathbb{R}^K | \beta^\top \mathbf{1}_K = 0\}$; in this sense, $\tilde{\beta}^*$ might not have a closed form. Here we consider using coordinate descent to solve the constrained lasso problem. For some fixed coordinate m, since $\beta^\top \mathbf{1}_K = 0$, we have that $b_m = -\sum_{l \neq m} b_l$. Rewriting the objective function of the lasso-constrained problem in a coordinate-wise form, we obtain

$$\sum_{l=1}^K \frac{1}{2}(b_l - b_{l*})^2 + \rho_1|b_l| = \left(b_k - \frac{(b_{k*} - b_{m*})}{2}\right)^2 + \frac{1}{2}\left(\sum_{l \neq k,m}^K b_l\right)^2 + \rho_1\left\{|b_k| + \left|b_k + \sum_{l \neq k,m}^K b_l\right|\right\}$$

$$+ \frac{1}{4}\left(b_{k*} + b_{m*} + \sum_{l \neq k,m}^K b_l\right)^2 + \sum_{l \neq k,m}^K \frac{1}{2}(b_l - b_{l*})^2 + \rho_1|b_l|. \tag{16}$$

Next, Theorem 4 provides the solution to the optimization problem (16).

Theorem 4. *Suppose that $t, s \in \mathbb{R}$ and $\varrho \geq 0$. Then the regularization problem*

$$\min_{b \in \mathbb{R}} \frac{1}{2}(b - t)^2 + \varrho\{|b| + |b + s|\}$$

has solution

$$b^* = \begin{cases} t, & |t| < C(s,t) \\ -C(s,t), & C(s,t) \leq |t| \leq C(s,t) + 2\varrho \\ \text{sign}(t)(|t| - 2\varrho), & |t| > C(s,t) + 2\varrho \end{cases}$$

$$= t - \mathcal{S}(t, C(s,t)) + \mathcal{S}\{\mathcal{S}(t, C(s,t)), 2\varrho\},$$

where $C(s,t) = \dfrac{1 - \text{sign}(s)\text{sign}(t)}{2}|s|$.

By Theorem 4, given some $m \in \{1, \ldots, K\}$, the coordinate-wise minimizer for any $k \neq m$ can be expressed as the proximal operator

$$b_k = \mathcal{T}(t, s, \rho_1) = t - \mathcal{S}(t, C(s,t)) + \mathcal{S}\{\mathcal{S}(t, C(s,t)), \rho_1\}, \tag{17}$$

with $s = \sum_{l \neq k,m} b_l$ and $t = (b_{k*} - b_{m*} - s)/2$. If we fix m during iteration, then the shrinkage of b_m will be indirectly reflected in the other b_k. We propose that m change with k in the coordinate-wise minimization process to ensure that every coordinate can be equally shrunk. We summarize our proposed algorithm in Algorithm 1.

Algorithm 1 Proximal gradient descent algorithm for SGL-MgDWD.

Input: λ_1, λ_2.
Initialization: $\boldsymbol{\alpha}^{(0)} = \mathbf{0}_K$, $\mathbf{B}^{(0)} = \mathbf{O}_{p \times K}$, $l = 0$.
1: **repeat**
2:     Update $\boldsymbol{\alpha}$ according to (10) and (12):

$$\boldsymbol{\alpha}^{(l+1)} = \mathbf{P}_K\{\boldsymbol{\alpha}^{(l)} - L_{\boldsymbol{\alpha}}^{-1} S(\boldsymbol{\alpha}^{(l)})\}.$$

3:     Update $\tilde{\mathbf{B}}$ according to (11):

$$\tilde{\mathbf{B}} = \mathbf{B}^{(l)} - L_{\mathbf{B}}^{-1} S(\mathbf{B}^{(l)}).$$

4:     Set $\mathbf{B}^{(l+1)} \leftarrow \tilde{\mathbf{B}}$.
5:     **repeat**
6:         **for** $m = 1$ to K **do**
7:             **for** k in $\{1, \ldots, K\} \setminus m$ **do**
8:                 Update (t, s):

$$t = \tilde{b}_k - \tilde{b}_m, \quad s = \sum_{r=1}^{K} b_r^{(l+1)} - b_k^{(l+1)} - b_m^{(l+1)}.$$

9:                 Update $b_k^{(l+1)}$ according to (17) and $b_m^{(l+1)}$ by constraint:

$$b_k^{(l+1)} = \mathcal{T}(t, s, L_{\mathbf{B}}^{-1}\lambda_1), \quad b_m^{(l+1)} = -s - b_k^{(l+1)}.$$

10:             **end for**
11:         **end for**
12:     **until** $\mathbf{B}^{(l+1)}$ convergence.
13:     Update $\mathbf{B}^{(l+1)}$ according to (15):

$$\mathbf{B}^{(l+1)} = \mathcal{U}(\mathbf{B}^{(l+1)}, L_{\mathbf{B}}^{-1}\lambda_2).$$

14:     Set $l \leftarrow l + 1$.
15: **until** some condition is met.
Output: $\boldsymbol{\alpha}^{(l)}$ and $\mathbf{B}^{(l)}$.

---

3. Numerical Analysis

In the following section, we use both simulated and real data sets to evaluate the finite sample properties of our proposed method. We compare the finite sample performance of SGL-MgDWD with L_1-regularized multinomial logistic regression (L_1-logistic).

3.1. Simulation Studies

The data is generated from the following model. Consider the K-category classification problem where $\pi_k = K^{-1}$ and $g_k(X)$ is the density function of a normal distribution with mean vector $\mu_k = (\mu_{1k}, \mu_{2k}, \mathbf{0}_{p-2}^\top)^\top$ and covariance matrix $\mathbf{I}_p$, where $(\mu_{1k}, \mu_{2k}) = (2\cos(\pi r_k), 2\sin(\pi r_k))$ with $r_k = \frac{2(k-1)}{K}$, for $k = 1, \ldots, K$. In this model, only the first two variables contribute to the classification and their corresponding parameter vectors β_1 and β_2 form two groups of coefficients. The true model has the sparsity structure $(s_e, s_g) = (2K, 2)$ for a total of $K(p+1)$ coefficients. We set the sample size for each category to $n_k = 50, 100, 200$ and 400, and the number of classes to $K = 5$ and 11. We consider dimensionality $p = 100$ and 1000.

In what follows, we compare the proposed SGL-MgDWD method with the OVR method based on SGL-MgDWD with $K = 2$ (OVR-SGL-gDWD). For SGL-MgDWD, logistic regression and OVR, the tuning parameter λ is optimized over a discrete set by minimizing the prediction error using 5-fold cross validation. In each simulation, we conduct 100 runs and use a testing set of equal size to evaluate each method's performance using the following criteria:

- Testing set accuracy, measuring the rate of correct classification;
- Signal, as the average number of correctly-selected element-wise and group-wise signals, that is, with $\hat{b}_{jk} \neq 0$ and $\hat{\beta}_j \neq \mathbf{0}$, respectively, denoted by the pair (s_e^+, s_g^+);
- Noise, as the average number of incorrectly-selected element-wise and group-wise components, that is, with $\hat{b}_{jk} = 0$ and $\hat{\beta}_j = 0$, respectively, denoted by the pair (n_e^+, n_g^+).

Simulation results are summarized in Tables 1 and 2.

As shown in Tables 1 and 2, the proposed SGL-MgDWD method performs better than the L_1-logistic and OVR methods. Specifically, in each scenario, predictions from the SGL-MgDWD method had higher accuracy relative to the other two methods. Similarly, the SGL-MgDWD method correctly selected the signal components of the model with fewer incorrectly-selected noise components, again relative to the L_1-logistic and OVR methods. These simulation results also demonstrate that test accuracy increases with increasing sample size n_k and that test accuracy decreases with higher dimension p at fixed n_k. This is consistent with the derived theoretical properties. All computations were performed on a Tensorflow 2.3 CPU on Threadripper 2950X at 4.1 Ghz.

3.2. HIV Data Analysis

Symptomatic distal sensory polyneuropathy (sDSP) is a common debilitating condition among people with HIV. This condition leads to neuropathic pain and is associated with substantial comorbidities and increased health care costs. Plasma miRNA profiles show differences between HIV patients with and without sDSP, and several miRNA biomarkers are reported to be associated with the presence of sDSP in HIV patients (see [30]). The corresponding binary classification problem was analyzed in [30] using random forest classifiers. However, the HIV data set can be further classified into four classes. The HIV data set has 1715 miRNA measures for 40 patients and is partitioned into four groups ($K = 4$) with $n_k = 10$ patients each category: non-HIV, HIV with no brain damage (HIVNBD), HIV with brain damage but stable (HIVBDS) and HIV with brain damage and unstable (HIVBDU). In the following analysis, we apply our proposed method to this classification problem. The primary aim was to identify critical miRNA biomarkers for each of the four groups. Beyond achieving a finer classification, this analysis is helpful in assessing related pathogenic effects for each patient group.

Given the small sample size of $N = 40$, we chose the tuning parameter λ by maximizing leave-one-out cross validation accuracy. We fixed $(q, \tau) = (1, 0.1)$. Table 3 shows the signal for coefficient estimates obtained from the SGL-MgDWD method using the selected λ. We conclude that there are 22 critical miRNA biomarkers important to the classification problem. In particular, the biomarkers miR-25-star, miR-3171, miR-3924 and miR-4307 are not relevant to the non-HIV group; miR-4641, miR-4655-3p and miR-660 are not relevant to the HIVNBD group; miR-217 and miR-4683 are not relevant to the HIVBDS group; and miR-217 and miR-4307 are not relevant to the HIVBDU group.

Table 1. Simulation results for the SGL-MgDWD, L_1-logistic, and OVR methods with $K = 5$. Time is measured relative to a baseline logistic regression model with $K = 5$, $p = 100$, and $N = 50$. Numbers in parentheses denote standard deviations.

n_k	p	Method	Test Accuracy	Signal (s_e^+, s_g^+)	Noise (n_e^+, n_g^+)	Time (SD)
50	100	SGL-MgDWD	0.980	(9.99, 2)	(0, 0)	1.150 (0.173)
		L_1-logistic	0.979	(9.00, 2)	(116.98, 26.17)	1.000 (0.153)
		OVR-SGL-gDWD	0.912	-	-	-
	1000	SGL-MgDWD	0.979	(10, 2)	(6.96, 1.94)	5.290 (0.166)
		L_1-logistic	0.966	(10, 2)	(2793.65, 722.38)	5.130 (0.063)
		OVR-SGL-gDWD	0.740	-	-	-
100	100	SGL-MgDWD	0.981	(10, 2)	(0.07, 0.03)	1.453 (0.155)
		L_1-logistic	0.980	(8.82, 2)	(35.18, 3.98)	1.258 (0.127)
		OVR-SGL-gDWD	0.828	-	-	-
	1000	SGL-MgDWD	0.980	(10, 2)	(1.01, 0.25)	4.863 (0.150)
		L_1-logistic	0.978	(9.93, 2)	(1380.38, 192.37)	4.703 (0.061)
		OVR-SGL-gDWD	0.546	-	-	-
200	100	SGL-MgDWD	0.980	(10, 2)	(7.67, 2.08)	1.776 (0.164)
		L_1-logistic	0.980	(9.39, 2)	(13.1, 0.72)	1.709 (0.175)
		OVR-SGL-gDWD	0.934	-	-	-
	1000	SGL-MgDWD	0.982	(10, 2)	(1.09, 0.29)	8.641 (0.186)
		L_1-logistic	0.981	(9.79, 2)	(199.02, 2.51)	2.505 (0.121)
		OVR-SGL-gDWD	0.950	-	-	-
400	100	SGL-MgDWD	0.981	(10, 2)	(0.02, 0)	2.792 (0.159)
		L_1-logistic	0.981	(10, 2)	(4.72, 3.95)	2.828 (0.115)
		OVR-SGL-gDWD	0.979	-	-	-
	1000	SGL-MgDWD	0.981	(10, 2)	(4.72, 3.95)	15.800 (0.221)
		L_1-logistic	0.981	(9.6, 2)	(16.17, 0.02)	17.915 (1.585)
		OVR-SGL-gDWD	0.964	-	-	-

Table 2. Simulation results for the SGL-MgDWD, L_1-logistic, and OVR methods with $K = 11$. Time is measured relative to a baseline logistic regression model with $K = 5$, $p = 100$, and $N = 50$. Numbers in parentheses denote standard deviations.

n_k	p	Method	Test Accuracy	Signal (s_e^+, s_g^+)	Noise (n_e^+, n_g^+)	Time (SD)
50	100	SGL-MgDWD	0.735	(21.41, 2)	(0.14, 0.02)	1.661 (0.143)
		L_1-logistic	0.735	(20.13, 2)	(337.77, 22.07)	1.610 (0.110)
		OVR-SGL-gDWD	0.647	-	-	-
	1000	SGL-MgDWD	0.733	(21.25, 2)	(0, 0)	7.105 (0.205)
		L_1-logistic	0.566	(20.67, 2)	(3805.97, 265.82)	6.933 (0.205)
		OVR-SGL-gDWD	0.382	-	-	-
100	100	SGL-MgDWD	0.737	(21.82, 2)	(0.06, 0.01)	2.518 (0.099)
		L_1-logistic	0.721	(20, 2)	(173.17, 5.81)	2.418 (0.103)
		OVR-SGL-gDWD	0.609	-	-	-
	1000	SGL-MgDWD	0.737	(21.88, 2)	(5.4, 0.77)	12.371 (0.109)
		L_1-logistic	0.697	(20.15, 2)	(1859.51, 9.04)	12.279 (0.114)
		OVR-SGL-gDWD	0.214	-	-	-
200	100	SGL-MgDWD	0.738	(22, 2)	(0, 0)	5.191 (0.079)
		L_1-logistic	0.730	(20, 2)	(50.7, 0.08)	4.246 (0.100)
		OVR-SGL-gDWD	0.609	-	-	-
	1000	SGL-MgDWD	0.738	(21.98, 2)	(0.23, 0.04)	21.950 (0.241)
		L_1-logistic	0.730	(20, 2)	(523.08, 1.07)	22.158 (0.163)
		OVR-SGL-gDWD	0.490	-	-	-
400	100	SGL-MgDWD	0.740	(22, 2)	(0, 0)	7.025 (0.172)
		L_1-logistic	0.738	(20, 2)	(3.71, 3.48)	7.997 (0.122)
		OVR-SGL-gDWD	0.709	-	-	-
	1000	SGL-MgDWD	0.738	(22, 2)	(0.68, 0.11)	38.301 (0.200)
		L_1-logistic	0.734	(20, 2)	(38.84, 35.37)	41.059 (2.064)
		OVR-SGL-gDWD	0.556	-	-	-

Table 3. Signal for the coefficient estimates obtained from the SGL-MgDWD method with $(q, \tau) = (1, 0.1)$ for the HIV data set. The symbols "+" and "-" denote positive and negative coefficient estimates, respectively, while "0" denotes a zero coefficient (i.e., an irrelevant variable).

	Non-HIV	HIVNBD	HIVBDS	HIVBDU
interception	+	+	-	+
miR-255b	-	+	-	+
miR-217	+	-	0	0
miR-25-star	0	+	+	-
miR-3136-5p	-	-	+	-
miR-3152-3p	+	-	-	+
miR-3159	-	-	-	+
miR-3171	0	+	-	-

Table 3. Cont.

	Non-HIV	HIVNBD	HIVBDS	HIVBDU
miR-33b	-	-	-	+
miR-34c-3p	-	-	+	+
miR-3545-5p	-	+	-	+
miR-3654	-	-	-	+
miR-3924	0	-	+	-
miR-4307	0	-	+	0
miR-4474-5p	-	+	+	+
miR-4526	+	-	-	-
miR-4641	+	0	-	-
miR-4655-3p	+	0	-	-
miR-4680-5p	-	-	+	-
miR-4683	-	-	0	+
miR-589	-	+	+	-
miR-619	+	-	-	+
miR-660	+	0	-	+

Author Contributions: Conceptualization, L.K. and N.T.; Methodology, T.S., L.K. and N.T.; Formal Analysis, Y.L.; Data Curation, W.G.B., E.A. and C.P.; Writing—Review & Editing, Y.W., B.J. and L.K.; Supervision, B.J., L.K. and N.T.. All authors have read and agreed to the published version of the manuscript.

Funding: A Canadian Institutes of Health Research Team Grant and Canadian HIV-Ageing Multidisciplinary Programmatic Strategy (CHAMPS) in NeuroHIV (Christopher Power) supported these studies. Bei Jiang and Linglong Kong were supported by the Natural Sciences and Engineering Research Council of Canada (NSERC). Christopher Power and Linglong Kong were supported by Canada Research Chairs in Neurological Infection and Immunity and Statistical Learning, respectively. Niansheng Tang was supported by grants from the National Natural Science Foundation of China (grant number: 11671349) and the Key Projects of the National Natural Science Foundation of China (grant number: 11731011).

Acknowledgments: The authors are thankful for the invitation of the two guest editors, Farouk Nathoo and Ejaz Ahmed. This work has also benefited from two anonymous reviewers' constructive comments and valuable feedback. The authors also thank the great help of Matthew Pietrosanu with editing.

Conflicts of Interest: The authors declare no conflict of interest.

Appendix A. Proofs

Appendix A.1. Proof of Lemma 1

Proof. For simplicity, we write $p_j = P(y = j|X)$ and $f_k = f_k(X)$. Using the Lagrange multiplier method, we define

$$L(\boldsymbol{F}) = \mathbb{E}\left\{\sum_{k=1}^{K} \mathbb{1}\{y = k\}\phi_q\{\boldsymbol{F}(\boldsymbol{X})\}\Big|\boldsymbol{X} = \boldsymbol{u}\right\} + \mu \mathbf{1}_K^\top \boldsymbol{F}(\boldsymbol{X}) = \sum_{k=1}^{K} p_k \phi_q(f_k) + \mu f_k.$$

Then for each k,

$$\frac{\partial L(\boldsymbol{F})}{\partial f_k} = \phi_q'(f_k)p_k + \mu = 0 \tag{A1}$$

with

$$\phi_q'(f_j) = \begin{cases} -1, & f_k \leq Q \\ -(Qf_k^{-1})^q, & f_k > Q. \end{cases}$$

Without loss of generality, assume that $p_1 > p_2 \geq p_3 \geq \cdots \geq p_{K-1} > p_K$. Note that $-1 \leq \phi_q' < 0$, and so $p_j \geq -\phi_q'(f_k)p_k = \mu > 0$ and $\mu = p_k$ if and only if $f_k \leq Q$.

If $\mu < p_K < p_k$, then $p_K \neq \mu$ when $f_K > Q$, which implies that $f_k > f_K > Q$ for all $1 \leq k \leq K$. Hence, substituting $\phi'_q(f_k) = -(Qf_k^{-1})^q$ into (A1) yields

$$f_k = Q\sqrt[q]{p_k\mu^{-1}} > Q > 0.$$

However, $\sum_{k=1}^{K} f_k > 0$, contradicting the sum-to-zero constraint. Therefore, $\mu = p_K < p_k$ for $k < K$ and the result follows. □

Appendix A.2. Proof of Theorem 1

Lemma A1. *Under (C1), $\mathcal{L}(\boldsymbol{\vartheta})$ exists, and it is convex on $\boldsymbol{\vartheta}$.*

Proof. The existence of $\mathcal{L}(\boldsymbol{\vartheta})$ will be satisfied if

$$\mathbb{E}_{\mathbf{X}|y}\{|\phi_q(\mathbf{Z}^\top\boldsymbol{\theta}_k)| \mid y = k\} = \int_{\mathscr{X}} |\phi_q(\mathbf{Z}^\top\boldsymbol{\theta}_k)|g_k(\mathbf{X})\mathrm{d}\mathbf{X} < \infty.$$

We divide $\mathscr{X}$ into two disjoint subsets. Defining $\mathscr{X}_k = \{\mathbf{X} \in \mathscr{X} \mid \mathbf{Z}^\top\boldsymbol{\theta}_k > Q\}$, it is clear that

$$\int_{\mathscr{X}_k} |\phi_q(\mathbf{Z}^\top\boldsymbol{\theta}_k)|g_k(\mathbf{X})\mathrm{d}\mathbf{X} \leq (q+1)^{-1}\int_{\mathscr{X}_k} g_k(\mathbf{X})\mathrm{d}\mathbf{X} < \infty.$$

Note that $0 < \phi_q(u) < (1+q)^{-1} < 1$ when $u > Q$. On the other hand, for $\mathscr{X}_k^c = \{\mathbf{X} \in \mathscr{X} \mid \mathbf{Z}^\top\boldsymbol{\theta}_k \leq Q\}$,

$$\int_{\mathscr{X}_k^c} |\phi_q(\mathbf{Z}^\top\boldsymbol{\theta}_k)|g_k(\mathbf{X})\mathrm{d}\mathbf{X} \leq |1 - a_k| + \sum_{j=1}^{p} b_{jk}\int_{\mathscr{X}} |x_j|g_k(\mathbf{X})\mathrm{d}\mathbf{X} < \infty,$$

if $\mathbb{E}_{\mathbf{X}|y}\{|x_j| \mid y = k\} < \infty$ for all $k \in \mathscr{Y}$. This completes the proof of the existence of $\mathcal{L}(\boldsymbol{\vartheta})$.

Recall that

$$\mathcal{L}(\boldsymbol{\vartheta}) = \sum_{k=1}^{K} \pi_k \int_{\mathscr{X}} \phi_q(\mathbf{Z}^\top\boldsymbol{\theta}_k)g_k(\mathbf{X})\mathrm{d}\mathbf{X},$$

where $\phi_q(u)$ is a convex function of u, so its composition with the affine mapping $u = \mathbf{Z}^\top\boldsymbol{\theta}_k$ is still convex in $\boldsymbol{\theta}_k$. Clearly, $g_k(\mathbf{X})$, $\pi_k > 0$, so the non-negatively-weighted integral and sum both preserve convexity. □

Lemma A2. *Existence of minimizers of $\mathcal{L}(\boldsymbol{\vartheta})$ on $\mathscr{C} = \{\boldsymbol{\vartheta} \in \mathbb{R}^{K(p+1)} \mid \mathbf{C}\boldsymbol{\vartheta} = \mathbf{0}_K\}$, where $\mathbf{C} = \mathbf{1}_K^\top \otimes \mathbf{I}_{p+1}$.*

Proof. By Jensen's inequality, for any $\boldsymbol{\vartheta} \in \mathscr{C}$, we have that

$$\mathcal{L}(\boldsymbol{\vartheta}) \geq \phi_q\Big(\sum_{k=1}^{K} \pi_k \mathbb{E}\{\mathbf{Z}^\top\boldsymbol{\theta}_k | y = k\}\Big).$$

Let $\mu = \text{vec}\{(\pi_k \mathbb{E}\{z_j | y = k\})_{jk}\}$, where $\|\mu\|_2 \geq (\sum_{k=1}^K \pi_k^2)^{\frac{1}{2}} \geq K^{-\frac{1}{2}} > 0$. For some $C > 0$, we have that

$$\begin{aligned}
\mathcal{L}(\vartheta) &\geq \phi_q(\mu^\top \vartheta) = \mathbb{1}\{\mu^\top \vartheta < Q\}(1 - \mu^\top \vartheta) + \mathbb{1}\{\mu^\top \vartheta \geq Q\}\varphi_q(\mu^\top \vartheta) \\
&\geq \mathbb{1}\{\mu^\top \vartheta < Q\}|1 - |\mu^\top \vartheta|| \\
&= \mathbb{1}\{\mu^\top \vartheta < -(C+1)\}(|\mu^\top \vartheta| - 1) + \mathbb{1}\{-(C+1) < \mu^\top \vartheta < -1\}(|\mu^\top \vartheta| - 1) \\
&\quad + \mathbb{1}\{-1 < \mu^\top \vartheta < Q\}(1 - |\mu^\top \vartheta|) \\
&> \mathbb{1}\{\|\mu\|_2 \|\vartheta\|_2 > C+1\}C \\
&= \mathbb{1}\left\{\|\vartheta\|_2 > \frac{C+1}{\|\mu\|_2}\right\}C.
\end{aligned}$$

Note that $1 - \mu^\top \vartheta > 1 - Q > 0$ when $\mu^\top \vartheta < Q$. By the Cauchy–Schwarz inequality, $-\mu^\top \vartheta = |\mu^\top \vartheta| \leq \|\mu\|_2 \|\vartheta\|_2$.

Hence, if $\|\vartheta\|_2 > \frac{C+1}{\|\mu\|_2} > 0$, then $\mathcal{L}(\vartheta) > C > 0$. The contrapositive of this result implies the existence of a minimizer in the unconstrained problem. That is, the closed set $\{\vartheta \in \mathscr{C} \mid \mathcal{L}(\vartheta) \leq C\}$ is bounded for some large enough C. This guarantees the existence of a solution, as desired. □

Lemma A3. *Under (C1), $\mathcal{S}(\vartheta)$ exists and*

$$\frac{\partial \mathcal{L}(\vartheta)}{\partial \vartheta} = \mathcal{S}(\vartheta).$$

Proof. The existence of $\mathcal{S}(\vartheta)$ will follow if

$$\int_{\mathscr{X}} |\phi_q'(\mathbf{Z}^\top \theta_k) z_j| \pi_k g_k(\mathbf{X}) d\mathbf{X} \leq \pi_k \int_{\mathscr{X}} |z_j| g_k(\mathbf{X}) d\mathbf{X} < \infty$$

for $j = 1, \ldots, p+1$. Note that $|\phi_q'(u)| \leq 1$ when $u > Q$.

For every $\theta_{kj} \in \mathbb{R}$, $\phi_q(\mathbf{Z}^\top \theta_k)$ is a Lebesgue integrable function of $\mathbf{X}$. For any $u \in \mathbb{R}$, $\phi_q'(u)$ exists and $|\phi_q'(u)| \leq 1$. Hence, by the Leibniz integral rule, we have that

$$\begin{aligned}
\frac{\partial}{\partial \theta_{jk}} \int_{\mathscr{X}} \phi_q(\mathbf{Z}^\top \theta_k) \pi_k g_k(\mathbf{X}) d\mathbf{X} &= \int_{\mathscr{X}} \frac{\partial \phi_q(\mathbf{Z}^\top \theta_k)}{\partial \theta_{jk}} \pi_k g_k(\mathbf{X}) d\mathbf{X} \\
&= \int_{\mathscr{X}} \phi_q'(\mathbf{Z}^\top \theta_k) z_j \pi_k g_k(\mathbf{X}) d\mathbf{X}
\end{aligned}$$

and for any $l \neq k$,

$$\frac{\partial}{\partial \theta_{jl}} \int_{\mathscr{X}} \phi_q(\mathbf{Z}^\top \theta_k) \pi_k g_k(\mathbf{X}) d\mathbf{X} = 0,$$

which is sufficient to show that

$$\frac{\partial \mathcal{L}(\vartheta)}{\partial \vartheta} = \mathcal{S}(\vartheta).$$

□

Lemma A4. *Suppose (C1) is satisfied. Then (C2) implies that $b_k^* \neq 0$.*

Proof. We can rewrite $\phi_q(u)$ as

$$\begin{aligned}\phi_q(u) &= \mathbb{1}\{u \leq Q\}(1-u) + \mathbb{1}\{u > Q\}(1-Q)\left(\frac{Q}{u}\right)^q \\ &= \left\{-\mathbb{1}\{u \leq Q\} - \mathbb{1}\{u > Q\}\left(\frac{Q}{u}\right)^{q+1}\right\}u + \mathbb{1}\{u \leq Q\} + \mathbb{1}\{u > Q\}\left(\frac{Q}{u}\right)^q \\ &= \phi_q'(u)u + \mathbb{1}\{u \leq Q\} + \mathbb{1}\{u > Q\}\left(\frac{Q}{u}\right)^q.\end{aligned}$$

Then for any $\gamma \in \mathbb{R}^{p+1}$ and its corresponding $\mathcal{X}_k = \{X \in \mathcal{X} | Z^\top \gamma > Q\}$, we have that

$$\begin{aligned}&\mathbb{E}\{\mathbb{1}\{y=k\}\phi_q(Z^\top \gamma)\} \\ &= \mathbb{E}\{\mathbb{1}\{y=k\}\phi_q'(Z^\top \gamma)Z^\top \gamma\} + \mathbb{E}\{\mathbb{1}\{y=k, Z^\top \gamma \leq Q\}\} \\ &\quad + \mathbb{E}\left\{\mathbb{1}\{y=k, Z^\top \gamma > Q\}\left(\frac{Q}{Z^\top \gamma}\right)^q\right\} \\ &= S_k^\top(\gamma)\gamma + \Pr\{y=k, X \notin \mathcal{X}_k\} + \mathbb{E}\left\{\mathbb{1}\{y=k, X \in \mathcal{X}_k\}\left(\frac{Q}{Z^\top \gamma}\right)^q\right\} \\ &= S_k^\top(\gamma)\gamma + \pi_k\left(1 - \mathbb{E}\left\{\mathbb{1}\{X \in \mathcal{X}_k\}\left\{1 - \left(\frac{Q}{Z^\top \gamma}\right)^q\right\}\Big|y=k\right\}\right).\end{aligned}$$

Let $\vartheta^* \in \mathscr{C}$ be a local minimizer. It follows that $P\mathcal{S}(\vartheta^*) = 0$ and $\sum_{k=1}^K S_k^\top(\theta_k^*)\theta_k^* = \mathcal{S}^\top(\vartheta^*)\vartheta^* = 0$ since $\vartheta^* = P\vartheta^*$ and $P = (I_K - K^{-1}1_K 1_K^\top) \otimes I_{p+1}$. Therefore,

$$\begin{aligned}\mathcal{L}(\vartheta^*) &= \mathbb{E}\{\mathbb{1}\{y=k\}\phi_q(Z^\top \theta_k^*)\} \\ &= \sum_{k=1}^K \pi_k\left(1 - \mathbb{E}\left\{\mathbb{1}\{X \in \mathcal{X}_k^*\}\left\{1 - \left(\frac{Q}{Z^\top \theta_k^*}\right)^q\right\}\Big|y=k\right\}\right) \\ &= \sum_{k=1}^K \pi_k\left(1 - \Pr\{X \in \mathcal{X}_k^*|y=k\}\mathbb{E}\left\{1 - \left(\frac{Q}{Z^\top \theta_k^*}\right)^q \Big| y=k, X \in \mathcal{X}_k^*\right\}\right).\end{aligned} \tag{A2}$$

For any $\gamma \in \mathbb{R}^{p+1}$ and its corresponding $\mathcal{X}_k = \{X \in \mathcal{X} | Z^\top \gamma > Q\}$, we always have that

$$0 < \mathbb{E}\left\{\left(\frac{Q}{Z^\top \gamma}\right)^q \Big| y=k, X \in \mathcal{X}_k\right\} < 1.$$

If $\gamma = 0_{p+1}$, then $\mathcal{X}_k = \varnothing$ so that $\Pr\{y=k, X \notin \mathcal{X}_k\} = \pi_k$ and $\Pr\{y=k, X \in \mathcal{X}_k\} = 0$. If $\gamma_1 \leq Q$ and $\gamma_{/1} = 0_p$, then $\mathcal{X}_k = \varnothing$, giving the same conclusions as the previous case. If $\gamma_1 > Q$ and $\gamma_{/1} = 0_p$, then $\mathcal{X}_k = \mathcal{X}$ so that $\Pr\{y=k, X \notin \mathcal{X}_k\} = 0$ and $\Pr\{y=k, X \in \mathcal{X}_k\} = \pi_k$. Consequently, when $0 < \Pr\{X \in \mathcal{X}_k|y=k\} < 1$, then neither $\mathcal{X}_k$ nor $\mathcal{X}$ equal $\varnothing$, so $b_k \neq 0$ follows.

Note that $\Pr\{X \notin \mathcal{X}_k|y=k\} > 0$ implies that $\Pr\{0 < Z^\top \gamma \leq Q|y=k\} > 0$ or $\Pr\{Z^\top \gamma \leq 0|y=k\} > 0$, and so special attention should be paid to bounded random variables. □

Lemma A5. *Under (C1), $\mathcal{H}(\vartheta)$ exists and*

$$\frac{\partial^2 \mathcal{L}(\vartheta)}{\partial \vartheta \partial \vartheta^\top} = \mathcal{H}(\vartheta).$$

Furthermore, $\mathcal{H}(\vartheta^*) \succ O_{K(p+1)}$ when (C2) and (C3) hold.

Proof. The existence of $\mathcal{H}(\vartheta)$ follows if its all entries are absolutely integrable, that is, for any $j, k = 1, \ldots, p+1$,

$$\int_{\mathscr{X}} |\mathbb{1}\{\mathbf{Z}^\top \boldsymbol{\theta}_k > Q\} \varphi_q''(\mathbf{Z}^\top \boldsymbol{\theta}_k) z_j z_l| \pi_k g_k(\mathbf{X}) \mathrm{d}\mathbf{X}$$
$$\leq (q + q^{-1} + 2) \int_{\mathscr{X}_k^c} |z_j z_l| g_k(\mathbf{X}) \mathrm{d}\mathbf{X}$$
$$< \infty.$$

Equivalently, the result follows if $\mathbb{E}_{\mathbf{X}|y}\{|z_j z_l| \mid y = k\} < \infty$ for all $k \in \mathscr{Y}$. Note that $0 < \varphi_q''(u) \leq q + q^{-1} + 2$ when $u > Q$.

Let η be a test function belonging to the Schwartz space $\mathscr{D}$. Then $\eta' \in \mathscr{D}$ with some support denoted by $\operatorname{supp}(\eta')$.

Clearly, $\phi_q'(u)$ is not differentiable at Q but is Lipschitz continuous. Therefore, the measurable function $S_k(\boldsymbol{\theta}_k)$ is a locally integrable function of $\boldsymbol{\theta}_k$. Then the (regular) generalized functions $S_k(\boldsymbol{\theta}_k)$ belong to the dual space of $\mathscr{D}$.

For the distributional derivative of $S_k(\boldsymbol{\theta}_k)$ with respect to θ_{jk}, we have that

$$\left| \left\langle \frac{\partial S_k(\boldsymbol{\theta}_k)}{\partial \theta_{jk}}, \eta(\theta_{jk}) \right\rangle \right| = \left| -\left\langle S_k(\boldsymbol{\theta}_k), \frac{\mathrm{d}\eta(\theta_{jk})}{\mathrm{d}\theta_{jk}} \right\rangle \right|$$
$$\leq \int_\mathbb{R} \left| S_k(\boldsymbol{\theta}_k) \eta'(\theta_{jk}) \right| \mathrm{d}\theta_{jk}$$
$$\leq \max_{\theta_{jk} \in \operatorname{supp}(\eta')} |\eta'(\theta_{jk})| \int_{\operatorname{supp}(\eta')} |S_k(\boldsymbol{\theta}_k)| \mathrm{d}\theta_{jk}$$
$$< \infty$$

implying that the function $f(\theta_{jk}, \mathbf{X}) = \phi_q'(\mathbf{Z}^\top \boldsymbol{\theta}_k) \mathbf{Z} \pi_k g_k(\mathbf{X}) \eta'(\theta_{jk})$ is integrable on $\mathbb{R} \times \mathscr{X}$. Therefore, by Fubini's Theorem,

$$\left\langle \frac{\partial S_k(\boldsymbol{\theta}_k)}{\partial \theta_{jk}}, \eta(\theta_{jk}) \right\rangle = -\left\langle S_k(\boldsymbol{\theta}_k), \frac{\mathrm{d}\eta(\theta_{jk})}{\mathrm{d}\theta_{jk}} \right\rangle$$
$$= \int_\mathscr{X} -\left\langle \phi_q'(\mathbf{Z}^\top \boldsymbol{\theta}_k) \mathbf{Z} \pi_k g_k(\mathbf{X}), \frac{\mathrm{d}\eta(\theta_{jk})}{\mathrm{d}\theta_{jk}} \right\rangle \mathrm{d}\mathbf{X}$$
$$= \int_\mathscr{X} \left\langle \frac{\partial \phi_q'(\mathbf{Z}^\top \boldsymbol{\theta}_k)}{\partial \theta_{jk}} \mathbf{Z} \pi_k g_k(\mathbf{X}), \eta(\theta_{jk}) \right\rangle \mathrm{d}\mathbf{X}$$
$$= \left\langle \mathbb{E}\left\{ \frac{\partial \phi_q'(\mathbf{Z}^\top \boldsymbol{\theta}_k)}{\partial \theta_{jk}} \mathbf{Z} \mathbb{1}\{y = k\} \right\}, \eta(\theta_{jk}) \right\rangle,$$

which implies that

$$\frac{\partial S_k(\boldsymbol{\theta}_k)}{\partial \theta_{jk}} = \mathbb{E}\left\{ \frac{\partial \phi_q'(\mathbf{Z}^\top \boldsymbol{\theta}_k)}{\partial \theta_{jk}} \mathbf{Z} \mathbb{1}\{y = k\} \right\}.$$

Recall that ϕ_q' can be written as

$$\phi_q'(u) = \varphi_q'(u) \mathbb{1}\{u > Q\} + (-1) \mathbb{1}\{u \leq Q\} = (\varphi_q'(u) + 1) \mathbb{1}\{u > Q\} - 1,$$

which contains a Schwartz product between the differentiable function $\varphi'_q(u)$ and the generalized function $\mathbb{1}\{u > Q\}$. Note that

$$\begin{aligned}\mathbb{1}\{\mathbf{Z}^\top\boldsymbol{\theta}_k > Q\} &= \mathbb{1}\{z_j > 0,\, \theta_{jk} > c_{jk}\} + \mathbb{1}\{z_j \leq 0,\, \theta_{jk} \leq c_{jk}\} \\ &= (2\mathbb{1}\{z_j > 0\} - 1)\mathbb{1}\{\theta_{jk} > c_{jk}\} + (1 - \mathbb{1}\{z_j > 0\}) \\ &= \operatorname{sign}(z_j)\mathbb{1}\{\theta_{jk} > c_{jk}\} + \mathbb{1}\{z_j \leq 0\},\end{aligned}$$

where $c_{jk} = (Q - \sum_{l\neq j} z_l \theta_{lk})/z_j$ and

$$\begin{aligned}\frac{\partial \mathbb{1}\{\mathbf{Z}^\top\boldsymbol{\theta}_k > Q\}}{\partial \theta_{jk}} + 0 &= \operatorname{sign}(z_j)\delta(\theta_{jk} - c_{jk}) \\ &= \operatorname{sign}(z_j)|z_j|\delta(\mathbf{Z}^\top\boldsymbol{\theta}_k - Q) \\ &= z_j \delta(\mathbf{Z}^\top\boldsymbol{\theta}_k - Q),\end{aligned}$$

where $\delta(x)$ is the Dirac delta function and the distributional derivative of $\mathbb{1}\{x > 0\}$. Recall that $\delta(cx) = \delta(x)/|c|$ and $f(x)\delta(x-c) = f(c)\delta(x-c)$ for some constant c and function f.

Thus, by the product rule for the distributional derivative of the Schwartz product,

$$\begin{aligned}\frac{\partial \varphi'_q(\mathbf{Z}^\top\boldsymbol{\theta}_k)}{\partial \theta_{jk}} &= \frac{\partial (\varphi'_q(\mathbf{Z}^\top\boldsymbol{\theta}_k) + 1)}{\partial \theta_{jk}}\mathbb{1}\{\mathbf{Z}^\top\boldsymbol{\theta}_k > Q\} + (\varphi'_q(\mathbf{Z}^\top\boldsymbol{\theta}_k) + 1)\frac{\partial \mathbb{1}\{\mathbf{Z}^\top\boldsymbol{\theta}_k > Q\}}{\partial \theta_{jk}} \\ &= \varphi''_q(\mathbf{Z}^\top\boldsymbol{\theta}_k)z_j\mathbb{1}\{\mathbf{Z}^\top\boldsymbol{\theta}_k > Q\} + (\varphi'_q(\mathbf{Z}^\top\boldsymbol{\theta}_k) + 1)z_j\delta(\mathbf{Z}^\top\boldsymbol{\theta}_k - Q) \\ &= \varphi''_q(\mathbf{Z}^\top\boldsymbol{\theta}_k)z_j\mathbb{1}\{\mathbf{Z}^\top\boldsymbol{\theta}_k > Q\}.\end{aligned}$$

Substituting the above expression, we obtain

$$\frac{\partial S_k(\boldsymbol{\theta}_k)}{\partial \theta_{jk}} = \mathbb{E}\left\{\varphi''_q(\mathbf{Z}^\top\boldsymbol{\theta}_k)\mathbf{Z}z_j\mathbb{1}\{\mathbf{Z}^\top\boldsymbol{\theta}_k > Q\}\mathbb{1}\{y = k\}\right\}.$$

Similarly, for $l \neq k$, we have the distributional derivative

$$\frac{\partial S_k(\boldsymbol{\theta}_k)}{\partial \theta_{jl}} = 0.$$

Recall that the distributional derivative does not depend on the order of differentiation and agrees with the classical derivative whenever the latter exists. To summarize, we have that

$$H_k(\boldsymbol{\theta}_k) = \frac{\partial^2 \mathcal{L}(\boldsymbol{\vartheta})}{\partial \boldsymbol{\theta}_k \partial \boldsymbol{\theta}_k^\top} = \frac{\partial S_k(\boldsymbol{\theta}_k)}{\partial \boldsymbol{\theta}_k^\top},\quad \mathcal{H}(\boldsymbol{\vartheta}) = \bigoplus_{k=1}^{K} H_k(\boldsymbol{\theta}_k).$$

The $H_k(\boldsymbol{\theta}_k)$ are symmetric matrices, so $\mathcal{H}(\boldsymbol{\vartheta})$ is also symmetric.

In the sense of generalized functions, differentiation is a continuous operation with respect to convergence in $\mathscr{D}'$. Therefore, $\phi'_0 = \lim_{q\to 0} \phi'_q = -\mathbb{1}\{u \leq 0\}$ and $\phi''_0 = \lim_{q\to 0} \phi''_q = \delta(u)$; $\phi'_\infty = \lim_{q\to\infty} \phi'_q = -\mathbb{1}\{u \leq 1\}$ and $\phi''_\infty = \lim_{q\to\infty} \phi''_q = \delta(u - 1)$, which coincides with results from the hinge loss.

Next, $\mathcal{H}(\boldsymbol{\vartheta}) \succ \mathbf{O}_{K(p+1)}$ if and only if both $H_1(\boldsymbol{\theta}_1)$ and its Schur complement $\bigoplus_{k=2}^{K} H_k(\boldsymbol{\theta}_k)$ are both symmetric and positive definite. We can deduce that $\mathcal{H}(\boldsymbol{\vartheta}) \succ \mathbf{O}_{K(p+1)}$ if and only if $H_k(\boldsymbol{\theta}_k) \succ \mathbf{O}_{p+1}$ for all k.

Note that there exists $c > 0$ such that $\varphi_q''(Z^\top \theta_k) \geq c$ on $\mathscr{X}_k$. Then for any $\gamma \in \mathbb{R}^{p+1}$,

$$\begin{aligned}
\gamma^\top H_k(\theta_k)\gamma &= \pi_k \int_{\mathscr{X}_k} \varphi_q''(Z^\top \theta_k)(Z^\top \gamma)^2 g_k(X) dX \\
&\geq c \Pr\{X \in \mathscr{X}_k, y = k\} \mathbb{E}\{(Z^\top \gamma)^2 | X \in \mathscr{X}_k, y = k\} \\
&\geq c \Pr\{X \in \mathscr{X}_k, y = k\} (\gamma_0^2 + \gamma_1^\top \text{Var}\{X | X \in \mathscr{X}_k, y = k\}\gamma_1),
\end{aligned}$$

which implies that $\gamma^\top H_k(\theta_k)\gamma = 0$ if and only if $\gamma = 0_{p+1}$ when $\text{Var}\{X | X \in \mathscr{X}_k, y = k\}$ is assumed to be non-singular. Assuming that $\text{Var}\{X | y = k\} \succ O$ implies that $\text{Var}\{X | X \in \mathscr{X}_k, y = k\} \succeq O$. □

Proof of Theorem 1. By Lemma A2, a minimizer $\vartheta^* \in \mathscr{C}$ exists with $b_k^* \neq 0_p$ (by Lemma A4) and $\mathcal{H}(\vartheta^*) \succ O_{K(p+1)}$ (by Lemma A5). By the second-order Lagrange condition and the convexity of $\mathcal{L}(\vartheta)$ (by Lemma A1), a minimizer of the population MgDWD loss is unique.

Recall from (A2) that

$$\begin{aligned}
\mathcal{L}(\vartheta^*) &= \mathbb{E}\{\mathbb{1}\{y = k\} \phi_q(Z^\top \theta_k^*)\} \\
&= \sum_{k=1}^K \pi_k \left(1 - \mathbb{E}\left\{\mathbb{1}\{X \in \mathscr{X}_k^*\}\left\{1 - \left(\frac{Q}{Z^\top \theta_k^*}\right)^q\right\} \bigg| y = k\right\}\right) \\
&= \sum_{k=1}^K A(k,q) \pi_k.
\end{aligned}$$

It follows that

$$\begin{aligned}
0 &\leq \mathbb{E}\left\{\mathbb{1}\{X \in \mathscr{X}_k^*\}\left\{1 - \left(\frac{Q}{Z^\top \theta_k^*}\right)^q\right\} \bigg| y = k\right\} \\
&< \mathbb{E}\left\{\mathbb{1}\{Z^\top \gamma > 1 + q^{-1}\} + \mathbb{1}\{Q < Z^\top \gamma \leq 1 + q^{-1}\}\left\{1 - \left(\frac{Q}{1 + q^{-1}}\right)^q\right\} \bigg| y = m\right\} \\
&= \Pr\{Z^\top \gamma > Q | y = m\} - \Pr\{Q < Z^\top \gamma \leq Q^{-1} | y = m\} Q^{2q} \\
&\leq 1
\end{aligned}$$

and

$$\begin{aligned}
1 &\geq \mathbb{E}\left\{\mathbb{1}\{X \in \mathscr{X}_k^*\}\left\{1 - \left(\frac{Q}{Z^\top \theta_k^*}\right)^q\right\} \bigg| y = k\right\} \\
&> \mathbb{E}\left\{\mathbb{1}\{Z^\top \theta_k^* > 1 + \epsilon\}\left\{1 - \left(\frac{Q}{1 + \epsilon}\right)^q\right\} \bigg| y = k\right\} \\
&\geq \sup_{\epsilon > 0}\left\{1 - \left(\frac{Q}{1 + \epsilon}\right)^q\right\} \Pr\{Z^\top \theta_k^* > 1 + \epsilon | y = m\} \\
&\geq 0.
\end{aligned}$$

Consequently, $0 \leq u(k,q) \leq A(k,q) \leq v(k,q) \leq 1$.

Note that $\lim_{q \to \infty}(1 + \epsilon)^{-q} Q^q = e^{-1}$ when $\epsilon = 0$ and $\lim_{q \to \infty}(1 + \epsilon)^{-q} Q^q = 0$ when $\epsilon > 0$. The difference between these two results is attributed to pointwise convergence.

Let $f_m = 1 - A(k,m) \in \mathscr{D}'$ with $m = 1, 2, \ldots$ and $\eta \in \mathscr{D}$. By Fubini's theorem and the dominated convergence theorem,

$$\lim_{m \to \infty} \langle f_m, \eta \rangle = \lim_{m \to \infty} \Big\langle \mathbb{E}\Big\{ \mathbb{1}\{X \in \mathscr{X}_k^*\} \Big(\frac{Q}{Z^\top \theta_k^*} \Big)^q \Big| y = k \Big\}, \eta(\gamma) \Big\rangle$$

$$= \lim_{m \to \infty} \mathbb{E}\Big\{ \Big\langle \mathbb{1}\{X \in \mathscr{X}_k^*\} \Big(\frac{Q}{Z^\top \theta_k^*} \Big)^q, \eta(\theta_k^*) \Big\rangle \Big| y = k \Big\}$$

$$= \mathbb{E}\Big\{ \lim_{m \to \infty} \Big\langle \mathbb{1}\{X \in \mathscr{X}_k^*\} \Big(\frac{Q}{Z^\top \theta_k^*} \Big)^q, \eta(\theta_k^*) \Big\rangle \Big| y = k \Big\}$$

$$= 0 = \langle 0, \eta(\theta_k^*) \rangle.$$

Similarly,

$$\lim_{m \to 0} \langle f_m, \eta \rangle = \mathbb{E}\Big\{ \lim_{m \to 0} \Big\langle \mathbb{1}\{X \in \mathscr{X}_k^*\} \Big(\frac{Q}{Z^\top \theta_k^*} \Big)^q, \eta(\gamma) \Big\rangle \Big| y = k \Big\}$$

$$= \mathbb{E}\Big\{ \langle \mathbb{1}\{Z^\top \theta_k^* > 0\}, \eta(\gamma) \rangle \Big| y = k \Big\}$$

$$= \Big\langle \mathbb{E}\{ \mathbb{1}\{Z^\top \theta_k^* > 0\} | y = k \}, \eta(\theta_k^*) \Big\rangle$$

$$= \Big\langle \Pr\{Z^\top \theta_k^* > 0 \mid y = k\}, \eta(\theta_k^*) \Big\rangle,$$

hence

$$A(k, \infty) = \lim_{q \to \infty} A(k, q) = \Pr\{X \notin \mathscr{X}_k^* \mid y = k\}, \text{ and } A(k, 0) = \lim_{q \to 0} A(k, q) = 1.$$

As a result, $A(k, \infty)$ coincides with the population hinge/SVM loss and $A(k, 0)$ is independent of θ_k^*. □

Appendix A.3. Proof of Lemma 2

Proof. By the definition of $\tilde{P}$,

$$\tilde{P}\{PS(\vartheta^*)\} = \tau \|PS(\vartheta^*)\|_\infty + (1 - \tau) \max_j \Big\{ \|P_K S(\alpha^*)\|_2, \|P_K S(\beta_j^*)\|_2 \Big\},$$

where

$$P_K S(\alpha^*) = P_K (E \circ \phi_q'\{F(\vartheta^*)\})^\top \mathbf{1}_K = \frac{1}{N} \sum_{i=1}^N P_K \text{diag}\{E_i\} \phi_q'(F_i^*),$$

$$P_K S(\beta_j^*) = P_K (E \circ \phi_q'\{F(\vartheta^*)\})^\top x_j = \frac{1}{N} \sum_{i=1}^N x_{ij} P_K \text{diag}\{E_i\} \phi_q'(F_i^*),$$

$P_K = (p_1, \ldots, p_K)$ with $p_k = (p_{lk}) = \mathbb{1}\{l = k\} - K^{-1}$, and

$$\mathbb{E}\{P_K S(\alpha^*)\} = P_K \mathcal{S}(\alpha^*) = 0_K, \quad \mathbb{E}\{P_K S(\beta_j^*)\} = P_K \mathcal{S}(\beta_j^*) = 0_K.$$

Denoting

$$d_{ik} = \{p_k^\top \text{diag}\{E_i\} \phi_q'(F_i^*)\} = \sum_{l=1}^K \Big(\mathbb{1}\{y_i = k\} - \frac{1}{K} \Big) e_{il} \phi_q'(f_{il}^*),$$

we have that $|d_{ik}| \leq 1 - K^{-1}$. Note that the d_{ik} are N i.i.d. random variables with

$$\frac{1}{N}\sum_{i=1}^{N}\mathbb{E}(d_{ik}) = \boldsymbol{p}_k^\top \mathcal{S}(\boldsymbol{\alpha}^*) = 0 \text{ and } \frac{1}{N}\sum_{i=1}^{N}\mathbb{E}(d_{ik}x_{ij}) = \boldsymbol{p}_k^\top \mathcal{S}(\boldsymbol{\beta}_j^*) = 0.$$

By Hoeffding's inequality, we have that

$$\Pr\left\{|\boldsymbol{p}_k^\top \mathcal{S}(\boldsymbol{\alpha}^*)| > c_1\left(1 - \frac{1}{K}\right)\sqrt{\frac{2\log(pK)}{N}}\right\} \leq 2(pK)^{-c_1^2}, \tag{A3}$$

where $c_1 > 1$.

Regarding the $d_{ik}x_{ij}$, we have that

$$\mathbb{E}\exp\{d_{ik}x_{ij}\} \leq \mathbb{E}\exp\{(1-K^{-1})|x_{ij}|\} \leq \exp\{4(1-K^{-1})^2\varsigma_1^2\kappa^2\},$$

which implies that the $d_{ik}x_{ij}$ are N independent sub-Gaussian random variables with variance proxy $(1-K^{-1})^2\varsigma_1^2\kappa^2$. Taking $c_1 > 1$, we have that

$$\Pr\left\{|\boldsymbol{p}_k^\top \mathcal{S}(\boldsymbol{\beta}_j^*)| > c_1\varsigma_1\kappa\left(1 - \frac{1}{K}\right)\sqrt{\frac{2\log(pK)}{N}}\right\} \leq 2(pK)^{-c_1^2}. \tag{A4}$$

Then by (A3) and (A4),

$$\Pr\left\{\max_j\{|\boldsymbol{p}_k^\top \mathcal{S}(\boldsymbol{\alpha}^*)|, |\boldsymbol{p}_k^\top \mathcal{S}(\boldsymbol{\beta}_j^*)|\} > \Lambda_1\right\} \leq 2(pK)^{-c_1^2} \tag{A5}$$

with

$$\Lambda_1 = \max\{\varsigma_1\kappa, 1\}c_1\left(1 - \frac{1}{K}\right)\sqrt{\frac{2\log(pK)}{N}}.$$

Taking a union bound over the Kp entries of $\mathbf{PS}(\boldsymbol{\beta}^*)$ yields that

$$\Pr\{\|\mathbf{PS}(\boldsymbol{\vartheta}^*)\|_\infty \geq \Lambda_1\} = \Pr\left\{\max_{j,k}\left\{\left|\frac{1}{N}\sum_{i=1}^{N}\boldsymbol{p}_k^\top \mathcal{S}(\boldsymbol{\alpha}^*)\right|, \left|\frac{1}{N}\sum_{i=1}^{N}\boldsymbol{p}_k^\top \mathcal{S}(\boldsymbol{\beta}_j^*)\right|\right\} \geq \Lambda_1\right\}$$
$$\leq 2K(p+1)(Kp)^{-c_1^2}.$$

On one hand,

$$\|\mathbf{P}\mathrm{diag}\{E_i\}\phi_q'(F_i^*)\|_2^2 = \|(E_i - K^{-1}) \circ \phi_q'(F_i^*)\|_2^2 \leq \sum_{l=1}^{K}(e_{il} - K^{-1})^2 \cdot 1 = 1 - K^{-1},$$

so for any $\gamma \in \mathbb{R}^K$,

$$|\gamma^\top \mathbf{P}\mathrm{diag}\{E_i\}\phi_q'(F_i^*)| \leq \|\gamma\|_2\sqrt{1 - \frac{1}{K}}$$

and $\mathbb{E}\{\gamma^\top \mathbf{P}\mathrm{diag}\{E_i\}\phi_q'(F_i^*)\} = 0$. Applying Hoeffding's lemma,

$$\mathbb{E}\exp\{\gamma^\top \mathbf{P}_K \mathcal{S}(\boldsymbol{\alpha}^*)\} = \prod_{i=1}^{N}\mathbb{E}\exp\left\{\frac{1}{N}\gamma^\top \mathbf{P}_K\mathrm{diag}\{E_i\}\phi_q'(F_i^*)\right\} \leq \exp\left\{\frac{\|\gamma\|_2^2}{2N}\left(1 - \frac{1}{K}\right)\right\}.$$

Applying a square root to Theorem 2.1 of [31] with $c_2 > 1$, we have that

$$\Pr\left\{\|\mathbf{P}S(\boldsymbol{\alpha}^*)\|_2 \geq \sqrt{\frac{K-1}{N}} + c_2\sqrt{\left(1 - \frac{1}{K}\right)\frac{2\log(p)}{N}}\right\} \leq p^{-c_2^2}. \tag{A6}$$

On the other hand, since the x_{ij} are N independent sub-Gaussian random variables with variance proxy $\varsigma_1^2 \kappa^2$,

$$\mathbb{E}\exp\{\boldsymbol{\gamma}^\top \mathbf{P}S(\boldsymbol{\beta}_j^*)\} = \prod_{i=1}^N \mathbb{E}\exp\left\{\frac{x_{ij}}{N}\{\boldsymbol{\gamma}^\top \mathbf{P}\mathrm{diag}\{E_i\}\phi_q'(F_i^*)\}\right\}$$

$$\leq \prod_{i=1}^N \mathbb{E}\exp\left\{\sqrt{1 - \frac{1}{K}}\frac{\|\boldsymbol{\gamma}\|_2}{N}|x_{ij}|\right\}$$

$$\leq = \exp\left\{\frac{\|\boldsymbol{\gamma}\|_2^2}{2}\left(1 - \frac{1}{K}\right)\frac{8\varsigma_1^2 \kappa^2}{N}\right\}$$

and $\mathbb{E}\{\mathbf{P}_K S(\boldsymbol{\beta}_j^*)\} = \mathbf{0}_K$. Similarly, we have that

$$\Pr\left\{\|\mathbf{P}S(\boldsymbol{\beta}_j^*)\|_2 \geq 2\sqrt{2}\varsigma_1 \kappa \left\{\sqrt{\frac{K-1}{N}} + c_2\sqrt{\left(1 - \frac{1}{K}\right)\frac{2\log(p)}{N}}\right\}\right\} \leq p^{-c_2^2} \tag{A7}$$

for a constant $c_2 > 1$.

Therefore, by (A6) and (A7),

$$\Pr\left\{\max_j\{\|\mathbf{P}S(\boldsymbol{\alpha}^*)\|_2, \|\mathbf{P}S(\boldsymbol{\beta}_j^*)\|_2\} \geq \Lambda_2\right\} \leq p^{-c_2^2}$$

with

$$\Lambda_2 = \max\{2\sqrt{2}\varsigma_1 \kappa, 1\}\left\{\sqrt{\frac{K-1}{N}} + c_2\sqrt{\left(1 - \frac{1}{K}\right)\frac{2\log(p)}{N}}\right\}.$$

Applying the union bound to (A5), it follows that

$$\Pr\left\{\tilde{P}\{\mathbf{P}S(\boldsymbol{\vartheta}^*)\} \geq \tau\Lambda_1 + (1-\tau)\Lambda_2\right\} \leq 2K(p+1)(pK)^{1-c_1^2} + p^{1-c_2^2},$$

and the desired result follows. □

Appendix A.4. Proof of Theorem 2

Lemma A6. *Suppose that* $\lambda = c_0 \sqrt{\frac{\log(pK)}{N}}$. *Then* $\hat{\boldsymbol{\vartheta}} - \boldsymbol{\vartheta}^* \in \mathscr{U}$, *where*

$$\mathscr{U} = \left\{\boldsymbol{\delta} \in \mathbb{R}^{K(p+1)} \;\middle|\; \frac{\tau}{1-\tau}\|\boldsymbol{\delta}_{\mathscr{E}_+}\|_1 + \sum_{j \in \mathscr{G}_+} \|\boldsymbol{\delta}_j\|_2 \geq C_0\left(\frac{\tau}{1-\tau}\|\boldsymbol{\delta}_{\mathscr{E}^c}\|_1 + \sum_{j \notin \mathscr{G}} \|\boldsymbol{\delta}_j\|_2\right)\right\},$$

$C_0 = \frac{(c_0-1)}{(c_0+1)}$, $\mathscr{E}^c$ *denotes the complement of* $\mathscr{E}$, $\mathscr{E}_+ = \mathscr{E} \cup \{l = 1 + (k-1)(p+1) | k = 1, \ldots, K\}$, *and* $\mathscr{G}_+ = \mathscr{G} \cup \{0\}$.

Proof. Since $\hat{\boldsymbol{\vartheta}} = \boldsymbol{\vartheta}^* + \boldsymbol{\delta}$ is the minimizer, we have that

$$\begin{aligned} L(\boldsymbol{\vartheta}^*) + \lambda P(\boldsymbol{\beta}^*) &\geq L(\hat{\boldsymbol{\vartheta}}) + \lambda P(\hat{\boldsymbol{\beta}}) \\ \lambda\{P(\boldsymbol{\beta}^*) - P(\boldsymbol{\beta}^* + \tilde{\boldsymbol{\delta}})\} &\geq L(\boldsymbol{\vartheta}^* + \boldsymbol{\delta}) - L(\boldsymbol{\vartheta}^*), \end{aligned} \tag{A8}$$

where β^* is the vector ϑ^* without the a_k components, replacing $\tilde{\delta}$ for δ. Then

$$\begin{aligned}P(\beta^*) - P(\beta^* + \tilde{\delta}) &= \tau(\|\beta^*_{\mathscr{E}}\|_1 - \|\beta^*_{\mathscr{E}} + \tilde{\delta}_{\mathscr{E}}\|_1 - \|\tilde{\delta}_{\mathscr{E}^c}\|_1) \\ &\quad + (1-\tau)\Big(\sum_{j\in\mathscr{G}}\|\beta^*_j\|_2 - \sum_{j\in\mathscr{G}}\|\beta^*_j + \delta_j\|_2 - \sum_{j\notin\mathscr{G}}\|\delta^*_j\|_2\Big) \\ &\leq \tau(\|\tilde{\delta}_{\mathscr{E}}\|_1 - \|\tilde{\delta}_{\mathscr{E}^c}\|_1) + (1-\tau)\Big(\sum_{j\in\mathscr{G}}\|\delta_j\|_2 - \sum_{j\notin\mathscr{G}}\|\delta^*_j\|_2\Big) \\ &\leq \tau(\|\delta_{\mathscr{E}_+}\|_1 - \|\delta_{\mathscr{E}^c}\|_1) + (1-\tau)\Big(\sum_{j\in\mathscr{G}_+}\|\delta_j\|_2 - \sum_{j\notin\mathscr{G}}\|\delta_j\|_2\Big).\end{aligned}$$

By the convexity of L,

$$L(\vartheta^* + \delta) - L(\vartheta^*) \geq \langle S(\vartheta^*), \delta \rangle \geq -\bar{P}\{\mathbf{PS}(\vartheta^*)\}P(\delta) \geq -\frac{\lambda}{c_0}P(\delta).$$

Note that

$$P(\delta) = \tau(\|\delta_{\mathscr{E}_+}\|_1 + \|\delta_{\mathscr{E}^c}\|_1) + (1-\tau)\Big(\sum_{j\in\mathscr{G}_+}\|\delta_j\|_2 + \sum_{j\notin\mathscr{G}}\|\delta_j\|_2\Big).$$

Combining the above results, we have that

$$\lambda\{P(\vartheta^*) - P(\vartheta^* + \delta)\} \geq \{L(\vartheta^* + \delta) - L(\vartheta^*)\}$$
$$(c+1)\tau\|\delta_{\mathscr{E}_+}\|_1 + (1-\tau)\sum_{j\in\mathscr{G}_+}\|\delta_j\|_2 \geq (c-1)\tau\|\delta_{\mathscr{E}^c}\|_1 + (1-\tau)\sum_{j\notin\mathscr{G}}\|\delta_j\|_2$$
$$\frac{\tau}{1-\tau}\|\delta_{\mathscr{E}_+}\|_1 + \sum_{j\in\mathscr{G}_+}\|\delta_j\|_2 \geq C_0\Big(\frac{\tau}{1-\tau}\|\delta_{\mathscr{E}^c}\|_1 + \sum_{j\notin\mathscr{G}}\|\delta_j\|_2\Big).$$

□

Lemma A7. *Assume that conditions (A1)-(A3) are satisfied. Then*

$$\sup_{v\in\mathscr{V}} \frac{|\Delta L(u,v) - \mathbb{E}\{\Delta L(u,v)\}|}{\|v\|_2} > \Lambda_3$$

with probability at most $2(Kp)^{2(s_e+K)(1-c_3^2)}$, *where*

$$\Lambda_3 = (1+\sqrt{2}c_3)\varsigma_2\sqrt{\frac{2(s_e+K)\log(pK)}{N}}$$

and $\Delta L(u,v) = L(u+v) - L(u)$ *for any* $u, v \in \mathbb{R}^{K(p+1)}$ *and for some constant* $c_3 > 1$.

Proof. Given any $u \in \mathbb{R}^{K(p+1)}$ and $v \in \mathscr{V}$ with $\mathscr{V} = \{v \in \mathbb{R}^{K(p+1)} | 0 < \|v\|_0 \leq s_e + K\}$,

$$\begin{aligned}\Delta L(u,v) &= \frac{1}{N}\sum_{i=1}^N E_i^\top\Big(\phi_q\{(\mathbf{U}+\mathbf{V})^\top Z_i\} - \phi_q\{\mathbf{U}^\top Z_i\}\Big) \\ &= \frac{1}{N}\sum_{i=1}^N\sum_{k=1}^K e_{ik}\Big(\phi_q\{Z_i^\top(u_k+v_k)\} - \phi_q\{Z_i^\top(u_k)\}\Big) \\ &= \frac{1}{N}\sum_{i=1}^N d_i(u,v),\end{aligned}$$

where $u = \text{vec}\{\mathbf{U}\}, v = \text{vec}\{\mathbf{V}\}$.

The bounded gradient implies the Lipschitz continuity of ϕ_q so that $|\phi_q(u+v) - \phi_q(u)| \le |v|$. Since $e_{ik} \in \{0,1\}$, we have that

$$|d_i(u,v)| \le \sum_{k=1}^{K} \left|e_{ik}\{\phi_q\{Z_i^\top(u_k+v_k)\} - \phi_q(Z_i^\top u_k)\}\right|$$
$$\le \sum_{k=1}^{K} |e_{ik}Z_i^\top v_k| \le E_i^\top \text{vec}\{V^\top Z_i\}$$
$$= v^\top(Z_i \otimes I_K)E_i.$$

Note that

$$\sum_{i=1}^{N} \left(v^\top(Z_i \otimes I_K)E_i\right)^2 = \|\text{diag}\{\text{vec}\{E^\top\}\}(Z \otimes I_K)v\|_2^2.$$

By Hoeffding's inequality, we have that

$$\Pr\left\{\left|\frac{1}{N}\sum_{i=1}^{N} d_i(u,v) - \mathbb{E}\left(\frac{1}{N}\sum_{i=1}^{N} d_i(u,v)\right)\right| > t\right\}$$
$$\le 2\exp\left\{-\frac{2N^2t^2}{4\|\text{diag}\{\text{vec}\{E^\top\}\}(Z \otimes I_K)v\|_2^2}\right\}$$
$$\le 2\exp\left\{-\frac{Nt^2}{2\varsigma_2^2\|v\|_2^2}\right\}.$$

Thus $\Pr\{R(v) > \Lambda_3\} \le 2(Kp)^{-(s_e+K)c_3^2}$ with

$$R(v) = \frac{|\Delta L(u,v) - \mathbb{E}\{\Delta L(u,v)\}|}{\|v\|_2} \quad \text{and} \quad \Lambda_3 = c_3\varsigma_2\sqrt{\frac{2(s_e+K)\log(pK)}{N}}.$$

Next, we consider covering $\mathscr{V}$ with ϵ-balls such that for any v_1 and v_2 in the same ball, $\left|\frac{v_1}{\|v_1\|_2} - \frac{v_1}{\|v_1\|_2}\right| \le \epsilon$, where ϵ is a small positive number. The number of ϵ-balls required to cover a m-dimensional unit ball is bounded by $(\frac{2}{\epsilon}+1)^m$. Then for those $\frac{v}{\|v\|_2}$, we require a covering number of at most $(3(Kp)/\epsilon)^{s_e+K}$. Let $\mathscr{N}$ denote such an ϵ-net. We have that

$$\Pr\left\{\sup_{v \in \mathscr{N}} R(v) > \Lambda_3\right\} \le \left(\frac{3Kp}{\epsilon}\right)^{s_e+K} 2(Kp)^{-(s_e+K)c_3^2} = 2\left\{\frac{3}{\epsilon}(Kp)^{1-c_3^2}\right\}^{s_e+K}.$$

Furthermore, for any $v_1, v_2 \in \mathscr{V}$,

$$|R(v_1) - R(v_2)| \le \frac{2}{N}\left\|\text{diag}\{\text{vec}\{E^\top\}\}(Z \otimes I_K)\left(\frac{v_1}{\|v_1\|_2} - \frac{v_1}{\|v_1\|_2}\right)\right\|_1$$
$$\le \frac{2}{\sqrt{N}}\left\|\text{diag}\{\text{vec}\{E^\top\}\}(Z \otimes I_K)\left(\frac{v_1}{\|v_1\|_2} - \frac{v_1}{\|v_1\|_2}\right)\right\|_2$$
$$\le 2\varsigma_2\epsilon.$$

Therefore $\sup_{v \in \mathscr{V}} R(v) \leq \sup_{v \in \mathscr{N}} R(v) + 2\varsigma_2 \epsilon$. Taking $\epsilon = \sqrt{\dfrac{(s_e + K) \log(pK)}{2N}}$, we have that

$$\Pr\left\{\sup_{v \in \mathscr{V}} R(v) > \Lambda_3\right\} \leq \Pr\left\{\sup_{v \in \mathscr{N}} R(v) > (c_3 - 1)\varsigma_1 \sqrt{\dfrac{2(s_e + K)\log(pK)}{N}}\right\}$$

$$\leq 2\left\{\sqrt{\dfrac{2N}{(s_e + K)\log(pK)}} 3(Kp)^{1-(c_3-1)^2}\right\}^{s_e+K}$$

$$\leq 2\left\{(Kp)^{2-(c_3-1)^2}\right\}^{s_e+K}.$$

Setting $c_3 = 1 + \sqrt{2}c_4$ and $c_4 > 1$, we obtain the desired result that

$$\Pr\left\{\sup_{v \in \mathscr{V}} R(v) > (1 + \sqrt{2}c_4)\varsigma_2 \sqrt{\dfrac{2(s_e + K)\log(pK)}{N}}\right\} \leq 2(Kp)^{2(s_e+K)(1-c_4^2)}.$$

□

Proof of Theorem 2. Consider a disjoint partition on the coordinate set $\delta = \hat{\vartheta} - \vartheta^*$, that is, $\delta = \sum_{m=1}^M v_m$ with $v_m \in \mathscr{V}$. Note that, each subvector v_m has at most $s_e + K$ non-zero coordinates. Denote $v_0 = 0$ and $u_m = \vartheta^* + \sum_{l=0}^{m-1} v_l$ so that $u_1 = \vartheta^*$ and $u_M + v_M = \vartheta^* + \delta$. We have the decomposition

$$\Delta L(\vartheta^*, \delta) = L\left(\vartheta^* + \sum_{m=1}^M v_m\right) - L(\vartheta^*) = \sum_{m=1}^M L\left(\vartheta^* + \sum_{l=0}^m v_l\right) - L\left(\vartheta^* + \sum_{l=0}^{m-1} v_l\right)$$

$$= \sum_{m=1}^M L(u_m + v_m) - L(u_m) = \sum_{m=1}^M \Delta L(u_m, v_m).$$

By Lemma A7,

$$\sum_{m=1}^M \Delta L(u_m, v_m) \geq \sum_{m=1}^M \mathbb{E}\{\Delta L(u_m, v_m)\} - \Lambda_3 \|v_m\|_2 = \mathbb{E}\{\Delta L(\vartheta^*, \delta)\} - \Lambda_3 \|\delta\|_2$$

with high probability. By Lemma A5, $\mathcal{L}$ is twice differentiable so that

$$\mathbb{E}\{\Delta L(\vartheta^*, \delta)\} = \dfrac{1}{N} \sum_{i=1}^N \mathbb{E}\left(E_i^\top \phi_q\{F_i(\vartheta^* + \delta)\}\right) - \mathbb{E}\left(E_i^\top \phi_q\{F_i(\vartheta^*)\}\right)$$

$$= \mathcal{L}(\vartheta^* + \delta) - \mathcal{L}(\vartheta^*)$$

$$= \mathcal{S}(\vartheta^*)^\top \delta + \dfrac{1}{2}\delta^\top \mathcal{H}(\vartheta^*)\delta + o(\|\delta\|_2^2)$$

$$\geq 0 + \dfrac{\varsigma_3^2}{2}\|\delta\|_2^2 + o(\|\delta\|_2^2).$$

Consequently, $\Delta L(\vartheta^*, \delta)$ is bounded below by $\dfrac{\varsigma_3^2}{2}\|\delta\|_2^2 - \Lambda_3 \|\delta\|_2$ with high probability. Note that

$$P(\beta^*) - P(\beta^* + \tilde{\delta}) \leq \tau(\|\delta_{\mathcal{E}_+}\|_1 - \|\delta_{\mathcal{E}^c}\|_1) + (1-\tau)\left(\sum_{j \in \mathcal{G}_+} \|\delta_j\|_2 - \sum_{j \notin \mathcal{G}} \|\delta_j\|_2\right)$$

$$\leq \left(\tau\|\delta_{\mathcal{E}_+}\|_1 + (1-\tau)\sum_{j \in \mathcal{G}_+}\|\delta_j\|_2\right).$$

From (A8),

$$L(\boldsymbol{\vartheta}^*) + \lambda P(\boldsymbol{\beta}^*) \geq L(\hat{\boldsymbol{\vartheta}}) + \lambda P(\hat{\boldsymbol{\beta}})$$
$$\lambda\{P(\boldsymbol{\beta}^*) - P(\boldsymbol{\beta}^* + \tilde{\boldsymbol{\delta}})\} \geq L(\boldsymbol{\vartheta}^* + \boldsymbol{\delta}) - L(\boldsymbol{\vartheta}^*)$$
$$\lambda\left(\tau\|\boldsymbol{\delta}_{\mathcal{E}_+}\|_1 + (1-\tau)\sum_{j\in\mathcal{G}_+}\|\boldsymbol{\delta}_j\|_2\right) \geq \frac{\varsigma_3^2}{2}\|\boldsymbol{\delta}\|_2^2 - \Lambda_3\|\boldsymbol{\delta}\|_2.$$

Clearly, $\|\boldsymbol{\delta}_{\mathcal{E}_+}\|_1 \leq \sqrt{s_e+K}\|\boldsymbol{\delta}_{\mathcal{E}_+}\|_2 \leq \sqrt{s_e+K}\|\boldsymbol{\delta}\|_2$ and $\sum_{j\in\mathcal{G}_+}\|\boldsymbol{\delta}_j\|_2 \leq \sqrt{s_g+1}\|\boldsymbol{\delta}\|_2$. We conclude that

$$\frac{\varsigma_3^2}{2}\|\boldsymbol{\delta}\|_2^2 \leq \lambda\left(\tau\|\boldsymbol{\delta}_{\mathcal{E}_+}\|_1 + (1-\tau)\sum_{j\in\mathcal{G}_+}\|\boldsymbol{\delta}_j\|_2\right) + \Lambda_3\|\boldsymbol{\delta}\|_2$$
$$\|\boldsymbol{\delta}\|_2^2 \leq 2\varsigma_3^{-2}\left\{\lambda\left(\tau\sqrt{s_e+K} + (1-\tau)\sqrt{s_g+1}\right) + \Lambda_3\right\}\|\boldsymbol{\delta}\|_2,$$

after which the desired result follows from straightforward algebraic manipulation. □

Appendix A.5. Proof of Lemma 3

Proof. Since

$$\text{vec}(\mathbf{F}^\top)^\top = \text{vec}\{(\mathbf{1}_N\boldsymbol{\alpha}^\top + \mathbf{X}\mathbf{B})^\top\}^\top = \boldsymbol{\alpha}^\top(\mathbf{1}_N^\top\otimes\mathbf{I}_K) + \text{vec}(\mathbf{B}^\top)^\top(\mathbf{X}^\top\otimes\mathbf{I}_K),$$

we have that

$$\begin{cases}\dfrac{\partial\text{vec}(\mathbf{F}^\top)^\top}{\partial\boldsymbol{\alpha}} = \dfrac{\partial\boldsymbol{\alpha}^\top(\mathbf{1}_N^\top\otimes\mathbf{I}_K)}{\partial\boldsymbol{\alpha}} = \dfrac{\partial\boldsymbol{\alpha}^\top}{\partial\boldsymbol{\alpha}}(\mathbf{1}_N^\top\otimes\mathbf{I}_K) = \mathbf{I}_K(\mathbf{1}_N^\top\otimes\mathbf{I}_K) = \mathbf{1}_N^\top\otimes\mathbf{I}_K \\[2ex] \dfrac{\partial\text{vec}(\mathbf{F}^\top)^\top}{\partial\text{vec}(\mathbf{B}^\top)} = \dfrac{\partial\text{vec}(\mathbf{B}^\top)^\top(\mathbf{X}^\top\otimes\mathbf{I}_K)}{\partial\text{vec}(\mathbf{B}^\top)} = \mathbf{I}_{pK}(\mathbf{X}^\top\otimes\mathbf{I}_K) = \mathbf{X}^\top\otimes\mathbf{I}_K.\end{cases}$$

The derivative with respect to $\boldsymbol{\alpha}$ is

$$\begin{aligned}NS(\boldsymbol{\alpha}) &= N\frac{\partial L(\boldsymbol{\theta})}{\partial\boldsymbol{\alpha}} = \frac{\partial}{\partial\boldsymbol{\alpha}}\text{vec}\{\mathbf{E}^\top\}^\top\text{vec}\{\phi_q(\mathbf{F}^\top)\}\\ &= \frac{\partial\text{vec}(\mathbf{F}^\top)^\top}{\partial\boldsymbol{\alpha}}\frac{\partial\phi_q\{\text{vec}(\mathbf{F}^\top)^\top\}}{\partial\text{vec}(\mathbf{F}^\top)}\text{vec}\{\mathbf{E}^\top\}\\ &= (\mathbf{1}_N^\top\otimes\mathbf{I}_K)\text{diag}(\text{vec}\{\phi_q'(\mathbf{F}^\top)\})\text{vec}\{\mathbf{E}^\top\}\\ &= \text{vec}(\mathbf{I}_K\left\{\mathbf{E}\circ\phi_q'(\mathbf{F})\right\}^\top\mathbf{1}_N)\\ &= \left\{\mathbf{E}\circ\phi_q'(\mathbf{F})\right\}^\top\mathbf{1}_N.\end{aligned}$$

Thus,

$$\begin{aligned}
\|S(\boldsymbol{\alpha})|_v^u\|_2^2 &= \|S(\boldsymbol{u}) - S(\boldsymbol{v})\|_2^2 = N^{-2}\|(\mathbf{1}_N^\top \otimes \mathbf{I}_K)\mathrm{vec}\{(\mathbf{E} \circ \phi_q'\{\mathbf{F}(\boldsymbol{\alpha})\}|_v^u)^\top\}\|_2^2 \\
&\leq N^{-2}\|\mathbf{1}_N^\top \otimes \mathbf{I}_K\|_2^2 \|\mathrm{vec}\{(\mathbf{E} \circ \phi_q'\{\mathbf{F}(\boldsymbol{\alpha})\}|_v^u)^\top\}\|_2^2 \\
&= N^{-1}\sum_{k=1}^K \sum_{i=1}^N e_{ik}^2(\phi_q'\{f_{ik}(u_k)\} - \phi_q'\{f_{ik}(v_k)\})^2 \\
&\leq N^{-1}\sum_{k=1}^K \Big(\sum_{i=1}^N e_{ik}\Big) L_q^2 (u_k - v_k)^2 \\
&\leq N^{-1} n_{\max} L_q^2 \|\boldsymbol{u} - \boldsymbol{v}\|_2^2,
\end{aligned}$$

where $L_q = \frac{(q+1)^2}{q}$ is the Lipschitz constant of ϕ_q'. We have that $L_\alpha = \sqrt{\frac{n_{\max}}{N}} L_q$.
The derivative with respect to $\mathrm{vec}(\mathbf{B}^\top)$ is

$$\begin{aligned}
N\frac{\partial L(\boldsymbol{\theta})}{\partial \mathrm{vec}(\mathbf{B}^\top)} &= \frac{\partial}{\partial \mathrm{vec}(\mathbf{B}^\top)}\mathrm{vec}\{\mathbf{E}^\top\}^\top \mathrm{vec}\{\phi_q(\mathbf{F}^\top)\} \\
&= \frac{\partial \mathrm{vec}(\mathbf{F}^\top)^\top}{\partial \mathrm{vec}(\mathbf{B}^\top)} \frac{\partial \phi_q\{\mathrm{vec}(\mathbf{F}^\top)^\top\}}{\partial \mathrm{vec}(\mathbf{F}^\top)} \mathrm{vec}\{\mathbf{E}^\top\} \\
&= (\mathbf{X}^\top \otimes \mathbf{I}_K)\mathrm{diag}(\mathrm{vec}\{\phi_q'(\mathbf{F}^\top)\})\mathrm{vec}\{\mathbf{E}^\top\} \\
&= \mathrm{vec}\Big(\mathbf{I}_K\{\mathbf{E} \circ \phi_q'(\mathbf{F})\}^\top \mathbf{X}\Big) \\
&= \mathrm{vec}\Big(\{\mathbf{E} \circ \phi_q'(\mathbf{F})\}^\top \mathbf{X}\Big).
\end{aligned}$$

Therefore, the derivative with respect to $\mathbf{B}$ is $S(\mathbf{B}) = N^{-1}\mathbf{X}^\top\{\mathbf{E} \circ \phi_q'(\mathbf{F})\}$. Note that

$$\begin{aligned}
\mathrm{vec}\Big(\mathbf{X}^\top\{\mathbf{E} \circ \phi_q'(\mathbf{F})\}\Big) &= (\mathbf{I}_K \otimes \mathbf{X}^\top)\mathrm{diag}\{\mathrm{vec}(\mathbf{E})\}\mathrm{vec}\{\phi_q'(\mathbf{F})\} \\
&= \Big\{\bigoplus_{k=1}^K \mathbf{X}^\top \mathrm{diag}(e_k)\Big\} \mathrm{vec}\{\phi_q'(\mathbf{F})\}
\end{aligned}$$

and

$$\sum_{i=1}^N \{e_{ik}\mathbf{X}_i^\top(u_k - v_k)\}^2 = \|\mathrm{diag}(e_k)\mathbf{X}(u_k - v_k)\|_2^2 \leq \|\mathrm{diag}(e_k)\mathbf{X}\|_2^2 \|u_k - v_k\|_2^2;$$

thus

$$N^2\|\text{vec}\{S(\mathbf{U}) - S(\mathbf{V})\}\|_2^2 = \sum_{k=1}^{K} \left\|\mathbf{X}^\top \text{diag}(e_k)\phi_q\{f_k(b_k)\}|_{v_k}^{u_k}\right\|_2^2$$

$$\leq \sum_{k=1}^{K} \|\mathbf{X}^\top \text{diag}(e_k)\|_2^2 \|\text{diag}(e_k)\phi_q\{f_k(b_k)\}|_{v_k}^{u_k}\|_2^2$$

$$\leq \sum_{k=1}^{K} \|\text{diag}(e_k)\mathbf{X}\|_2^2 \sum_{i=1}^{N} e_{ik} \left(\phi_q\{f_{ik}(u_k)\} - \phi_q\{f_{ik}(v_k)\}\right)^2$$

$$\leq L_q^2 \sum_{k=1}^{K} \|\text{diag}(e_k)\mathbf{X}\|_2^2 \sum_{i=1}^{N} \left\{e_{ik}\mathbf{X}_i^\top(u_k - v_k)\right\}^2$$

$$\leq L_q^2 \sum_{k=1}^{K} \|\text{diag}(e_k)\mathbf{X}\|_2^4 \|u_k - v_k\|_2^2$$

$$\leq \max_k \left\{\|\text{diag}(e_k)\mathbf{X}\|_2^2\right\}^2 \|\text{vec}(\mathbf{U} - \mathbf{V})\|_2^2.$$

We conclude that $L_\mathbf{B} = L_q N^{-1} \max_k \|\text{diag}(e_k)\mathbf{X}\|_2^2$. □

Appendix A.6. Proof of Theorem 3

Lemma A8. *The indicator function*

$$\delta_{\mathcal{R}}(x) = \begin{cases} 0, & \text{if } x \in \mathcal{R} \\ \infty, & \text{if } x \notin \mathcal{R}, \end{cases}$$

where $\mathcal{R} = \{x \in \mathbb{R}^p \mid \mathbf{1}_p^\top x = 0\}$, has subdifferential

$$\partial \delta_{\mathcal{R}}(x) = \begin{cases} \{g \in \mathbb{R}^p \mid g = s\mathbf{1}_p, s \in \mathbb{R}\}, & \text{if } x \in \mathcal{R} \\ \varnothing, & \text{if } x \notin \mathcal{R}. \end{cases}$$

Proof. Suppose that $x \in \mathcal{R}$. Then $g \in \partial \delta_{\mathcal{R}}(x)$ if and only if both

$$\delta_{\mathcal{R}}(y) \geq \delta_{\mathcal{R}}(x) + \langle g, y - x \rangle \text{ for all } y \in \mathcal{R} \text{ and}$$
$$\omega^\top(y - x) \leq 0.$$

Let $z = y - x$. Then $z \in \mathcal{R}$ since $\mathbf{1}_p^\top(y - x) = 0$. Thus, $g^\top z \leq 0$. If $g^\top z = 0$, then $g \in \{g \in \mathbb{R}^p \mid g = s\mathbf{1}_p, s \in \mathbb{R}\}$. If there exists $g \in \partial \delta_{\mathcal{R}}(x)$ satisfying $g^\top z < 0$ for some $z \in \mathcal{R}$, then $-z \in \mathcal{R}$, so we must have that $g^\top z > 0$. This is a contradiction.

Now, for any $x \notin \mathcal{R}$, we have that $g \in \partial \delta_{\mathcal{R}}(x)$ if and only if both

$$\delta_{\mathcal{R}}(y) \geq \delta_{\mathcal{R}}(x) + \langle g, y - x \rangle \text{ for all } y \in \mathcal{R} \text{ and}$$
$$\omega^\top(x - y) \geq \infty.$$

For $x \notin \mathcal{R}$ and $y \in \mathcal{R}$, since $z = x - y \in \mathbb{R}^p$ and $g^\top z \geq \infty$, it must be that $g \in \varnothing$. □

Proof of Theorem 3. It is sufficient to minimize the objective function

$$G(\beta) = \frac{1}{2}\|\beta - \beta_*\|_2^2 + \rho_1\|\beta\|_1 + \rho_2\|\beta\|_2 + \delta_{\mathcal{R}}(\beta),$$

where $\mathcal{R} = \{x \in \mathbb{R}^K \mid \mathbf{1}_K^\top \beta = 0\}$. Then the subdifferential of $G(\beta)$ is

$$\partial G(\beta) = \beta - \beta_* + \rho_1 \partial \|\beta\|_1 + \rho_2 \partial \|\beta\|_2 + \partial \delta_\mathcal{R}(\beta).$$

For an optimal solution $\beta^* \in \mathcal{R}$, we have that $\mathbf{0}_p \in \partial G(\beta^*)$ if and only if there exist $u \in \partial \|\beta\|_1$, $v \in \partial \|\beta\|_2$ and $s \in \mathbb{R}$ such that $\beta^* = \beta_* - \rho_1 u - \rho_2 v - s\mathbf{1}_p$. Since $\mathbf{1}^\top \beta^* = 0$, we have that $s = p^{-1}\mathbf{1}_p^\top(\beta_* - \rho_1 u - \rho_2 v)$, so

$$\beta^* = \mathbf{P}_K(\beta_* - \rho_1 u - \rho_2 v).$$

If $\beta^* = \mathbf{0}_p$, then $|u_j| < 1$ for $j = 1, \ldots, p$, $\|v\|_2 \leq 1$ and

$$\|\mathbf{P}_K(\beta_* - \rho_1 u)\|_2 = \rho_2 \|\mathbf{P}_K v\|_2 \leq \rho_2 \|\mathbf{P}_K\|_2 \|v\|_2 = \rho_2 \|v\|_2 \leq \rho_2;$$

If $\beta^* \neq \mathbf{0}_K$, then $u \in \partial \|x\|_1$, $v = \frac{\beta^*}{\|\beta^*\|_2}$ and

$$\beta^* = \mathbf{P}_K\left(\beta_* - \rho_1 u - \rho_2 \frac{\beta^*}{\|\beta^*\|_2}\right)$$

$$\left(1 + \frac{\rho_2}{\|\beta^*\|_2}\right)\beta^* = \mathbf{P}_K(\beta_* - \rho_1 u).$$

Note that $\beta^* = \mathbf{P}_K \beta^* \in \mathcal{R}$. Taking the norm of both sides, we see that

$$\left(1 + \frac{\rho_2}{\|\beta^*\|_2}\right)\|\beta^*\|_2 = \|\mathbf{P}_K(\beta_* - \rho_1 u)\|_2$$

$$\|\beta^*\|_2 = \|\mathbf{P}_K(\beta_* - \rho_1 u)\|_2 - \rho_2 > 0.$$

Substituting this result back into the $\beta^* \neq \mathbf{0}_K$ case, we have that

$$\beta^* = \left\{1 - \frac{\rho_2}{\|\mathbf{P}_K(\beta_* - \rho_1 u)\|_2}\right\} \mathbf{P}_K(\beta_* - \rho_1 u).$$

Combining the above two cases gives the desired result. □

Appendix A.7. Proof of Theorem 4

Proof. Denote the objective function by

$$G(b) = \frac{1}{2}(b - t)^2 + \varrho\{|b| + |b + s|\}.$$

When $s = 0$, we obtain a lasso problem with

$$b^* = \underset{b \in \mathbb{R}}{\arg\min} \, \frac{1}{2}(b - t)^2 + 2\varrho|x| = \mathcal{S}(t, 2\varrho).$$

When $s \neq 0$, the subdifferential of $G(b)$ is

$$\partial G(b) = b - t + \varrho\{\partial |x| + \partial |x + s|\}.$$

We see that $0 \in \partial G(b^*)$ if and only if there exist $u \in \partial |b|$ and $v \in \partial |b + s|$ with

$$b^* = b - \varrho(u + v).$$

If $b^* = 0$, then $|u| < 1$ and $v = \text{sign}(s)$, hence

$$b^* = 0 \text{ if } |t - \varrho\text{sign}(s)| \leq \varrho.$$

If $s > 0$, then $\text{sign}(s) = 1$ and $0 \leq t \leq 2\varrho$. If $s < 0$, then $\text{sign}(s) = -1$, and $-2\varrho \leq t \leq 0$. Note that if $t \neq 0$, then $\text{sign}(s) = \text{sign}(t)$ or $\text{sign}(s)\text{sign}(t) = 1$.

When $b^* = -s$, then $u = -\text{sign}(s)$ and $|v| < 1$, hence

$$b^* = -s \text{ if } |t + s + \varrho\text{sign}(s)| \leq \varrho.$$

If $s > 0$, then $\text{sign}(s) = 1$ and $-(s + 2\lambda) \leq t \leq -s < 0$. If $s < 0$, then $\text{sign}(s) = -1$ and $0 < -s \leq t \leq -(s - 2\lambda)$. Note that $\text{sign}(s) = -\text{sign}(t)$ is equivalent to $\text{sign}(s)\text{sign}(t) = -1$.

Let $C(s,t) = \dfrac{1 - \text{sign}(s)\text{sign}(t)}{2}|s| \geq 0$. We can summarize the two cases above as

$$b^* = -C(s,t) \text{ if } 0 \leq C(s,t) \leq |t| \leq C(s,t) + 2\varrho. \tag{A9}$$

If $b^* \neq 0, -s$, then $u = \text{sign}(b^*)$ and $v = \text{sign}(b^* + s)$, thus

$$b^* = t - \varrho\{\text{sign}(b^*) + \text{sign}(b^* + s)\}$$
$$b^* + s = t + s - \varrho\{\text{sign}(b^*) + \text{sign}(b^* + s)\}.$$

If $\text{sign}(b^*) = -\text{sign}(b^* + s) = 1$, then $b^*(b^* + s) < 0$ or $0 < t < -s$. Thus $b^* = t > 0$ if $0 < t < -s$. If $\text{sign}(b^*) = -\text{sign}(b^* + s) = -1$, then $b^*(b^* + s) < 0$ or $-s < t < 0$. Thus $b^* = t < 0$ if $-s < t < 0$. Rewriting the two cases above, we have that

$$b^* = t \quad \text{if } 0 < |t| < C(s,t). \tag{A10}$$

If $\text{sign}(b^*) = \text{sign}(b^* + s) = 1$, then

$$\min\{b^*, b^* + s\} > 0$$
$$t - 2\varrho + \frac{s - |s|}{2} > 0$$
$$\text{sign}(t)|t| > \text{sign}(t)(\frac{|s|}{2} + 2\varrho) - \frac{s}{2} > 0.$$

Note that $t > 0$ and $\text{sign}(x) = \text{sign}(t)$. If $\text{sign}(b^*) = \text{sign}(b^* + s) = -1$, then

$$\max\{b^*, b^* + s\} < 0$$
$$t + 2\varrho + \frac{s + |s|}{2} > 0$$
$$\text{sign}(t)|t| < \text{sign}(t)(\frac{|s|}{2} + 2\varrho) - \frac{s}{2} < 0.$$

Note that $t < 0$ and $\text{sign}(x) = \text{sign}(t)$. Rewriting the two cases above, we have that

$$b^* = t - 2\varrho\text{sign}(t) \text{ if } |t| > 2\varrho + C(s,t). \tag{A11}$$

Summarizing (A9)–(A11),

$$b^* = \begin{cases} t, & |t| < C(s,t), \\ -C(s,t), & C(s,t) \leq |t| \leq C(s,t) + 2\varrho, \\ \text{sign}(t)(|t| - 2\varrho), & |t| > C(s,t) + 2\varrho, \end{cases}$$

with $C(s,t) = \dfrac{1 - \text{sign}(s)\text{sign}(t)}{2}|s| \geq 0$. On one hand, when $s \neq 0$,

$$b^* = t - \mathcal{S}(t, C(s,t)) + \mathcal{S}\{\mathcal{S}(t, C(s,t)), 2\varrho\}.$$

On the other hand, when $s = 0$, it follows that $b^* = \mathcal{S}(t, 2\varrho)$ since $\mathcal{S}(z, 0) = z$. □

References

1. Haralick, R.M.; Shanmugam, K.; Dinstein, I.H. Textural features for image classification. *IEEE Trans. Syst. Man Cybern.* **1973**, *SMC-3*, 610–621.
2. Wang, X.; Zhang, H.H.; Wu, Y. Multiclass probability estimation with support vector machines. *J. Comput. Graph. Stat.* **2019**, *28*, 586–595.
3. Hansen, J.H.; Hasan, T. Speaker recognition by machines and humans: A tutorial review. *IEEE Signal Process. Mag.* **2015**, *32*, 74–99.
4. Duda, R.O.; Hart, P.E.; Stork, D.G. *Pattern Classification*; John Wiley & Sons: New York, NY, USA 2012.
5. Hastie, T.; Tibshirani, R.; Friedman, J. *The Elements of Statistical Learning: Data Mining, Inference, and Prediction*; Springer Science & Business Media: New York, NY, USA, 2009.
6. Cortes, C.; Vapnik, V. Support-vector networks. *Mach. Learn.* **1995**, *20*, 273–297.
7. Cristianini, N.; Shawe-Taylor, J. *An Introduction to Support Vector Machines and Other Kernel-Based Learning Methods*; Cambridge University Press: Cambridge, UK, 2000.
8. Marron, J.S.; Todd, M.J.; Ahn, J. Distance-weighted discrimination. *J. Am. Stat. Assoc.* **2007**, *102*, 1267–1271.
9. Qiao, X.; Zhang, H.H.; Liu, Y.; Todd, M.J.; Marron, J.S. Weighted distance weighted discrimination and its asymptotic properties. *J. Am. Stat. Assoc.* **2010**, *105*, 401–414.
10. Marron, J. Distance-weighted discrimination. *Wiley Interdiscip. Rev. Comput. Stat.* **2015**, *7*, 109–114.
11. Zhang, L.; Lin, X. Some considerations of classification for high dimension low-sample size data. *Stat. Methods Med. Res.* **2013**, *22*, 537–550.
12. Wang, B.; Zou, H. Another look at distance-weighted discrimination. *J. R. Stat. Soc. Ser. B (Stat. Methodol.)* **2018**, *80*, 177–198.
13. Liu, Y.; Zhang, H.H.; Wu, Y. Hard or soft classification? Large-margin unified machines. *J. Am. Stat. Assoc.* **2011**, *106*, 166–177.
14. Huang, H.; Liu, Y.; Du, Y.; Perou, C.M.; Hayes, D.N.; Todd, M.J.; Marron, J.S. Multiclass distance-weighted discrimination. *J. Comput. Graph. Stat.* **2013**, *22*, 953–969.
15. Wang, B.; Zou, H. A multicategory kernel distance weighted discrimination method for multiclass classification. *Technometrics* **2019**, *61*, 396–408.
16. Wang, B.; Zou, H. Sparse distance weighted discrimination. *J. Comput. Graph. Stat.* **2016**, *25*, 826–838.
17. Wang, L.; Shen, X. On L1-norm multiclass support vector machines: Methodology and theory. *J. Am. Stat. Assoc.* **2007**, *102*, 583–594.
18. Zhang, X.; Wu, Y.; Wang, L.; Li, R. Variable selection for support vector machines in moderately high dimensions. *J. R. Stat. Soc. Ser. B (Stat. Methodol.)* **2016**, *78*, 53–76.
19. Peng, B.; Wang, L.; Wu, Y. An error bound for L1-norm support vector machine coefficients in ultra-high dimension. *J. Mach. Learn. Res.* **2016**, *17*, 8279–8304.
20. Simon, N.; Friedman, J.; Hastie, T.; Tibshirani, R. A sparse-group lasso. *J. Comput. Graph. Stat.* **2013**, *22*, 231–245.
21. Friedman, J.; Hastie, T.; Tibshirani, R. A note on the group lasso and a sparse group lasso. *arXiv* **2010**, arXiv:1001.0736.
22. Cai, T.T.; Zhang, A.; Zhou, Y. Sparse group lasso: Optimal sample complexity, convergence rate, and statistical inference. *arXiv* **2019**, arXiv:1909.09851.
23. Yu, D.; Zhang, L.; Mizera, I.; Jiang, B.; Kong, L. Sparse wavelet estimation in quantile regression with multiple functional predictors. *Comput. Stat. Data Anal.* **2019**, *136*, 12–29.
24. He, Q.; Kong, L.; Wang, Y.; Wang, S.; Chan, T.A.; Holland, E. Regularized quantile regression under heterogeneous sparsity with application to quantitative genetic traits. *Comput. Stat. Data Anal.* **2016**, *95*, 222–239.

25. Huang, H. Large dimensional analysis of general margin based classification methods. *arXiv* **2019**, arXiv:1901.08057.
26. Huang, H.; Yang, Q. Large scale analysis of generalization error in learning using margin based classification methods. *arXiv* **2020**, arXiv:2007.10112.
27. Lam, X.Y.; Marron, J.; Sun, D.; Toh, K.C. Fast algorithms for large-scale generalized distance weighted discrimination. *J. Comput. Graph. Stat.* **2018**, *27*, 368–379.
28. Sun, D.; Toh, K.C.; Yang, L. A convergent 3-block semiproximal alternating direction method of multipliers for conic programming with 4-type constraints. *SIAM J. Optim.* **2015**, *25*, 882–915.
29. Parikh, N.; Boyd, S. Proximal algorithms. *Found. Trends Optim.* **2014**, *1*, 127–239.
30. Asahchop, E.L.; Branton, W.G.; Krishnan, A.; Chen, P.A.; Yang, D.; Kong, L.; Zochodne, D.W.; Brew, B.J.; Gill, M.J.; Power, C. HIV-associated sensory polyneuropathy and neuronal injury are associated with miRNA–455-3p induction. *JCI Insight* **2018**, *3*, e122450.
31. Hsu, D.; Kakade, S.; Zhang, T. A tail inequality for quadratic forms of subgaussian random vectors. *Electron. Commun. Probab.* **2012**, *17*, 52.

Publisher's Note: MDPI stays neutral with regard to jurisdictional claims in published maps and institutional affiliations.

© 2020 by the authors. Licensee MDPI, Basel, Switzerland. This article is an open access article distributed under the terms and conditions of the Creative Commons Attribution (CC BY) license (http://creativecommons.org/licenses/by/4.0/).

Article

Ensemble Linear Subspace Analysis of High-Dimensional Data

S. Ejaz Ahmed [1], Saeid Amiri [2,*] and Kjell Doksum [3]

1 Department of Mathematics and Statistics, Brock University, St. Catharines, ON L2S 3A1, Canada; sahmed5@brocku.ca
2 Department of Civil, Geologic and Mining Engineering Polytechnique Montreál, Montreál, QC H3T 1J4, Canada
3 Department of Statistics, University of Wisconsin, Madison, WI 53706, USA; doksum@cs.wisc.edu
* Correspondence: saeid.amiri1@gmail.com

Citation: Ahmed, S.E.; Amiri, S.; Doksum, K. Ensemble Linear Subspace Analysis of High-Dimensional Data. *Entropy* **2021**, *23*, 324. https://doi.org/10.3390/e23030324

Academic Editor: Alessandro Giuliani

Received: 14 January 2021
Accepted: 5 March 2021
Published: 9 March 2021

Publisher's Note: MDPI stays neutral with regard to jurisdictional claims in published maps and institutional affiliations.

Copyright: © 2021 by the authors. Licensee MDPI, Basel, Switzerland. This article is an open access article distributed under the terms and conditions of the Creative Commons Attribution (CC BY) license (https://creativecommons.org/licenses/by/4.0/).

Abstract: Regression models provide prediction frameworks for multivariate mutual information analysis that uses information concepts when choosing covariates (also called features) that are important for analysis and prediction. We consider a high dimensional regression framework where the number of covariates (p) exceed the sample size (n). Recent work in high dimensional regression analysis has embraced an ensemble subspace approach that consists of selecting random subsets of covariates with fewer than p covariates, doing statistical analysis on each subset, and then merging the results from the subsets. We examine conditions under which penalty methods such as Lasso perform better when used in the ensemble approach by computing mean squared prediction errors for simulations and a real data example. Linear models with both random and fixed designs are considered. We examine two versions of penalty methods: one where the tuning parameter is selected by cross-validation; and one where the final predictor is a trimmed average of individual predictors corresponding to the members of a set of fixed tuning parameters. We find that the ensemble approach improves on penalty methods for several important real data and model scenarios. The improvement occurs when covariates are strongly associated with the response, when the complexity of the model is high. In such cases, the trimmed average version of ensemble Lasso is often the best predictor.

Keywords: ensembling; high-dimensional data; Lasso; elastic net; penalty methods; prediction; random subspaces

1. Introduction

Recent research in statistical science has focused on developing effective and useful techniques for analyzing high-dimensional data where the number of variables substantially exceeds the number of cases or subjects. Examples of such data sets are genome or gene expression arrays, and other biomarkers based on RNA and proteins. The challenge is to find associations between such markers (X's) and phenotype (Y).

Regression models provide useful frameworks for multivariate mutual information analysis that uses information concepts when choosing covariates (also called features) that are important for the analysis and prediction. A recent article that includes both the concept of mutual information and the Lasso is [1]. This paper develops properties of methods that use the information in a vector X to reduce prediction error, that is, to reduce entropy. We consider regression experiments, that is, experiments with a response variable $Y \in \mathbb{R}$ and a covariate vector $(X_1, \ldots, X_p)^t$. The objective is to use a sample of i.i.d. vectors $(\mathbf{x}_i, y_i), 1 \leq i \leq n$, where $\mathbf{x}_i = (x_{i1}, \ldots, x_{ip})^t$ with $x_{ij} \in \mathbb{R}$, to construct a predictor $\widehat{Y}_0$ of a response Y_0 corresponding to a covariate vector $\mathbf{x}_0 = (x_{01}, \ldots, x_{0p})^t$ that is not part of the sample. Let $\mathbf{X} = (x_{ij})_{n \times p}$ be the design matrix of explanatory variables (covariates) and

$\mathbf{y} = (y_1, \ldots, y_n)^t$ be the vector of response variables. Denote $\mathbf{X}[, j]$ as the jth column vector of the design matrix. We will use the linear model

$$\mathbf{y} = \mathbf{X}\boldsymbol{\beta} + \boldsymbol{\epsilon}, \tag{1}$$

where $\boldsymbol{\beta} = (\beta_1, \ldots, \beta_p)^t$ is the vector of regression coefficients and $\boldsymbol{\epsilon} = (\epsilon_1, \ldots, \epsilon_n)^t \sim N(0, \sigma^2 I)$ is the residual error term. In this model, predictors $\widehat{Y}_0$ take the form

$$\widehat{Y}_0 = \sum_{j=1}^{p} \widehat{\beta}_j x_{0,j},$$

where $\widehat{\beta}_j$ is an estimator based on the i.i.d. sample $(\mathbf{x}_i, y_i), 1 \leq i \leq n$.

Under $n \geq p$, the ordinary least square (OLS) estimator of $\boldsymbol{\beta}$ can be used. When $n < p$ a unique OLS estimate does not exist. However, for sparse models where most of the β's are zero, we can use the Lasso [2] criteria that forces many of the estimated β's to be set to zero. For a given penalty level $\lambda \geq 0$, the Lasso estimate of $\boldsymbol{\beta}$ is

$$\widehat{\boldsymbol{\beta}} = \mathrm{argmin}_{\beta} \{ \frac{1}{2} \|\mathbf{y} - \mathbf{x}\boldsymbol{\beta}\|_2^2 + \lambda \|\boldsymbol{\beta}\|_1 \},$$

where $\|.\|_2$ is the Euclidean distance and $\|\boldsymbol{\beta}\|_1 = \sum |\beta_j|$ is the ℓ_1-norm. The Lasso not only sets a subset of β's to zero, it also shrinks OLS estimates of the remaining β's towards zero. It is an effective procedure for experiments when one can assume that the number r of covariates that are relevant for the response in the sense that their β coefficient is not zero, satisfies $r \leq n$. That is, for sparse models.

Other effective high-dimension methods that we consider are adaptive Lasso, ref. [3], smoothly clipped absolute deviation (SCAD), ref. [4], least angle regression (LARS), ref. [5], and elastic net, ref. [6]. The properties of Lasso, and its variants, are well studied to examine consistency of parameter estimates [7,8], and to assess the prediction error and the variable selection process [9,10] examined properties of the Lasso in partially linear models. Several variants of Lasso were introduced by [11] and more recently by [12]. See [13–15] for many of the extensions of the original Lasso.

In this paper, we examine properties of statistical methods based on Ensemble Linear Subspace Analysis (ELSA) for analyzing high-dimensional data. ELSA is based on repeated random selection of subsets of covariates, doing statistical inference on each of the subsets, and then combing the results from subsets to construct a final inference. One advantages of this ensemble subspace approach is that it makes the analysis of studies with a million or more covariates variables more manageable. Another advantage is that for many situations the ensemble approach is more efficient because it takes advantage of the high efficiency of statistical methods for the case where the number of covarites is less than or equal to the sample size.

Classical examples using sub-models whose results are pooled and aggregated into a final statistical analysis is the bagging method ([16]) and the random forests approach ([17]). Recent studies that use ensemble ideas include [18,19]. These papers focus on feature selection, that is, selecting the covariates that are associated with the response variable. This paper deals with using the selected covariates to construct efficient predictors of the response. We examine conditions under which penalty methods such as Lasso perform better when used in the ensemble approach by computing mean squared prediction errors for simulations and a real data example. Linear models with both random and fixed designs are considered. We examine two versions of penalty methods: one where the tuning parameter is selected by cross-validation; and one where the final predictor is a trimmed average of individual predictors corresponding to the members of a set of fixed tuning parameters. We find that the ensemble approach improves on penalty methods for several important real data and model scenarios. The improvement occurs when covariates are strongly associated with the response, when the complexity of the model (represented by r/p is high. In such cases, the trimmed average version of ensemble Lasso is often the best predictor.

The rest of this article is organized as follows. In Sections 2 and 3, we introduce six different approaches to subspace selection. Section 3 describes a new approach for dealing with tuning parameters λ. Instead of using the standard Lasso based on a $\widehat{\lambda}$ obtained by cross validation, it computes Lasso predictors for a fixed set of tuning parameters and uses the average of these predictors as the fixed predictors. Section 4 outlines other penalty-based ensemble methods for high dimensional data. Section 5 introduces the concepts of mean squared Prediction Error (MSPE) and efficiency (EFF) for fixed and random design experiments as well as for real data. Section 6 gives efficiency of various penalty methods with respect to CV Lasso, including efficiencies of ensemble subspace version of these penalty methods. The efficiency results show that when the model complexity r/p is moderately high, trimmed subspace method perform best in all but one case. Section 7 compares six ensemble subspace Lasso methods to the standard CV Lasso. For models with a mixture of strong and weak signals, the ensemble methods perform best except when the models are very sparse. The final section gives a summary of results.

2. Ensembling via Random Subspaces

The following three-step protocol provides the ensemble subspace approach:
- Divide the initial dataset $(\mathbf{X}, \mathbf{y})$, $\mathbf{X} = (x_{ij})_{n \times p}, \mathbf{y} \in R^n$ randomly into smaller sub-datasets by selecting at random subsets covariates. The sample size n remains the same.
- Construct predictors of the future response Y_0 within each sub dataset.
- Combine the results obtained from each sub dataset into a final analysis.

We consider three approaches to choosing subsets of $\mathbf{X}$-variables

1. Choose subspaces with p^* covariates, where p^* is the number of distinct covariates after randomly selecting p covariates with replacement from the collection of all covariates. Here the random variable p^* is known to have expected value approximately $0.63p$. Let $\mathbf{x}^*$ denote the distinct covariates and $\mathbf{X}^*$ denote the corresponding design matrix. The subspace data is $(\mathbf{X}^*, \mathbf{y})$ where $\mathbf{y} \in R^n$ and $\mathbf{X}^* = (x_{ij}^*)_{n \times p^*}$. By repeating this procedure B times independently and using a method such as Lasso we get predictors $\{\widehat{Y}_{0,1}, \ldots, \widehat{Y}_{0,B}\}$.

2. Choose n covariates without replacement from the p covariates, repeating B times independently and using a method such as Lasso thereby obtaining $\{\widehat{Y}_{0,1}, \ldots, \widehat{Y}_{0,B}\}$.

3. Same as 2., except choose $n/2$ covariates.

The final prediction of the response based on a covariate vector $\mathbf{x}_0$ is $\widehat{Y}_0(\mathbf{x}_0) = B^{-1} \sum_{b=1}^{B} \widehat{Y}_{0b}(\mathbf{x}_0)$. Note that the terms in the sum that defines $\widehat{Y}_{0b}(\mathbf{x}_0)$ are identically distributed, but not independent. Thus, with $\widehat{Y}_0 = \widehat{Y}_0(\mathbf{x}_0)$ and $\widehat{Y}_{0b} = \widehat{Y}_{0b}(\mathbf{x}_0)$

$$Var(\widehat{Y}_{0b}) = \frac{1}{B} Var(\widehat{Y}_{01}) + \frac{B-1}{B} Cov(\widehat{Y}_{01}, \widehat{Y}_{02}) = \rho \sigma^2 + \frac{1-\rho}{B} \sigma^2, \qquad (2)$$

where σ^2 is the variance of one predictor $\widehat{Y}_0$ and ρ is the pairwise correlation between two such predictors. By selecting B large, we can make the second term negligible. When ρ is sufficiently small $\rho \sigma^2$ can in many cases be smaller than the variance of the predictor based on all the covariates. When $\widehat{Y}_0$ is prediction unbiased, that is, $E(\widehat{Y}_0 - Y) = 0$, then $Var(\widehat{Y}_0)$ equals the prediction mean squared error (PMSE). When the subspace have n or fewer variables, OLS is prediction unbiased.

3. Prediction on Subspaces

We consider two approaches for dealing with Lasso tuning parameters: the cross-validated and the Trimmed Lasso. The same approaches will be applied to the other penalty methods. Let $\mathbf{X}^* = \{x^*_{ij}\}$ be the subspace design matrix. The Lasso estimate based on a linear model on the subspace is

$$\widehat{\beta} = \mathrm{argmin}_\beta \{\frac{1}{2}\|\mathbf{y} - \mathbf{X}^*\beta\|_2^2 + \lambda\|\beta\|_1\},$$

The standard procedure is to choose the tuning parameter λ using 10-fold cross-validation (CV), which denoted as CVLasso hereafter. Note, since the size of subspace design $\mathbf{X}^* = \{x^*_{ij}\}$ is changed, $\widehat{\beta}$ is changed as well and correspond the number variables in $\mathbf{X}^* = \{x^*_{ij}\}$. It is implemented in the library "glmnet" in R. Cross validation may sometimes lead to unfortunate choices of λ because the random choices of training and test sample may not yield a λ that represents a λ that will give a good predictor. Thus we will consider a method based on a collection of fixed λ's. This method, which we call the *Trimmed Lasso (TrLasso)*, uses as predictor the trimmed average (10% in each tails) of Lasso predictors computed from a path of 100 λ's. The path is generated using the library glmnet in R with option "nlambda". The largest lambda, λ_{MAX}, is the smallest value for which all beta coefficients are zero while $\lambda_{MIN} = \lambda_{MAX}e^{-6}$. The λ values are equally spaced on the log scale. We consider six versions of ensemble subspace methods. In the following, "approach j" for $j = 1, 2$ and 3 chooses subspace sizes p^*, n, and $n/2$, respectively.

ETrLasso (j): For $j = 1, 2$ and 3 use approach (j) to choose the number of variables in each subspace. Then apply TrLasso in each subspace.

ECVTLasso (j): For $j = 1, 2$ and 3 use approach (j) to choose the number of variables in each subspace. Then apply CVLasso in each subspace.

4. Competitors to Lasso

4.1. Elastic-Net

For highly correlated predictor variables the Lasso tends to select a few of them and shrink the rest to zero, see [6,15] for an extensive discussion. For such cases the Elastic Net, denoted ELNET hereafter, is suggested as a compromise between the ridge and the Lasso methods. The estimates of coefficients can be obtained from:

$$\widehat{\beta} = \mathrm{argmin}_\beta \left\{\frac{1}{2}\|Y - \mathbf{X}\beta\|_2^2 + \lambda\left(\frac{1}{2}(1-\alpha)\|\beta\|_2^2 + \alpha\|\beta\|_1\right)\right\}, \quad (3)$$

where $\alpha \in [0, 1]$. Here $\alpha = 1$ leads to the regular Lasso. The penalty parameters, λ and α, are two nonnegative tuning parameters.

We examine properties of ELNET using of $\alpha = 0.25, 0.5$, and 0.75, while λ is treated as for the Lasso. Thus we obtain TrELNET(α) and CVELNET(α). For ELNET the ensemble subspace method is also carried out as for the Lasso but only using the trimmed (10%) option, resulting in three methods for each α. We use the notation TrELNET(j, α) and ELNET(j, α), $j = 1, 2, 3$ for the trimmed and CV ensemble subspace option for subspace of size p^*, n, and $n/2$. The calculations of these ELNETs, including the Lasso where $\alpha = 1$, are done using the library glmnet in R.

4.2. Adaptive Lasso

Ref. [3] introduced the adaptive Lasso for linear regression. It uses a weighted penalty of the form $\sum_{j=1}^{p} w_j|\widehat{\beta}_j|$ where $w_j = 1/|\widehat{\beta}_j|$ and $\widehat{\beta}_j$ is a preliminary estimate of β_j and

$$\widehat{\beta} = \mathrm{argmin}_\beta \left\{\frac{1}{2}\|Y - \mathbf{X}\beta\|_2^2 + \lambda\|w\beta\|_1\right\}. \quad (4)$$

The preliminary beta estimate is typically the Ridge estimate. We use that in our simulation studies. The Adaptive Lasso is also computed as a 10% trimmed average of Lasso predictors for a sequenced of λ's and as the predictor obtained when λ is selected using CV. They are denoted as TrALasso and CVALasso, respectively. We consider these methods for the proposed ensembled subspace procedures and denote them as ETrAlasso(j) and ECVAlasso(j), $j = 1, 2, 3$.

4.3. Lars

Least angle regression, also called LARS, was developed in [5]. It uses a model selection algorithms based on forward selection that enables the procedure to select a parsimonious set of predictors to be used for the efficient prediction of a response variable from an available large collection of possible covariates. It improves computational efficiency compared to the Lasso. As in Section 3, LARS is considered with trimming and with CV in prediction. They are denoted as TrLARS and CVLARS, respectively. We consider the trimmed and CV versions of these methods for the proposed ensembled subspace procedure and denoted them as ETrLARS(j) and ECVLARS(j), $j = 1, 2, 3$. The calculation of LARS is done by using the library LAR in R.

4.4. Scad

Ref. [4] introduced the SCAD penalty for linear regression. It is a symmetric and quadratic spline on the reals whose first order derivative is

$$SCAD'_{\lambda,a}(x) = \lambda \left\{ I(|x| \leq \lambda) + \frac{(a\lambda - |x|)_+}{(a-1)\lambda} I(|x| > \lambda) \right\}, \tag{5}$$

where $\lambda > 0$ and $a = 3.7$ as recommended by [4]. The SCAD penalty is continuously differentiable and can produce sparse solutions and nearly unbiased estimates for sparce models with large beta coefficients. The CV and trimmed version of SCAD will be labeled as CVSCAD and TrSCAD, while the ensemble subspace methods will be ECVSCAD(j) and ETrSCAD(j), $j = 1, 2, 3$.

5. Mean Squared Prediction Error (MSPE)

5.1. (a) Random Covariates, Simulated Data

To examine prediction error, we generate a training set $\mathcal{D} = \{(\mathbf{x}_1, y_1), \ldots, (\mathbf{x}_n, y_n)\}$ using the simulation model under consideration, and for each method considered obtain a predictor of the form $\hat{y}_i = \sum_{j=1}^{p} \hat{\beta}_j x_{ij}$, $i = 1, \ldots, n$. To explore the performance of proposed methods on data not used in producing the prediction formula, we independently generate a test set $\mathcal{D}_0 = \{(\mathbf{x}_{01}, y_{01}), \ldots, (\mathbf{x}_{0n_0}, y_{0n_0})\}$ and compute

$$\text{MSPE} = \frac{1}{n_0} \sum_{i=1}^{n_0} (y_{0i} - \hat{y}_{0i})^2,$$

where

$$\hat{y}_{0i} = \sum_{j=1}^{p} \hat{\beta}_j x_{0ij}, \quad i = 1, \ldots, n_0,$$

is the predicted value of y_{0i} based on $\mathbf{x}_{0i}$. We use $n_0 = 0.3n$ in the simulation studies. We repeat the process of generating independent collections for training and test sets $M = 2000$ times, therby obtaining $\text{MSPE}_1, \ldots, \text{MSPE}_M$. We measure the efficiency of a predictor $\hat{Y}$ by comparing it to the standard method, Lasso with cross-validation

$$EFF(\hat{Y}) = \frac{1}{M} \sum_b \frac{MSPE_b(\text{CVLasso})}{MSPE_b(\hat{Y})}, \tag{6}$$

where the sum is over the simulation, and as mentioned earlier for the Lass the standard procedure is to choose the tuning parameter λ using 10-fold cross-validation (CV).

5.2. (b) Fixed Covariate, Simulated and Real Data

Let $\mathcal{D} = \{(\mathbf{x}_1, y_1), \ldots, (\mathbf{x}_n, y_n)\}$, $\mathbf{x} \in R^p$ and $y \in R$, denote a real or simulated data set with random y's and fixed $\mathbf{x}$'s. Split this set into a test set $\mathcal{D}_0$ with n_0 data vectors and a training set $\mathcal{D}_1$ with the remaining n_1 data vectors, where $n_0 = 0.3n$ and $n_1 = 0.7n$. For each of the discussed methods, the training set is used to produce a prediction algorithm that is used to predict the y's in the test set. The MSPE is then MSPE $= \frac{1}{n_0} \sum_{i=1}^{n_0} (\widehat{y}_{0i} - y_{0i})^2$, where $\widehat{y}_{0i}$ is the predicted value of y_{0i} based on $\mathbf{x}$'s in the test set. Next we compute the ratio with respect to CVLasso(MSPE). This procedure is repeated 2000 times and the average is the final EFF($\widehat{Y}$). For simulated experiments, an additional $M = 2000$ repetitions is carried out.

6. Efficiency Result for Lasso Competitors

In the following, we compare the accuracy of the methods presented in Sections 3 and 4. The results are presented with $B = 250$ subspaces; we also tried $B = 500$, but since the result were nearly the same, they are not presented here. We examine the relative performance of the methods as a function of the complexity index which is defined as the ratio r/p of the number of covariates that are relevant for the response y to the total number of covariates.

6.1. Syndrome Gene Data

Ref. [20] studied expression quantitative trait locus mapping in the laboratory rat to gain a broad perspective of gene regulation in the mammalian eye and to identify genetic variation relevant to human eye disease. The dataset which is from the `flare` library in R has $n = 120$ with $p = 200$ predictors, it includes the expression level of TRIM32 gene which can be considered as dependent variable. To compare the accuracy of the proposed methods on this dataset, we randomly select 30% of the data as a test set and consider the rest as a training set, and calculate the relative efficiency EFF($\widehat{Y}$) to CVLasso. We repeat the procedure of selecting training and test set 2000 times which provide good accuracy. The results are reported in Table 1.

Among the seven Lasso Type competitor to CVLasso, the most efficient in terms of EFF($\widehat{Y}$) is the one based on subspaces of sizes $n/2 = 60$ and based on a trimmed average of Lasso predictors computed for a sequence of λ tuning parameters. We found that it improves on CVLasso 83% of the time. However, the average of the mean square prediction error ratios is EFF($\widehat{Y}$) = 1.11, thus the improvement does not appear to be substantial.

Turning to the other procedures in Table 1, we see that, generally, the best performance is obtained for the trimmed ensemble versions based on subspaces of size $n/2$, expect for adaptive Lasso which is best for subspace size n. Generally, the improvement ensemble over CvLasso is about 1.1 in terms of EFF($\widehat{Y}$). Moreover, the performance of these methods are very close, including ELNET methods with different α. That is, using subspaces and a robust trimmed average of response predictors obtained from the path of glment lambdas is more efficient than using the predictor based on the lambda selected by glment cross validation. The improvement achieved by the trimmed ensemble versions of SCAD based on subspaces of size $n/2$ over the basic (CV and trimmed) versions of SCAD is striking.

Table 1. Efficiencies with respect to CVLasso for the Syndrome Gene data.

		Method		
CVLasso -	TrLasso 1.048(0.002)	ETrLasso(1) 1.059(0.002)	ETrLasso(2) 1.079(0.002)	ETrLasso(3) 1.102(0.002)
		ECVLasso(1) 1.056(0.001)	ECVLasso(2) 1.067(0.002)	ECVLasso(3) 1.059(0.002)
CVELNET(0.25) 1.028(0.001)	TrELNET(0.25) 1.057(0.002)	ETrELNET(1,0.25) 1.056(0.002)	ETrELNET(2,0.25) 1.092(0.002)	ETrELNET(3,0.25) 1.103(0.002)
	1.068(0.001)	ECVELNET(1,0.25) 1.071(0.002)	ECVELNET(2,0.25) 1.059(0.002)	ECVELNET(3,0.25)
CVELNET(0.50) 1.014(0.000)	TrELNET(0.50) 1.053(0.002)	ETrELNET(1,0.50) 1.059(0.002)	ETrELNET(2,0.50) 1.084(0.002)	ETrELNET(3,0.50) 1.103(0.002)
		ECVELNET(1,0.50) 1.062(0.001)	ECVELNET(2,0.50) 1.069(0.002)	ECVELNET(3,0.50) 1.060(0.002)
CVELNET(0.75) 1.006(0.000)	TrELNET(0.75) 1.049(0.002)	ETrELNET(1,0.75) 1.059(0.002)	ETrELNET(2,0.75) 1.081(0.002)	ETrELNET(3,0.75) 1.103(0.002)
		ECVELNET(1,0.75) 1.059(0.001)	ECVELNET(2,0.75) 1.067(0.002)	ECVELNET(3,0.75) 1.059(0.002)
CVLARS 0.963(0.002)	TrLARS 0.990(0.002)	ETrLARS(1) 1.076(0.002)	ETrLARS(2) 1.100(0.002)	ETrLARS(3) 1.083(0.002)
		ECVLARS(1) 1.067(0.001)	ECVLARS(2) 1.046(0.003)	ECVLARS(3) 0.775(0.005)
CVALasso 0.899(0.002)	TrAlasso 0.958(0.002)	ETrAlasso(1) 1.004(0.003)	ETrAlasso(2) 1.110(0.002)	ETrAlasso(3) 1.100(0.002)
		ECVALasso(1) 1.070(0.002)	ECVALasso(2) 1.086(0.002)	ECVALasso(3) 1.075(0.002)
CVSCAD 0.837(0.003)	TrSCAD 0.891(0.003)	ETrSCAD(1) 0.954(0.003)	ETrSCAD(2) 0.969(0.003)	ETrSCAD(3) 1.099(0.002)
		ECVSCAD(1) 0.986(0.001)	ECVSCAD(2) 1.014(0.002)	ECVSCAD(3) 1.033(0.002)

6.2. Simulation Efficiency Results

We next used a modification of a model set forth by [21]. We set $p = 1000$, and in contrast to the syndrome Gene inspired model, we now use i.i.d. random x's, as indicated in Model (7). The model provides a large range of β values corresponding to strong, moderate and weak covariate signals. The correlations between covariates renage from 0.28 and 0.94.

$$X \sim N(M, \Sigma), \tag{7}$$

$$M = (\mu_i)_{i=1,\ldots,p}, \quad \mu_i \stackrel{i.i.d}{\sim} N(5,2),$$

$$\Sigma = (\sigma_{i,j})_{i,j=1,\ldots,p}, \quad \sigma_{i,j} = \sigma_{j,i} \stackrel{i.i.d}{\sim} Unif(0.4, 0.6), i \neq j$$

$$\sigma_{i,i} \sim Unif(0.8, 1.2),$$

$$\beta_{j_0+1},\ldots,\beta_{j_0+r} \stackrel{i.i.d}{\sim} Unif(-2,2), \quad j_0 \in \{1,\ldots,p-r\},$$

$$\beta_j = 0, \text{ for all other } j,$$

$$y_i = \sum_{j=1}^{p} \beta_j x_{ij} + \epsilon_i, \text{ with } \epsilon_i \stackrel{i.i.d}{\sim} N(0, 0.15), i = 1,\ldots,n.$$

Using this model, we generate $(\mathbf{x}_1, y_1), \ldots, (\mathbf{x}_n, y_n)$, $n = 180$. Tables 2–5 give the mean of the efficiency criteria over $M = 2000 trials$. The numbers in parentheses are standard deviations (SD). We next discuss the result for the case with $r = 150$ relevant variables. Here k denotes the number of covariates in the subspaces, and p^* is the number of distinct variables in a bootstrap sample from the set of covariates.

Table 2. Efficiencies of trimmed mean methods with respect to the CVLasso for the model (7) with complexity index $r/p = 0.15$.

	Method		
TrLasso 1.021(0.002)	ETrLasso(1) 1.015(0.003)	ETrLasso(2) 0.841(0.004)	ETrLasso(3) 0.759(0.004)
TrELNET(0.25) 1.023(0.003)	ETrELNET(1,0.25) 0.978(0.004)	ETrELNET(2,0.25) 0.835(0.004)	ETrELNET(3,0.25) 0.754(0.004)
ETrELNET(0.50) 1.026(0.002)	ETrELNET(1,0.50) 1.001(0.003)	ETrELNET(2,0.50) 0.841(0.004)	ETrELNET(3,0.50) 0.756(0.004)
TrELNET(0.75) 1.023(0.002)	ETrELNET(1,0.75) 1.009(0.003)	ETrELNET(2,0.75) 0.841(0.004)	ETrELNET(3,0.75) 0.756(0.004)
TrLARS 0.998(0.002)	ETrLARS(1) 1.049(0.003)	ETrLARS(2) 0.880(0.004)	ETrLARS(3) 0.733(0.004)
TrAlasso 0.995(0.003)	ETrAlasso(1) 0.971(0.003)	ETrAlasso(2) 0.823(0.004)	ETrAlasso(3) 0.763(0.004)
TrSCAD 0.844(0.005)	ETrSCAD(1) 1.017(0.003)	ETrSCAD(2) 0.826(0.004)	ETrSCAD(3) 0.771(0.004)

Table 3. Efficiencies of cross validated methods with respect to the CVLasso for the model (7) with complexity index $r/p = 0.15$.

	Method		
CVLasso -	ECVLasso(1) 0.974(0.003)	ECVLasso(2) 0.727(0.004)	ECVLasso(3) 0.671(0.004)
CVELNET(0.25) 1.033(0.002)	ECVELNET(1,0.25) 0.971(0.003)	ECVELNET(2,0.25) 0.722(0.004)	ECVELNET(3,0.25) 0.668(0.004)
CVELNET(0.50) 1.016(0.001)	ECVELNET(1,0.50) 0.977(0.003)	ECVELNET(2,0.50) 0.725(0.004)	ECVELNET(3,0.50) 0.670(0.004)
CVELNET(0.75) 1.006(0.000)	ECVELNET(1,0.75) 0.976(0.003)	ECVELNET(2,0.75) 0.726(0.004)	ECVELNET(3,0.75) 0.671(0.004)
CVLARS 0.953(0.003)	ECVLARS(1) 1.040(0.003)	ECVLARS(2) 0.822(0.004)	ECVLARS(3) 0.680(0.004)
CVALasso 1.015(0.003)	ECVAlasso(1) 1.073(0.004)	ECVAlasso(2) 0.711(0.004)	ECVAlasso(3) 0.732(0.004)
CVSCAD 0.816(0.004)	ECVSCAD(1) 0.875(0.004)	ECVSCAD(2) 0.733(0.004)	ECVSCAD(3) 0.682(0.004)

6.2.1. Results for $r/p = 0.15$

(a) Lasso Based Methods

Trimmed Lasso based on all $p = 1000$ covariates performs best, with ensemble trimmed Lasso with $k = p^*$, a close second. Ensemble CVLasso performs poorly for all k. The trimming approach dominates the cross validation approach.

(b) ELNET Based Methods

CV and trimmed ELNET based on all $p = 1000$ covariates are close and better than the ensemble methods and CVLasso. The value α in ELNET does not make much difference. Among ensemble methods, the trimmed version with $k = p^*$ and $\alpha = 0.75$ is the best, it is slightly better than CVLasso.

(c) LARS Based Methods

The trimmed and CV ensemble subspace methods with $k = p^*$ are best with the trimmed version slightly better. Both are better than CV Lasso.

(d) Adaptive Lasso Based Methods

CV ensemble adaptive Lasso based on subspaces with $k = p^*$ is best among all methods.

(e) SCAD Based Methods

For this model, SCAD does poorly for all but one version, presumably because it produces poor predictors for β's that are close to zero. The one version that does well is the trimmed ensemble method with $k = p^*$ variables.

6.2.2. Results for $r/p = 0.30$

(a) Lasso Based Methods

Trimmed ensemble Lasso based on p^* covariates in the subspaces performs best. The trimming approach outperforms the CV approach for each of k.

(b) ELNET Based Methods

Trimmed ensemble ELNET based on p^* covariates performs best. The trimming approach outperforms the CV approach for each k. The value of α does not make much difference.

(c) LARS Based Methods

Trimmed ensemble LARS based on p^* covariates is best among all LARS methods. Trimmed methods outperform CV methods.

(d) Adaptive Lasso Based Methods

CV Adaptive ensemble Lasso based on subspaces with p^* covariates is best among all methods. Trimmed methods outperform CV methods except when $k = p^*$.

(e) SCAD Based Methods

Trimmed ensemble SCAD with p^* covariates in the supspaces does well. Trimmed ensemble versions outperform CV version and the $k = 1000$ version.

Table 4. Efficiencies of trimmed methods with respect to the CVLasso for the model (7) with $r/p = 0.3$.

	Method		
TrLasso 1.056(0.002)	ETrLasso(1) 1.135(0.003)	ETrLasso(2) 1.092(0.005)	ETrLasso(3) 1.002(0.004)
TrELNET(0.25) 1.095(0.002)	ETrELNET(1,0.25) 1.130(0.003)	ETrELNET(2,0.25) 1.087(0.004)	ETrELNET(3,0.25) 0.997(0.004)
TrELNET(0.50) 1.073(0.002)	ETrELNET(1,0.50) 1.133(0.003)	ETrELNET(2,0.50) 1.092(0.005)	ETrELNET(3,0.50) 1.000(0.004)
TrELNET(0.75) 1.062(0.002)	ETrELNET(1,0.75) 1.133(0.003)	ETrELNET(2,0.75) 1.096(0.005)	ETrELNET(3,0.75) 1.003(0.004)
TrLARS 1.037(0.002)	ETrLARS(1) 1.146(0.003)	ETrLARS(2) 1.121(0.005)	ETrLARS(3) 0.957(0.004)
TrAlasso 1.055(0.002)	ETrAlasso(1) 1.104(0.003)	ETrAlasso(2) 1.072(0.004)	ETrAlasso(3) 1.006(0.004)
TrSCAD 0.836(0.004)	ETrSCAD(1) 1.098(0.003)	ETrSCAD(2) 1.054(0.004)	ETrSCAD(3) 1.021(0.004)

Table 5. Efficiencies of cross validated methods with respect to the CVLasso for the model (7) with $r/p = 0.3$.

	Method		
-	ECVLasso(1) 1.050(0.002)	ECVLasso(2) 0.914(0.004)	ECVLasso(3) 0.873(0.004)
CVELNET(0.25) 1.060(0.002)	ECVELNET(1,0.25) 1.082(0.003)	ECVELNET(2,0.25) 0.920(0.004)	ECVELNET(3,0.25) 0.875(0.004)
CVELNET(0.50) 1.024(0.001)	ECVELNET(1,0.50) 1.063(0.002)	ECVELNET(2,0.50) 0.915(0.004)	ECVELNET(3,0.50) 0.874(0.004)
ECVELNET(0.75) 1.008(0.000)	ECVELNET(1,0.75) 1.055(0.002)	ECVELNET(2,0.75) 0.915(0.004)	ECVELNET(3,0.75) 0.873(0.004)
CVLARS 0.964(0.003)	ECVLARS(1) 1.106(0.002)	ECVLARS(2) 1.029(0.004)	ECVLARS(3) 0.883(0.004)
CVALasso 1.004(0.003)	ECVAlasso(1) 1.178(0.004)	ECVAlasso(2) 0.913(0.004)	ECVAlasso(3) 0.948(0.004)
CVSCAD 0.888(0.003)	ECVSCAD(1) 0.936(0.003)	ECVSCAD(2) 0.899(0.004)	ECVSCAD(3) 0.874(0.004)

6.2.3. Overall Summary

Tables 2–5 show that the ensemble and trimming methods can improve on the CV Lasso. Overall, the CV esnsemble Adaptive Lasso based on subspaces with p^* covariates performs best. For $r/p = 0.30$, that is, 30% complexity, ensemble subsace with p^* covariates does best overall and the trimmed approach is best except for the Adaptive Lasso. When $r/p = 0.15$, the results are less clear, except the ensemble subspaces with p^* covariates yields the overall best result when coupled with the Adaptive Lasso. The overall superior performance of ensemble subspace methods based on p^* can in part be explained by formula (2) because the p^* methods produce predictors that are weakly correlated.

7. Comparison of Cv and Trimmed Lasso Methods

7.1. Syndrome Gene Data Inspired Simulation Model

Simulation based on real data is very important from an application perspective, because the structure of the underlying population is often unknown. In this subsection, we use **x** from [20] as described in Section 6.1. That is we use non-random covariates to compare the efficiencies of the proposed Lasso-based methods on this dataset as a function of the complexity index r/p. We randomly selected r predictor variables from $p = 200$ predictors, where r/p ranges from 0 to and 0.5, and used the following models with r covariates relevant to the response Y.

$$\beta_{j_0+1},\ldots,\beta_{j_0+r} \overset{i.i.d}{\sim} Unif(-2,2), \ j_0 \in \{1,\ldots,200-r\}, \quad (8)$$
$$\beta_j = 0, \text{ for all other } j,$$

$$y_i = \sum_{j=1}^{p} \beta_j x_{ij} + \epsilon_i, \text{ with } \epsilon \overset{i.i.d}{\sim} N(0,0.4).$$

The average of the standard deviations of the predictors is 0.28, so we considered $\epsilon \sim N(0,0.4)$. We then calculated the discussed efficiencies of the proposed methods using $M = 2000$. The result are reported in Figure 1. It shows that for r/p less than 0.29 the Lasso cross validated method has the best performance. For r/p larger than 0.29, the trimmed subspace version with n variables in the subspaces is best with cross validatioed ensemble Lasso with p^* covariates a close second. This CV ensemble Lasso is also second best for $r/p < 0.29$. For $r/p < 0.29$, the performance of subspace methods are poor.

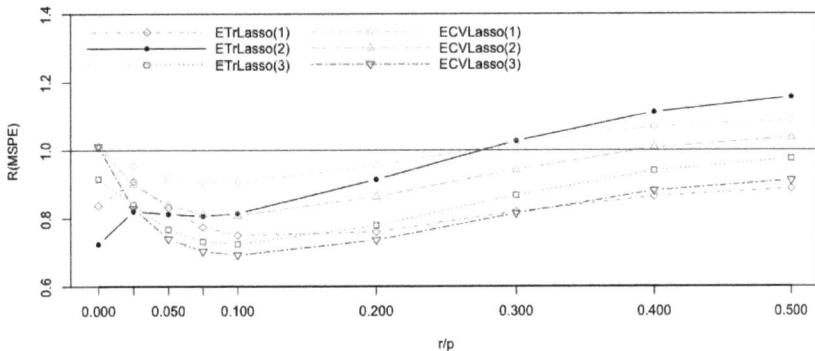

Figure 1. Efficiencies of the Lasso ensemble subspace methods with respect to the CVLasso for the Syndrome Gene inspired simulation model, with different complexity indices r/p.

To summarize, in terms of predictor error, for sparse models, the cross validated lasso based on all covariates performs best, while for the model with r/p larger than 0.29, the trimmed ensemble lasso based on subspaces of size n performs best.

7.2. Simulated Models with Random Covariates

7.2.1. (a) Strong and Weak Signals. Strong Covariate Correlations

We consider model (7) with values of r/p ranging from 0 to 0.5. The results in Figure 2 show that the ensemble CV Lasso based on subspaces with p^* covariates improves on the CV Lasso for all values of the complexity index r/p. The ensemble trimmed Lasso with p^* covariates is for best $0.07 < r/p < 0.3$ while the ensemble trimmed Lasso with n covariates in each subspace is best for $r/p > 0.3$. The ensemble CV Lasso's with n and $n/2$ covariates are slightly worse than CV Lasso.

To summarize, the ensemble methods with p^* covariates in the subspaces perform very well when compared to the CV Lasso. The ensemble trimmed Lasso versions are best for values of r/p larger than 0.2. This shows that when there are many covariates with strong and weak signals cross validation may lead to a poor choice of the trimming parameter λ.

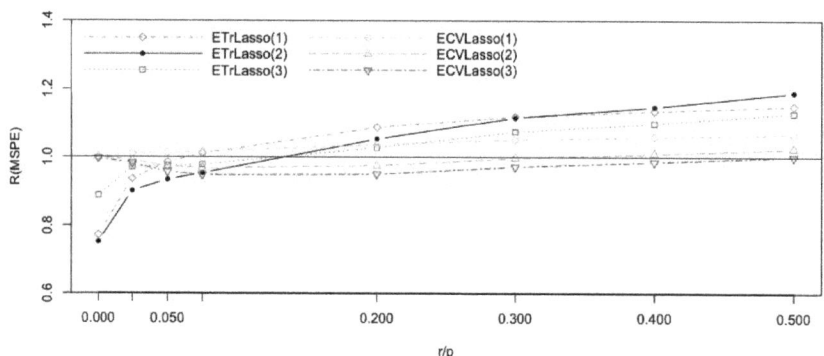

Figure 2. Efficiencies of the Lasso ensemble subspace methods with respect to the CVLasso for the model (7), with different complexity indices r/p.

7.2.2. (b) Strong and Weak Signals. Weak Covariate Correlations

We consider model (7) with σ_{ij} replaced by

$$\sigma_{ij} \sim Unif(0.0, 0.2). \tag{9}$$

Figure 3 shows that the dominance of the ensemble trimmed Lasso methods holds for $r/p > 0.09$. In other words, when there is weak correlations between the covariates, and the complexity of the model is more than 0.09, it is better to use the trimmed average of ensemble predictors based on a sequence of fixed trimming parameters than using trimming parameters obtained by cross validation.

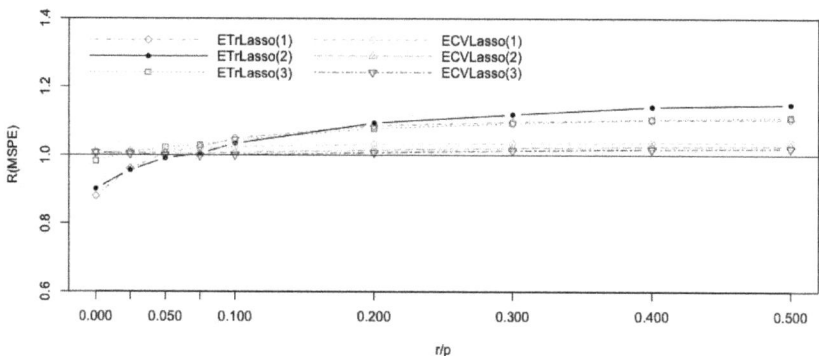

Figure 3. Efficiencies of the Lasso ensemble subspace methods with respect to the CVLasso for the model (9), with different complexity indices r/p.

7.2.3. (c) Strong Signals. Weak Covariate Correlations

We consider model (9) with β replaced by

$$\beta \sim Unif(2, 3). \tag{10}$$

Figure 4 shows that for very small complexity ($r/p \leq 0.020$), CV Lasso is best, while for $r/p > 0.020$, the ensemble trimmed Lasso with p^* covariates in the subspaces improves an

CV Lasso and does very well overall. For $r/p > 0.15$, the ensemble trimmed Lasso with n covariates in the subspaces is best. The trimmed ensemble versions do better than the CV ensemble versions for $r/p > 0.025$.

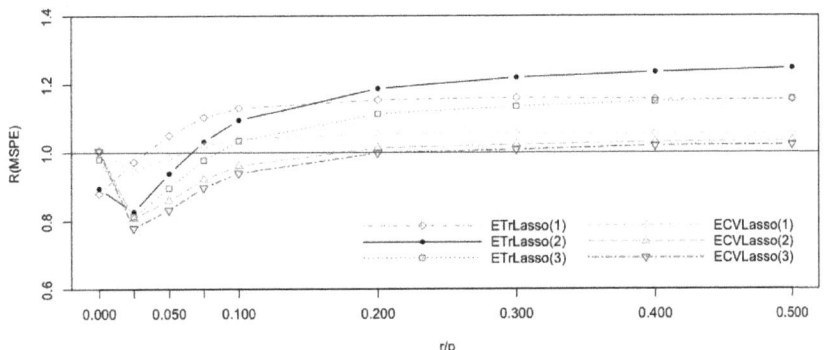

Figure 4. Efficiencies of the Lasso ensemble subspace methods with respect to the CVLasso for the model (10), with different complexity indices r/p.

7.2.4. (d) Weak Signal. Weak and Strong Correlation between Covariates

These two cases had very similar results. Here we give only the case where we use model (9) with

$$\beta \sim Unif(-0.2, 0.2). \tag{11}$$

Figure 5 shows that in this case the ensemble trimmed Lasso methods with p^* and with n covariates in the subspaces do poorly. The ensemble CV Lasso methods performs at the same level as CV Lasso, as does the ensemble trimmed mean approach with $k = n/2$.

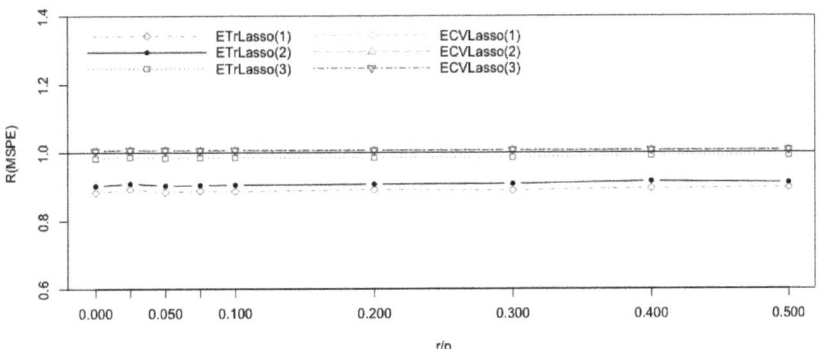

Figure 5. Efficiencies of the Lasso ensemble subspace methods with respect to the CVLasso for the model (11), with different complexity indices r/p.

8. Conclusions

This article explores the random ensemble subspace approach for high-dimensional data analysis. This technique splits the data into covariate subspaces and generates models and methods on each covariate subspace. Merging and assembling the methods provides a global solution to the high-dimensional data analysis challenge. Let n denote the sample size and p the member of covariates, under $p >> n$. We consider three different approaches of selecting subspaces: repeatedly select subspaces as follows (1) n covariates with replacement from p covariates, then use the distinct covariates to form subspaces, (2)

n covariates at random without replacement, and (3) $n/2$ covariates on random without replacement. This approach is applied to a variety of penalty methods and compared to cross-validation (CV) Lasso using mean squared predictor error (MSPE). We consider MSPE as a function of model complexity, which is defined as r/p where r is the number of covariates that are associated with the response and find that when r/p is moderate to large, the cross-validation ensemble subspace approach improves the CVLasso that uses all p covariates in one step. We also introduced an alternative to cross-validation that consists of computing predictors for a fixed set of data-based tuning parameters and using these predictors' trimmed mean. This approach works well when the ratio r/p is above 0.2.

To facilitate communication among researchers and provide possible collaborations between scientists across disciplines and as supporters of open-science, the codes are written in R according to the end-to-end protocol we implemented in this manuscript, which are available on request.

Author Contributions: Conceptualization, S.E.A., S.A. and K.D.; Formal analysis, S.E.A., S.A. and K.D.; Funding acquisition, S.E.A.; Investigation, K.D.; Methodology, S.E.A., S.A. and K.D.; Writing—review & editing, S.E.A., S.A. and K.D. All authors have read and agreed to the published version of the manuscript.

Funding: The research is supported by the Natural Sciences and the Engineering Research Council of Canada (NSERC).

Institutional Review Board Statement: Not applicable.

Informed Consent Statement: Not applicable.

Data Availability Statement: Not applicable.

Acknowledgments: We would like to thank the constructive comments by four anonymous referees and an associate editor which improved the quality and the presentation of our results.

Conflicts of Interest: The authors declare no conflict of interest.

References

1. Guo, H.; Yu, Z.; An, J.; Han, G.; Ma, Y.; Tang, R. A two-stage mutual information based Bayesian Lasso algorithm for multi-locus genome-wide association studies. *Entropy* **2020**, *22*, 329. [CrossRef] [PubMed]
2. Tibshirani, R. Regression shrinkage and selection via the Lasso. *J. R. Stat. Soc. Ser. B* **1996**, *267*–288. [CrossRef]
3. Zou, H. The adaptive lasso and its oracle properties. *J. Am. Stat. Assoc.* **2006**, *101*, 1418–1429. [CrossRef]
4. Fan, J.; Li, R. Variable selection via nonconcave penalized likelihood and its oracle properties. *J. Am. Stat. Assoc.* **2001**, *96*, 1348–1360. [CrossRef]
5. Efron, B.; Hastie, T.; Johnstone, I.; Tibshirani, R. Least angle regression. *Ann. Stat.* **2004**, *32*, 407–499.
6. Zou, H.; Hastie, T. Regularization and variable selection via the elastic net. *J. R. Stat. Soc.* **2005**, *67*, 301–320. [CrossRef]
7. Meinshausen, N.; Yu, B. Lasso-type recovery of sparse representations for high-dimensional data. *Ann. Stat.* **2009**, *37*, 246–270. [CrossRef]
8. Zhao, P.; Yu, B. On model selection consistency of Lasso. *J. Mach. Learn. Res.* **2006**, *7*, 2541–2563.
9. Raheem, S.E.; Ahmed, S.E.; Doksum, K.A. Absolute penalty and shrinkage estimation in partially linear models. *Comput. Stat. Data Anal.* **2012**, *56*, 874–891. [CrossRef]
10. Wainwright, M.J. Sharp thresholds for high-dimensional and noisy sparsity recovery using-constrained quadratic programming (Lasso). *Inf. Theory IEEE Trans.* **2009**, *55*, 2183–2202. [CrossRef]
11. Schelldorfer, J.; Meier, L.; Bühlmann, P. GlmmLasso: An algorithm for high-dimensional generalized linear mixed models using ℓ_1-penalization. *J. Comput. Graph. Stat.* **2014**, *23*, 460–477. [CrossRef]
12. Ranganai, E.; Mudhombo, I. Variable Selection and Regularization in Quantile Regression via Minimum Covariance Determinant Based Weights. *Entropy* **2021**, *23*, 33. [CrossRef] [PubMed]
13. Ahmed, S.E. *Penalty, Shrinkage and Pretest Strategies: Variable Selection and Estimation*; Springer: New York, NY, USA, 2014.
14. Bühlmann, P.; Van De Geer, S. *Statistics for High-Dimensional Data: Methods, Theory and Applications*; Springer Science & Business Media: Berlin/Heidelberg, Germany, 2011.
15. Hastie, T.; Tibshirani, R.; Wainwright, M. *Statistical Learning with Sparsity: The Lasso and Generalizations*; CRC Press: New York, NY, USA, 2015.
16. Breiman, L. Bagging predictors. *Mach. Learn.* **1996**, *24*, 123–140. [CrossRef]
17. Breiman, L. Random forests. *Mach. Learn.* **2001**, *45*, 5–32. [CrossRef]

18. Bolón-Canedo, V.; Alonso-Betanzos, A. Ensembles for feature selection: A review and future trends. *Inf. Fusion* **2019**, *52*, 1–12. [CrossRef]
19. Tu, W.; Yang, D.; Kong, L.; Che, M.; Shi, Q.; Li, G.; Tian, G. Ensemble-based Ultrahigh-dimensional Variable Screening. In *Proceedings of the* International Joint Conferences on Artificial Intelligence Organization, Macao, China, 10–16 August 2019; pp. 3613–3619. [CrossRef]
20. Scheetz, T.E.; Kim, K.Y.A.; Swiderski, R.E.; Philp, A.R.; Braun, T.A.; Knudtson, K.L.; Dorrance, A.M.; DiBona, G.F.; Huang, J.; Casavant, T.L.; et al. Regulation of gene expression in the mammalian eye and its relevance to eye disease. *Proc. Natl. Acad. Sci. USA* **2006**, *103*, 14429–14434. [CrossRef] [PubMed]
21. Lv, J.; Fan, Y. A unified approach to model selection and sparse recovery using regularized least squares. *Ann. Statist.* **2009**, *37*, 3498–3528. [CrossRef]

Article

A Nuisance-Free Inference Procedure Accounting for the Unknown Missingness with Application to Electronic Health Records

Jiwei Zhao [1,*] and Chi Chen [2]

1 Department of Biostatistics and Medical Informatics, University of Wisconsin-Madison, Madison, WI 53726, USA
2 Novartis Institutes for Biomedical Research, Shanghai 201203, China; chi-2.chen@novartis.com
* Correspondence: jiwei.zhao@wisc.edu

Received: 24 August 2020; Accepted: 12 October 2020; Published: 14 October 2020

Abstract: We study how to conduct statistical inference in a regression model where the outcome variable is prone to missing values and the missingness mechanism is unknown. The model we consider might be a traditional setting or a modern high-dimensional setting where the sparsity assumption is usually imposed and the regularization technique is popularly used. Motivated by the fact that the missingness mechanism, albeit usually treated as a nuisance, is difficult to specify correctly, we adopt the conditional likelihood approach so that the nuisance can be completely ignored throughout our procedure. We establish the asymptotic theory of the proposed estimator and develop an easy-to-implement algorithm via some data manipulation strategy. In particular, under the high-dimensional setting where regularization is needed, we propose a data perturbation method for the post-selection inference. The proposed methodology is especially appealing when the true missingness mechanism tends to be missing not at random, e.g., patient reported outcomes or real world data such as electronic health records. The performance of the proposed method is evaluated by comprehensive simulation experiments as well as a study of the albumin level in the MIMIC-III database.

Keywords: nuisance; post-selection inference; missingness mechanism; regularization; asymptotic theory; unconventional likelihood

1. Introduction

A major step towards scientific discovery is to identify useful associations from various features and to quantify their uncertainties. This usually warrants building a regression model for an outcome variable and estimating the coefficient associated with each feature as well as the precision of the estimator. Besides the traditional regression with a small dimensionality, with advances in biotechnology, the modern high-dimensional regression usually posits a sparse parameter in the model, and then applies regularization to select the significant features in order to recover the sparsity. In particular, the post-selection inference could be challenging in a regularized regression framework. In this paper, our main interest is to consider a regression model where the outcome variable is prone to missing values. We study both the traditional setting where regularization is not needed and the modern one with regularization.

The missing data issue is an inevitable concern for statistical analysis in various disciplines ranging from biomedical studies to social sciences. In many applications, the occurrence of missing data is usually not the investigator's primary interest but complicates the statistical analysis. The validity of any method devised for missing data heavily depends on the assumption of the missingness mechanism [1]. Unfortunately, those assumptions are largely unknown and difficult, if not infeasible, to be empirically tested. Therefore, one prefers to concentrate on analyzing the regression model for the outcome variable, while treating the mechanism model as a nuisance. A flexible assumption imposed at the minimum level on the mechanism would provide protection against model misspecification at this level.

While it is indeed promising to regard the missingness mechanism as a nuisance with a flexible assumption, a potential issue is the model identifiability problem if the mechanism contains missing-not-at-random cases, i.e., allowing the mechanism to depend on the missing values themselves. In the past few years, researchers have made great progress on this topic by introducing a so-called instrument. This instrument could be a shadow variable [2–7] or an instrumental variable [8,9]. Both approaches are reasonable and are suitable for different applications. In this paper, we adopt the shadow variable approach as it facilitates the interpretability of the regression model for the outcome. The details of the shadow variable approach will be articulated later throughout the paper.

Therefore, we proceed with a semiparametric framework where our primary interest is a parametric regression, e.g., a linear model, where the statistical task is to estimate the parameter of interest and conduct statistical inference (particularly post-selection inference for the setting with regularization). For the nuisance missingness mechanism, we only impose a nonparametric assumption without specifying a concrete form. We encode the shadow variable as Z, which is one component of the covariate $\mathbf{X}$. In general, a shadow variable with a smaller dimensionality allows more flexibility of the missingness mechanism. Therefore, although it could be multidimensional, we only consider univariate Z throughout the paper. With all of these ingredients, we analyze a conditional likelihood approach which will eventually result in a nuisance-free procedure for parameter estimation and statistical inference.

There are at least two extra highlights of our proposed method that are worth mentioning. The first pertains to the algorithm and computation. Although it looks complicated at first sight, we show that, via some data manipulation strategy, the conditional likelihood function can be analytically written as the likelihood of a conventional logistic regression with some prespecified format. Therefore, our objective function can be readily optimized by many existing software packages. This greatly alleviates the computational burden of our procedure. Second, while the variance estimation under the traditional setting is straightforward following the asymptotic approximation, it is challenging for the setting with regularization. To resolve this problem, we present an easy-to-implement data-driven method to estimate the variance of the regularized estimator via a data perturbation technique. It is noted that the current literature on the inference procedure for regularized estimation in the presence of missing values is very scarce. The authors of [10–12] all considered the model selection problem under high dimensionality with missing data; however, none of them studied the post-selection inference in this context.

The remainder of the paper is structured as follows. In Section 2, we first layout our model formulation and introduce the shadow variable and the conditional likelihood. Section 3 details the traditional setting without regularization. We present our algorithm of how to maximize the conditional likelihood function, the theory of how to derive the asymptotic representation of our proposed estimator and how to estimate its variance. In Section 4, we devote ourselves to the modern setting where the sparsity assumption is imposed and the regularization technique is adopted. Both algorithm and theory as well as the variance estimation through the data perturbation technique are presented. In Section 5, we conduct comprehensive simulation studies to examine the finite sample performance of our proposed estimator as well as the comparison to some existing methods. Section 6 is the application of our method to the regression model for the albumin level which suffers from a large amount of missing values in the MIMIC-III study [13]. The paper is concluded with a discussion in Section 7.

2. Methodology

Denote the outcome variable as Y and covariate $\mathbf{X}$. We assume $\mathbf{X} = (\mathbf{U}^\mathrm{T}, Z)^\mathrm{T}$ where $\mathbf{U}$ is p-dimensional and Z univariate, with detailed interpretation later. We consider the linear model

$$Y = \alpha + \boldsymbol{\beta}^\mathrm{T}\mathbf{U} + \gamma Z + \epsilon, \tag{1}$$

where $\boldsymbol{\beta}$ is also p-dimensional, α and γ are scalars and the true value of γ, γ_0, is nonzero, $\epsilon \sim N(0, \sigma^2)$. We consider the situation that Y has missing values while $\mathbf{X}$ is fully observed. We introduce a binary variable R to indicate missingness: $R = 1$ if Y is observed and $R = 0$ if missing. To allow the greatest flexibility of the missingness mechanism model, we assume

$$\mathrm{pr}(R = 1 \mid Y, \mathbf{X}) = \mathrm{pr}(R = 1 \mid Y, \mathbf{U}) = s(Y, \mathbf{U}), \tag{2}$$

where $s(\cdot)$ merely represents an unknown and unspecified function not depending on Z. We reiterate that, as the assumption (2), in a nonparametric flavor, does not specify a concrete form of $s(\cdot)$, one does not need to be worrisome of the mechanism model misspecification. Moreover, as it allows the dependence on Y, besides missing-completely-at-random (MCAR) and many scenarios of missing-at-random (MAR), the assumption (2) also contains various situations of missing-not-at-random (MNAR).

We term Z the shadow variable following the works in [5–7,14]. Its existence depends on whether it is sensible that Z and R are conditionally independent (given Y and $\mathbf{U}$) and that Y heavily relies on Z (as $\gamma_0 \neq 0$). There are many examples in the literature documenting that the existence of Z is practically reasonable. In application, a surrogate or a proxy of the outcome variable Y, which would not synchronically affect the missingness mechanism, could be a good choice for the shadow variable Z.

We assume independent and identically distributed observations $\{r_i, y_i, \mathbf{u}_i, z_i\}$ for $i = 1, ..., N$ and the first n subjects are free of missing data. Now we present a $s(\cdot)$-free procedure via the use of the conditional likelihood. Denote $\mathbf{V} = (Y, \mathbf{U}^\mathrm{T})^\mathrm{T}$. We start with

$$\prod_{i=1}^{n} p(\mathbf{v}_i \mid z_i, r_i = 1) = \prod_{i=1}^{n} \frac{s(\mathbf{v}_i)}{g(z_i)} p(\mathbf{v}_i \mid z_i),$$

where $g(z_i) = \mathrm{pr}(r_i = 1 \mid z_i) = \int \mathrm{pr}(r_i = 1 \mid \mathbf{v}) p(\mathbf{v} \mid z_i) d\mathbf{v}$ and $p(\cdot \mid \cdot)$ is a generic notation for conditional probability density/mass function. If $\mathbf{V}$ were univariate, we denote $\mathcal{A}$ as the rank statistic of $\{v_1, ..., v_n\}$, then

$$\prod_{i=1}^{n} p(v_i \mid z_i, r_i = 1) = p(v_1, ..., v_n \mid z_1, ..., z_n, r_1 = \cdots = r_n = 1)$$
$$= p(\mathcal{A} \mid v_{(1)}, ..., v_{(n)}, z_1, ..., z_n, r_1 = \cdots = r_n = 1) p(v_{(1)}, ..., v_{(n)} \mid z_1, ..., z_n, r_1 = \cdots = r_n = 1). \tag{3}$$

The conditional likelihood that we use, the first term on the right hand side of (3), is exactly

$$p(\mathcal{A} \mid v_{(1)}, ..., v_{(n)}, z_1, ..., z_n, r_1 = \cdots = r_n = 1) = \frac{p(v_1, ..., v_n \mid z_1, ..., z_n, r_1 = \cdots = r_n = 1)}{p(v_{(1)}, ..., v_{(n)} \mid z_1, ..., z_n, r_1 = \cdots = r_n = 1)}$$

$$= \frac{\prod_{i=1}^{n} p(v_i \mid z_i, r_i = 1)}{\sum_{\omega \in \Omega} \prod_{i=1}^{n} p(v_{\omega(i)} \mid z_i, r_i = 1)} = \frac{\prod_{i=1}^{n} p(v_i \mid z_i)}{\sum_{\omega \in \Omega} \prod_{i=1}^{n} p(v_{\omega(i)} \mid z_i)}, \tag{4}$$

where Ω represents the collection of all one-to-one mappings from $\{1, ..., n\}$ to $\{1, ..., n\}$. Now (4) is nuisance-free and can be used to estimate the unknown parameters in $p(v_i \mid z_i)$.

Although $\mathbf{V}$ is multidimensional in our case, the idea presented above can still be applied and it leads to

$$\frac{\prod_{i=1}^{n} p(y_i, \mathbf{u}_i \mid z_i, r_i = 1)}{\sum_{\omega \in \Omega} \prod_{i=1}^{n} p(y_{\omega(i)}, \mathbf{u}_{\omega(i)} \mid z_i, r_i = 1)} = \frac{\prod_{i=1}^{n} p(y_i, \mathbf{u}_i \mid z_i)}{\sum_{\omega \in \Omega} \prod_{i=1}^{n} p(y_{\omega(i)}, \mathbf{u}_{\omega(i)} \mid z_i)}. \tag{5}$$

Furthermore, to simplify the computation, we adopt the pairwise fashion of (5) following the previous discussion on pairwise pseudo-likelihood in [15], which results

$$\prod_{1 \leq i < j \leq n} \frac{p(y_i, \mathbf{u}_i \mid z_i) p(y_j, \mathbf{u}_j \mid z_j)}{p(y_i, \mathbf{u}_i \mid z_i) p(y_j, \mathbf{u}_j \mid z_j) + p(y_i, \mathbf{u}_i \mid z_j) p(y_j, \mathbf{u}_j \mid z_i)}.$$

After plugging in model (1) and some algebra, the objective eventually becomes to minimize

$$L(\boldsymbol{\theta}) = \binom{N}{2}^{-1} \sum_{1 \leq i < j \leq N} \phi_{ij}(\boldsymbol{\theta}) = \binom{N}{2}^{-1} \sum_{1 \leq i < j \leq N} r_i r_j \log\{1 + W_{ij} \exp(\boldsymbol{\theta}^\mathrm{T} \mathbf{d}_{ij})\}, \tag{6}$$

where $\boldsymbol{\theta} = (\widetilde{\gamma}, \widetilde{\boldsymbol{\beta}}^\mathrm{T})^\mathrm{T}$, $\widetilde{\gamma} = \gamma/\sigma^2$, $\widetilde{\boldsymbol{\beta}} = \widetilde{\gamma} \boldsymbol{\beta}$, $\mathbf{d}_{ij} = (-y_{i \backslash j} z_{i \backslash j}, \mathbf{u}_{i \backslash j}^\mathrm{T} z_{i \backslash j})^\mathrm{T}$, $y_{i \backslash j} = y_i - y_j$, $\mathbf{u}_{i \backslash j} = \mathbf{u}_i - \mathbf{u}_j$, $z_{i \backslash j} = z_i - z_j$ and $W_{ij} = p(z_i \mid \mathbf{u}_j) p(z_j \mid \mathbf{u}_i) / \{p(z_i \mid \mathbf{u}_i) p(z_j \mid \mathbf{u}_j)\}$.
Denote the minimizer of (6) as $\widehat{\boldsymbol{\theta}}$. By checking that

$$\frac{\partial^2 \phi_{ij}(\boldsymbol{\theta})}{\partial \boldsymbol{\theta} \partial \boldsymbol{\theta}^\mathrm{T}} = r_i r_j \{1 + W_{ij} \exp(\boldsymbol{\theta}^\mathrm{T} \mathbf{d}_{ij})\}^{-2} W_{ij} \exp(\boldsymbol{\theta}^\mathrm{T} \mathbf{d}_{ij}) \mathbf{d}_{ij} \mathbf{d}_{ij}^\mathrm{T}$$

is positive definite, $\widehat{\boldsymbol{\theta}}$ uniquely exists. To compute $\widehat{\boldsymbol{\theta}}$, one also needs a model for W_{ij}. Fortunately, this model only depends on fully observed data $\mathbf{x}_i$ and $\mathbf{x}_j$. Essentially any existing parametric, semiparametric, or nonparametric modeling technique for $p(z \mid \mathbf{u})$ can be used, and W_{ij} can be estimated accordingly. Throughout, we denote $\widehat{W}_{ij}$ as an available well-behaved estimator of W_{ij}. Although our procedure stems from $p(y, \mathbf{u} \mid z, r = 1)$, which only relies on the data $\{y_i, \mathbf{x}_i\}$ with $i = 1$, it can be seen that, not only the data $\{y_i, \mathbf{x}_i\}$ with $i = 1$ are used to compute $\widehat{\boldsymbol{\theta}}$, the data $\{\mathbf{x}_i\}$ with $i = 0$ are also used in the process of estimating W_{ij}. Therefore, all observed data, both from completely-observed subjects and from partially-observed subjects, are utilized in our procedure.

One can notice that, due to the assumption (2) which allows the greatest flexibility of the mechanism model and the adoption of the conditional likelihood, not all parameters α, β, γ, and σ^2 are estimable. Nevertheless, the parameter β, which quantifies the association between Y and $\mathbf{U}$ after adjusting for Z and is of primarily scientific interest, can be fully estimable. The remainder of the paper focuses on the estimation and inference of β, as well as the variable selection procedure based on β.

Before moving on, we give some comparison with the existing literature to underline the novel contributions we make in this paper. Based on a slightly different but more restrictive missingness mechanism assumption that $\mathrm{pr}(R = 1 \mid Y, \mathbf{X}) = a(Y) b(\mathbf{X})$, Refs. [16–18] used the similar idea to analyze non-ignorable missing data for a generalized linear model and a semiparametric proportional likelihood ratio model, respectively. They focused on different aspects of how to use the conditional likelihoods and their consequences such as the partial identifiability issue and the large bias issue. In this paper, we focus on the linear model (1) and we just showed that the parameter β is fully identifiable. It can be seen that the method presented in this paper can be applied to different models, but their identifiability problems or some other relevant issues have to be analyzed on a case-by-case basis. For instance, Ref. [19] studied the parameter estimation problem in a logistic regression model with a low dimensionality under assumption (2). They showed that, different from the current paper, all the unknown parameters are identifiable in their context. However, because of the complexity of their objective function, the algorithm studied in [19] is trivial and cannot be extended to a high dimensional setting.

3. Traditional Setting without Regularization

Computation. Directly minimizing $L(\boldsymbol{\theta})$ is feasible; however, it is very computationally involved. From rearranging the terms in $L(\boldsymbol{\theta})$, we realize that it can be rewritten as the negative log-likelihood function of a standard logistic regression model. To be more specific, let k be the index of pair (i,j) with $k = 1, ..., K$ and $K = \binom{n}{2}$. Then,

$$L(\boldsymbol{\theta}) = \frac{1}{K} \sum_{k=1}^{K} \log\left\{1 + \exp\left(s_k \boldsymbol{\theta}^\mathsf{T} \mathbf{t}_k + \log \widehat{W}_k\right)\right\}, \tag{7}$$

where $s_k = -\text{sign}(z_{i\backslash j})$, $\mathbf{t}_k = (|z_{i\backslash j}|y_{i\backslash j}, -|z_{i\backslash j}|\mathbf{u}_{i\backslash j}^\mathsf{T})^\mathsf{T}$. Denote $g_k = I\{z_{i\backslash j} > 0\}$, then one can show that the summand in (7), $\log\left\{1 + \exp\left(s_k \boldsymbol{\theta}^\mathsf{T} \mathbf{t}_k + \log \widehat{W}_k\right)\right\}$, equals,

$$-\left[g_k\left(\boldsymbol{\theta}^\mathsf{T} \mathbf{t}_k + s_k \log \widehat{W}_k\right) - \log\left\{1 + \exp\left(\boldsymbol{\theta}^\mathsf{T} \mathbf{t}_k + s_k \log \widehat{W}_k\right)\right\}\right],$$

which is the contribution of the k-th subject to the negative log-likelihood of a logistic regression with g_k as the response, $\boldsymbol{\theta}$ as the coefficient, $\mathbf{t}_k$ as the covariate, and $s_k \log \widehat{W}_k$ as the offset term, but without an intercept. Therefore, $\widehat{\boldsymbol{\theta}}$ can be obtained by fitting the aforementioned logistic regression model. Algorithm 1 describes the steps for data manipulation and model fitting to estimate $\boldsymbol{\theta}$ under this traditional setting.

Algorithm 1 Minimization of (6) without penalization

1: **Inputs:** $\{y_i, \mathbf{u}_i, z_i\}, \{y_j, \mathbf{u}_j, z_j\}, \widehat{W}_{ij}$, for $i = 1, ..., n$ and $j = 1, ..., n$
2: **Initialize:** $k \leftarrow 0$
3: **for** $j \in \{2 : n\}$ **do**
4:     **for** $i \in \{1 : (j-1)\}$ **do**
5:         $k \leftarrow k + 1$
6:         $y_{i\backslash j} \leftarrow y_i - y_j, \mathbf{u}_{i\backslash j} \leftarrow \mathbf{u}_i - \mathbf{u}_j, z_{i\backslash j} \leftarrow z_i - z_j, \widehat{W}_k \leftarrow \widehat{W}_{ij}$
7:         $g_k \leftarrow I\{z_{i\backslash j} > 0\}$
8:         $s_k \leftarrow -\text{sign}(z_{i\backslash j})$
9:         $\mathbf{t}_k \leftarrow (|z_{i\backslash j}|y_{i\backslash j}, -|z_{i\backslash j}|\mathbf{u}_{i\backslash j}^\mathsf{T})^\mathsf{T}$
10: Fit logistic regression with response $\mathbf{g}$, covariate $\mathbf{t}$, offset $\mathbf{s}^\mathsf{T} \log \widehat{\mathbf{W}}$, and no intercept.
11: **Outputs:** $\widehat{\boldsymbol{\theta}}$

Asymptotic Theory. The asymptotic theory of $\widehat{\boldsymbol{\theta}}$ involves a model of $p(z \mid \mathbf{u})$, which does not contain any missing values, and therefore any statistical model, either parametric, or semiparametric, or nonparametric, can be used. For simplicity, we only discuss the parametric case here, and any further elaborations will be rendered into Section 7. For a parametric model $p(z \mid \mathbf{u}; \boldsymbol{\eta})$, one can apply the standard maximum likelihood estimate $\widehat{\boldsymbol{\eta}}$. Here, we simply assume

$$\sqrt{N}\left(\widehat{\boldsymbol{\eta}} - \boldsymbol{\eta}_0\right) = -\mathbf{G}^{-1} \sqrt{N} \frac{1}{N} \sum_{i=1}^{N} \frac{\partial}{\partial \boldsymbol{\eta}} \log\left\{p(z_i \mid \mathbf{u}_i; \boldsymbol{\eta}_0)\right\} + o_p(1), \tag{8}$$

where $\mathbf{G} = E\left[\frac{\partial^2}{\partial \boldsymbol{\eta} \partial \boldsymbol{\eta}^\mathsf{T}} \log\{p(z \mid \mathbf{u}; \boldsymbol{\eta}_0)\}\right]$, $E\|\frac{\partial^2}{\partial \boldsymbol{\eta} \partial \boldsymbol{\eta}^\mathsf{T}} \log\{p(z \mid \mathbf{u}; \boldsymbol{\eta}_0)\}\|^2 < \infty$, $\boldsymbol{\eta}_0$ is the true value of $\boldsymbol{\eta}$, and $\|\mathbf{M}\| = \sqrt{\text{trace}(\mathbf{MM}^\mathsf{T})}$ for a matrix $\mathbf{M}$. With this prerequisite, we have the following result for $\widehat{\boldsymbol{\theta}}$, and its proof is provided in Appendix A.

Theorem 1. *Assume (8) as well as* $E\left\|\frac{\partial^2 \phi_{ij}(\boldsymbol{\theta}_0, \boldsymbol{\eta}_0)}{\partial \boldsymbol{\theta} \partial \boldsymbol{\theta}^\mathsf{T}}\right\|^2 < \infty$. *Denote $\boldsymbol{\theta}_0$ the true value of $\boldsymbol{\theta}$. Then*

$$\sqrt{N}\left(\widehat{\boldsymbol{\theta}} - \boldsymbol{\theta}_0\right) \xrightarrow{d} N\left(\mathbf{0}, \mathbf{A}^{-1} \boldsymbol{\Sigma} \mathbf{A}^{-1}\right),$$

where $\mathbf{A} = E\left\{\frac{\partial^2 \phi_{ij}(\theta_0, \eta_0)}{\partial \theta \partial \theta^T}\right\}$, $\Sigma = 4E\left\{\lambda_{12}(\theta_0, \eta_0)\lambda_{13}(\theta_0, \eta_0)^T\right\}$, $\lambda_{ij}(\theta_0, \eta_0) = \mathbf{B}\mathbf{G}^{-1}\mathbf{M}_{ij}(\eta_0) - \mathbf{N}_{ij}(\theta_0, \eta_0)$, $\mathbf{B} = E\left\{\frac{\partial^2 \phi_{ij}(\theta_0, \eta_0)}{\partial \theta \partial \eta^T}\right\}$, $\mathbf{M}_{ij}(\eta_0) = \frac{1}{2}\left\{\frac{\partial}{\partial \eta}\log p(z_i \mid \mathbf{u}_i; \eta_0) + \frac{\partial}{\partial \eta}\log p(z_j \mid \mathbf{u}_j; \eta_0)\right\}$, and $\mathbf{N}_{ij}(\theta_0, \eta_0) = \frac{\partial \phi_{ij}(\theta_0, \eta_0)}{\partial \theta}$.

If one prefers the asymptotic result of $\widehat{\beta}$, we have

Corollary 1. *Let $\mathbf{C}$ be a $p \times (p+1)$ matrix such that $\mathbf{C}\theta = \beta$, i.e.,*

$$\mathbf{C} = \begin{pmatrix} 0 & 1/\widetilde{\gamma}_0 & 0 & \cdots & 0 \\ 0 & 0 & 1/\widetilde{\gamma}_0 & \cdots & 0 \\ \vdots & \vdots & \vdots & \ddots & \vdots \\ 0 & 0 & 0 & \cdots & 1/\widetilde{\gamma}_0 \end{pmatrix}.$$

Denote β_0 the true value of β. Then, following Theorem 1, we have $\sqrt{N}\left(\widehat{\beta} - \beta_0\right) \xrightarrow{d} N\left(0, \mathbf{C}\mathbf{A}^{-1}\Sigma \mathbf{A}^{-1}\mathbf{C}^T\right)$.

Variance Estimation. With Theorem 1 and Corollary 1, the variance estimation is straightforward using the plugging in strategy. Note that $\mathrm{var}(\widehat{\theta}) = \frac{1}{N}\mathbf{A}^{-1}\Sigma \mathbf{A}^{-1}$, then one would have the estimate $\widehat{\mathrm{var}}(\widehat{\theta}) = \frac{1}{N}\widehat{\mathbf{A}}^{-1}\widehat{\Sigma}\widehat{\mathbf{A}}^{-1}$ where $\widehat{\mathbf{A}} = \binom{N}{2}^{-1}\sum_{1 \leq i < j \leq N}\frac{\partial^2 \phi_{ij}(\widehat{\theta},\widehat{\eta})}{\partial \theta \partial \theta^T}$, $\widehat{\Sigma} = \frac{4}{N-1}\sum_{i=1}^N\left[\frac{1}{N-1}\sum_{j=1, j\neq i}^N\left\{\widehat{\mathbf{B}}\widehat{\mathbf{G}}^{-1}\mathbf{M}_{ij}(\widehat{\eta}) - \mathbf{N}_{ij}(\widehat{\theta}, \widehat{\eta})\right\}\right]^{\otimes 2}$, $\widehat{\mathbf{B}} = \binom{N}{2}^{-1}\sum_{1 \leq i < j \leq N}\frac{\partial^2 \phi_{ij}(\widehat{\theta},\widehat{\eta})}{\partial \theta \partial \eta^T}$, and $\widehat{\mathbf{G}} = \frac{1}{N}\sum_{i=1}^N\frac{\partial^2}{\partial \eta \partial \eta^T}\log\{p(z_i \mid \mathbf{u}_i; \widehat{\eta})\}$.

4. Modern Setting with Regularization

In the past few decades, it has become a standard practice to consider the high-dimensional regression model, where one assumes the parameter β is sparse and often uses the regularization technique to recover the sparsity. While it is a prominent problem to analyze this type of model when the data are prone to missing values, the literature is quite scarce primarily because it is cumbersome to rigorously address the missingness under high dimensionality. Therefore, it is valuable to extend the nuisance-free likelihood procedure proposed in Section 3 to the setting with regularization.

Computation. Regularization is a powerful technique to identify the zero elements of a sparse parameter in a regression model. Various penalty functions have been extensively studied, such as LASSO [20], SCAD [21], and MCP [22]. In particular, we study the adaptive LASSO penalty [23] with the objective of minimizing the following function

$$L_\lambda(\theta) = L(\theta) + \sum_{j=1}^p \lambda \left|\widehat{\widetilde{\beta}}_j\right|^{-1}\left|\widetilde{\beta}_j\right|, \tag{9}$$

where $\lambda > 0$ is the tuning parameter. Following [23], $\widehat{\widetilde{\beta}}_j$ is a root-N-consistent estimator of $\widetilde{\beta}_j$; for example, one can use the estimator via minimizing the unregularized objective Function (6). Obviously, the penalty term in (9) does not alter the numerical characteristic of $L(\theta)$ that we presented in Section 3. The $L_\lambda(\theta)$ is essentially the regularized log-likelihood of a logistic regression model with the similar format as discussed in (7).

To choose the tuning parameter λ, one can follow either the cross-validation method or various information-based criteria. Fortunately, all of these approaches have been extensively studied in the literature. In this paper, we follow the Bayesian information criterion (BIC) to determine λ. Specifically, we choose λ to be the minimizer of the following BIC function

$$\mathrm{BIC}(\lambda) = 2L(\boldsymbol{\theta}) + p_\lambda \frac{\log(n)}{n},$$

where p_λ is the number of nonzero elements in $\widehat{\widetilde{\boldsymbol{\beta}}}_\lambda$ and the minimizer of (9) is encoded as $\widehat{\boldsymbol{\theta}}_\lambda = (\widehat{\widetilde{\gamma}}_\lambda, \widehat{\widetilde{\boldsymbol{\beta}}}_\lambda^\mathrm{T})^\mathrm{T}$. We summarize the whole computation pipeline as Algorithm 2 below.

Algorithm 2 Minimization of (9) with the ALASSO penalty

1: **Inputs:** $\{y_i, \mathbf{u}_i, z_i\}, \{y_j, \mathbf{u}_j, z_j\}, \widehat{W}_{ij}$, for $i = 1, \ldots, n$ and $j = 1, \ldots, n$
2: **Initialize:** $k \leftarrow 0$
3: **for** $j \in \{2 : n\}$ **do**
4:     **for** $i \in \{1 : (j-1)\}$ **do**
5:         $k \leftarrow k + 1$
6:         $y_{i \backslash j} \leftarrow y_i - y_j$, $\mathbf{u}_{i \backslash j} \leftarrow \mathbf{u}_i - \mathbf{u}_j$, $z_{i \backslash j} \leftarrow z_i - z_j$, $\widehat{W}_k \leftarrow \widehat{W}_{ij}$
7:         $g_k \leftarrow I\{z_{i \backslash j} > 0\}$
8:         $s_k \leftarrow -\mathrm{sign}(z_{i \backslash j})$
9:         $\mathbf{t}_k \leftarrow (|z_{i \backslash j}|y_{i \backslash j}, |z_{i \backslash j}|\mathbf{u}_{i \backslash j}^\mathrm{T})^\mathrm{T}$
10: Fit logistic regression with response $\mathbf{g}$, covariates $\mathbf{t}$, offset $\mathbf{s}^\mathrm{T} \log \mathbf{W}$, and no intercept.
11: Obtain $\widehat{\widetilde{\boldsymbol{\theta}}}$.
12: Fit logistic regression with ALASSO penalty.
13: Find $\lambda^\star$ which minimizes the BIC.
14: **Outputs:** $\widehat{\boldsymbol{\theta}}(\lambda^\star) = \widehat{\boldsymbol{\theta}}_\lambda$

Asymptotic Theory. Recall that $\boldsymbol{\theta} = (\widetilde{\gamma}, \widetilde{\boldsymbol{\beta}}^\mathrm{T})^\mathrm{T}$. Without loss of generality, we assume the first p_0 parameters in $\widetilde{\boldsymbol{\beta}}$ are nonzero, where $1 \le p_0 < p$. For simplicity, we denote $\boldsymbol{\theta}_T = (\widetilde{\gamma}, \widetilde{\beta}_1, \ldots, \widetilde{\beta}_{p_0})^\mathrm{T}$ as the vector of nonzero components and $\boldsymbol{\theta}_{T^c} = (\widetilde{\beta}_{p_0+1}, \ldots, \widetilde{\beta}_p)^\mathrm{T}$ as the vector of zeros.

In Theorem 1, we defined $\mathbf{A} = \mathrm{E}\left\{\frac{\partial^2 \phi_{ij}(\boldsymbol{\theta}_0, \boldsymbol{\eta}_0)}{\partial \boldsymbol{\theta} \partial \boldsymbol{\theta}^\mathrm{T}}\right\}$, a $(p+1) \times (p+1)$ matrix. Now we assume it can be partitioned as $\mathbf{A} = \begin{pmatrix} \mathbf{A}_1 & \mathbf{A}_2 \\ \mathbf{A}_2^\mathrm{T} & \mathbf{A}_3 \end{pmatrix}$, where $\mathbf{A}_1$ is a $(p_0+1) \times (p_0+1)$ submatrix corresponding to $\boldsymbol{\theta}_T$. Similarly, we defined $\boldsymbol{\Sigma} = 4\mathrm{E}\left\{\lambda_{12}(\boldsymbol{\theta}_0, \boldsymbol{\eta}_0)\lambda_{13}(\boldsymbol{\theta}_0, \boldsymbol{\eta}_0)^\mathrm{T}\right\}$, and we also assume it can be partitioned as $\boldsymbol{\Sigma} = \begin{pmatrix} \boldsymbol{\Sigma}_1 & \boldsymbol{\Sigma}_2 \\ \boldsymbol{\Sigma}_2^\mathrm{T} & \boldsymbol{\Sigma}_3 \end{pmatrix}$, where $\boldsymbol{\Sigma}_1$ is a $(p_0+1) \times (p_0+1)$ submatrix corresponding to $\boldsymbol{\theta}_T$ as well. We denote the minimizer of (9), $\widehat{\boldsymbol{\theta}}_\lambda$, as $\widehat{\boldsymbol{\theta}}_\lambda = (\widehat{\boldsymbol{\theta}}_{\lambda,T}^\mathrm{T}, \widehat{\boldsymbol{\theta}}_{\lambda,T^c}^\mathrm{T})^\mathrm{T}$, and its true value $\boldsymbol{\theta}_0 = (\boldsymbol{\theta}_{0,T}^\mathrm{T}, \boldsymbol{\theta}_{0,T^c}^\mathrm{T})^\mathrm{T}$.

Now, we present the oracle property pertaining to $\widehat{\boldsymbol{\theta}}_\lambda$, which includes the asymptotic normality for the nonzero components and the variable selection consistency. The proof is provided in Appendix B.

Theorem 2. *Assume (8), $\mathbf{A}_1$ is positive definite and $\mathrm{E}\|\frac{\partial \phi_{ij}(\boldsymbol{\theta}_0, \boldsymbol{\eta}_0)}{\partial \boldsymbol{\theta}}\|^2 < \infty$ for each $\boldsymbol{\theta}$ in a neighborhood of $\boldsymbol{\theta}_0$. We also assume $\sqrt{N}\lambda \to 0$ and $N\lambda \to \infty$. Then,*

$$\sqrt{N}\left(\widehat{\boldsymbol{\theta}}_{\lambda,T} - \boldsymbol{\theta}_{0,T}\right) \xrightarrow{d} N\left(\mathbf{0}, \mathbf{A}_1^{-1}\boldsymbol{\Sigma}_1 \mathbf{A}_1^{-1}\right).$$

In addition, let $T_N = \{j \in \{1, \ldots, p\} : \widehat{\widetilde{\beta}}_{j,\lambda} \ne 0\}$ and $T = \{j \in \{1, \ldots, p\} : \widetilde{\beta}_{j,0} \ne 0\}$, then

$$\lim_{N \to \infty} pr(T_N = T) = 1.$$

Variance Estimation. Although the above theory provides a rigorous justification for the asymptotic property of $\widehat{\boldsymbol{\theta}}_\lambda$, in practice, however, it does not guide the standard error estimation. Here, we propose a data perturbation approach for the variance estimation. Specifically, following [24], we generate a set of independent and identically distributed positive random variables $\Xi = \{\xi_i, i = 1, ..., N\}$ with $E(\xi_i) = 1$ and $var(\xi_i) = 1$, e.g., the standard exponential distribution. Since it is based on a U-statistic structure, we perturb our objective function by adding $\kappa_{ij} = \xi_i \xi_j$ to each of its pairwise terms. We first obtain the estimator $\hat{\theta}^\star$ by minimizing the perturbed version of (6):

$$L^\star(\theta) = \binom{N}{2}^{-1} \sum_{1 \leq i < j \leq N} \kappa_{ij} \phi_{ij}(\theta).$$

Then, we obtain the estimator $\hat{\theta}^\star_\lambda$ by minimizing the perturbed version of (9):

$$L^\star_\lambda(\theta) = \binom{N}{2}^{-1} \sum_{1 \leq i < j \leq N} \kappa_{ij} \phi_{ij}(\theta) + \sum_{j=1}^p \frac{\lambda}{\left|\widetilde{\beta}^\star_j\right|} \left|\widetilde{\beta}_j\right|,$$

where the optimal λ is also computed by the BIC. We repeat this data perturbation scheme a large number of times, say, M.

Following the theory in [25,26], under some regularity conditions, one can first show that $\sqrt{N}\left(\hat{\theta}^\star_{\lambda,T} - \theta_{0,T}\right)$ converges in distribution to $N(0, A_1^{-1} \Sigma_1 A_1^{-1})$, the same limiting distribution of $\sqrt{N}\left(\hat{\theta}_\lambda - \theta_0\right)$. Furthermore, one can also show $pr^*\left(\hat{\theta}^\star_{\lambda,T^c} = 0\right) \to 1$, where pr^* is the probability measure generated by the original data $\mathcal{X}$ and the perturbation data Ξ. In addition, one can show that the distribution of $\sqrt{N}\left(\hat{\theta}^\star_{\lambda,T} - \hat{\theta}_{\lambda,T}\right)$ conditional on the data can be used to approximate the unconditional distribution of $\sqrt{N}\left(\hat{\theta}_{\lambda,T} - \theta_{0,T}\right)$ and that $pr^*\left(\hat{\theta}^\star_{\lambda,T^c} = 0 \mid \mathcal{X}\right) \to 1$.

To achieve a confidence interval for θ_j, the j-th coordinate in θ, the lower and upper bounds can be formed by $\hat{\theta}^\star_{\lambda,j,\alpha/2}$ and $\hat{\theta}^\star_{\lambda,j,1-\alpha/2}$, respectively, where $\hat{\theta}^\star_{\lambda,j,q}$ represents the q-th quantile of $\left\{\hat{\theta}^\star_{\lambda,j,m}, m = 1, ..., M\right\}$.

5. Simulation Studies

We conduct comprehensive simulation studies to evaluate the finite sample performance of our proposed estimators and also compare with some currently existing methods. We first present the results under the model without regularization, then with regularization.

5.1. Scenarios without Regularization

For the proposed estimator studied in Section 3, we generate $\{R_i, Y_i, U_i^T, Z_i\}, i = 1, ..., N$, independent and identically distributed copies of (R, Y, U^T, Z), as follows. We first generate the random vector $U = (U_1, ..., U_p)^T$ with $U_i \sim N(0.5, 1)$ and $p = 4$, and then generate $Z = \alpha_z + \eta^T U + \epsilon_z$ with $\alpha_z = 0.5$, $\eta = (-0.5, 1, -1, 1.5)^T$, $\epsilon_z \sim N(0, 1)$. Afterwards, the outcome variable Y is generated following the model (1) with $\alpha = -1$, $\beta = (-0.5, 1, -1, 1.5)^T$, $\gamma = 0.5$, and $\epsilon \sim N(0, 1)$, and the missingness indicator R is generated following $pr(R = 1 \mid Y, U) = I(Y < 2.5, U_1 < 2, U_2 < 2, U_3 < 2, U_4 < 2)$ which results in around 40% missing values. We examine two situations with sample size $N = 500$ and $N = 1000$ respectively. Besides the estimator studied in Section 3 (Proposed), we also implement the estimator using all simulated data (FullData) and the estimator using completely observed subjects only (CC). Based on 1000 simulation replicates, for each of the three estimators, we summarize the sample bias, sample standard deviation, estimated standard error, and coverage probability of 95% confidence intervals in Table 1.

Table 1. In Section 5.1, sample bias (Bias), sample standard deviation (SD), estimated standard error (SE), and coverage probability (CP) of 95% confidence interval of the estimator of FullData (using all simulated data), CC (using only completely observed subjects), and of the proposed estimator studied in Section 3.

N	Parameter	Method	Bias	SD	SE	CP
500	$\tilde{\gamma}$	FullData	0.0026	0.0444	0.0450	0.9540
		CC	−0.0329	0.0564	0.0560	0.9100
		Proposed	0.0174	0.0829	0.0789	0.9450
	β_1	FullData	0.0022	0.0489	0.0503	0.9510
		CC	0.0376	0.0670	0.0699	0.9300
		Proposed	0.0164	0.1644	0.1607	0.9400
	β_2	FullData	−0.0017	0.0657	0.0635	0.9310
		CC	−0.0649	0.0851	0.0835	0.8680
		Proposed	−0.0399	0.2305	0.2239	0.9360
	β_3	FullData	0.0022	0.0616	0.0635	0.9540
		CC	0.0778	0.0871	0.0867	0.8430
		Proposed	0.0462	0.2323	0.2298	0.9410
	β_4	FullData	−0.0045	0.0792	0.0810	0.9530
		CC	−0.0988	0.1007	0.1043	0.8550
		Proposed	−0.0672	0.3081	0.3047	0.9380
1000	$\tilde{\gamma}$	FullData	−0.0012	0.0317	0.0317	0.9540
		CC	−0.0348	0.0396	0.0393	0.8510
		Proposed	0.0068	0.0573	0.0555	0.9350
	β_1	FullData	0.0011	0.0367	0.0355	0.9370
		CC	0.0399	0.0490	0.0494	0.8840
		Proposed	0.0154	0.1154	0.1138	0.9460
	β_2	Full Data	0.0020	0.0448	0.0448	0.9500
		CC	−0.0649	0.0577	0.0588	0.8110
		Proposed	−0.0153	0.1531	0.1591	0.9590
	β_3	Full Data	−0.0015	0.0458	0.0449	0.9460
		CC	0.0779	0.0605	0.0611	0.7490
		Proposed	0.0135	0.1598	0.1634	0.9480
	β_4	Full Data	0.0009	0.0564	0.0571	0.9540
		CC	−0.0949	0.0720	0.0734	0.7550
		Proposed	−0.0242	0.2091	0.2167	0.9430

Furthermore, we consider a similar simulation setting where the generation is the same as above except for a logistic missingness mechanism model with logit$\{pr(R = 1 \mid Y, \mathbf{U})\} = 3 - 2Y + 0.5U_1 - U_2 + U_3 - 1.5U_4$, which also results in around 40% missing values. We replicate the results, shown in Table 2.

We can reach the following conclusions from Tables 1 and 2. For the estimator Proposed, although its bias is slightly larger than the benchmark FullData, it is still very close to zero. The sample standard deviation and the estimated standard error are rather close to each other. The sample coverage probability of the estimated 95% confidence interval is also very close to the nominal level. This observation well matches our theoretical justification in Theorem 1. On the contrary, the estimator CC is clearly biased, resulting in empirical coverage far from the nominal level, and therefore is not recommended to use in practice. It is also clear that, compared to the benchmark FullData, the estimator Proposed has estimation efficiency loss to some extent. This is because the proposed method uses the conditional likelihood approach and it completely eliminates the effect of the nuisance.

Table 2. In Section 5.1, sample bias (Bias), sample standard deviation (SD), estimated standard error (SE), and coverage probability (CP) of 95% confidence interval of the estimator of FullData (using all simulated data), CC (using only completely observed subjects), and of the proposed estimator studied in Section 3, with a logistic missingness mechanism model.

N	Parameter	Method	Bias	SD	SE	CP
500	$\tilde{\gamma}$	FullData	−0.0011	0.0464	0.0451	0.9410
		CC	−0.0306	0.0567	0.0567	0.9200
		Proposed	0.0100	0.0822	0.0787	0.9380
	β_1	FullData	−0.0004	0.0509	0.0503	0.9520
		CC	0.0440	0.0636	0.0637	0.8930
		Proposed	0.0146	0.1308	0.1236	0.9420
	β_2	FullData	0.0013	0.0639	0.0637	0.9520
		CC	−0.0871	0.0828	0.0821	0.8190
		Proposed	−0.0173	0.1824	0.1753	0.9430
	β_3	FullData	−0.0030	0.0655	0.0636	0.9400
		CC	0.0876	0.0847	0.0821	0.8030
		Proposed	0.0214	0.1840	0.1756	0.9440
	β_4	FullData	0.0023	0.0845	0.0812	0.9390
		CC	−0.1307	0.1083	0.1061	0.7560
		Proposed	−0.0331	0.2533	0.2384	0.9360
1000	$\tilde{\gamma}$	FullData	0.0004	0.0315	0.0317	0.9490
		CC	−0.0286	0.0396	0.0398	0.8950
		Proposed	0.0060	0.0568	0.0555	0.9390
	β_1	FullData	0.0007	0.0362	0.0354	0.9420
		CC	0.0442	0.0451	0.0447	0.8410
		Proposed	0.0079	0.0910	0.0859	0.9290
	β_2	FullData	−0.0004	0.0450	0.0448	0.9390
		CC	−0.0879	0.0571	0.0576	0.6640
		Proposed	−0.0044	0.1277	0.1220	0.9420
	β_3	FullData	−0.0009	0.0450	0.0448	0.9450
		CC	0.0880	0.0588	0.0577	0.6660
		Proposed	0.0114	0.1309	0.1222	0.9380
	β_4	FullData	−0.0005	0.0576	0.0572	0.9510
		CC	−0.1342	0.0755	0.0745	0.5740
		Proposed	−0.0191	0.1757	0.1661	0.9370

5.2. Scenarios with Regularization

For the estimator studied in Section 4, the independent and identically distributed samples are generated as follows. The variable $\mathbf{U} = (U_1, \ldots, U_p)^T$ is generated from MVN$(\mathbf{0}, \Sigma_u)$ with $\Sigma_u = (0.5^{|i-j|})_{1 \leq i,j \leq p}$ and $p = 8$. Then, the shadow variable Z is generated following $Z = \alpha_z + \eta^T \mathbf{U} + \epsilon_z$ with $\alpha_z = 0$, $\eta = (-0.5, 0.5, -1, 1, -0.5, 0.5, -1, 1)^T$ and $\epsilon_z \sim N(0,1)$. The outcome variable Y is generated from model (1) with $\alpha = 0$, $\beta = (3, 1.5, 0, 0, 2, 0, 0, 0)^T$, $\gamma = 3$, $\epsilon \sim N(0, \sigma^2)$ and $\sigma = 3$. The distribution of the missingness indicator follows from logit$\{\text{pr}(R = 1 \mid Y, \mathbf{U})\} = 5 + 5Y + 0.2U_1 + 0.2U_7$, which results in about 45% missing values. Similar to Section 5.1, we also examine two situations with sample size $N = 500$ and $N = 1000$ respectively, and we implement three estimators FullData, CC, and Proposed. When the estimator Proposed is implemented, we perform $M = 500$ perturbations in order to obtain the confidence interval for the unknown parameter. The results summarized below are based on 1000 simulation replicates.

Figure 1 shows the L_1, L_2, and L_∞ norms of the bias for the three different estimators. As sample size increases, there is no doubt that the estimation bias is getting smaller for any method. It is also clear that the bias of the Proposed estimator is larger than the benchmark FullData, but much smaller than the method CC.

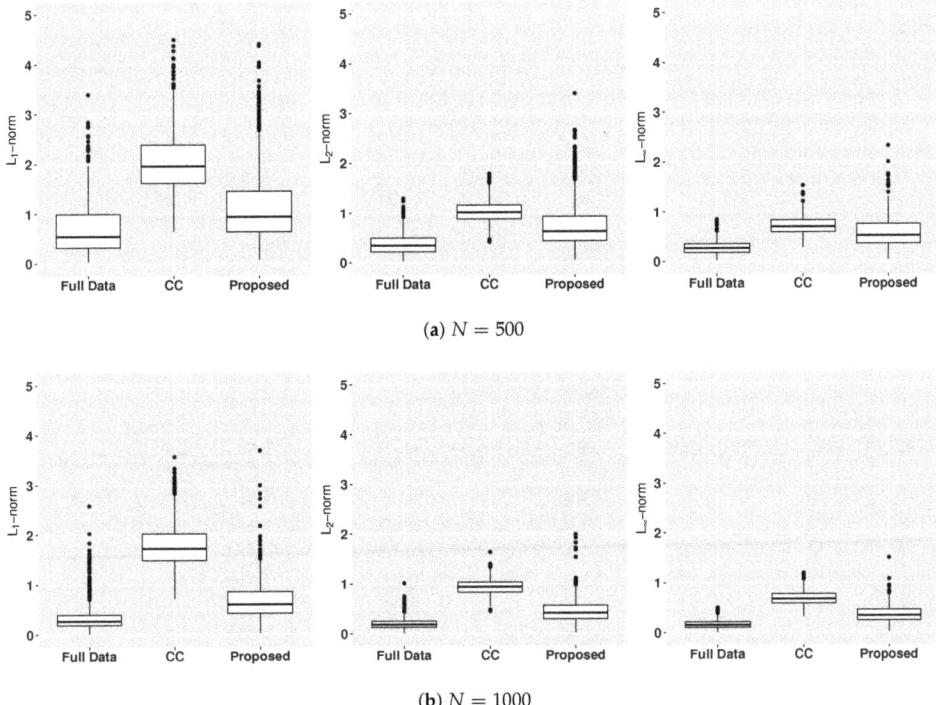

Figure 1. In Section 5.2, L_1 (1st column), L_2 (2nd column), and L_∞ (3rd column) norms of the estimation bias of the estimator of FullData (using all simulated data), CC (using only completely observed subjects), and of the proposed estimator studied in Section 4.

We present the statistical inference results in Table 3 for $N = 500$ and Table 4 for $N = 1000$, respectively, including sample bias, sample standard deviation, estimated standard error, coverage probability, and length of 95% confidence interval for the three different methods. For the nonzero β's as well as $\widetilde{\gamma}$, similar to Section 5.1, the method CC clearly prompts coverage probability far from the nominal level hence is not reliable. For the method Proposed, its estimation bias is quite close to zero, and its sample standard deviation and estimated standard error are quite close to each other. The coverage probability of the confidence interval converges to the nominal level 95% as the sample size gets larger. For the noisy zero β's, the coverage probabilities in the three methods are all close to 1, reflecting the variable selection consistency in the oracle property, even for the CC method. Furthermore, a very nice finite sample property of our proposed estimator is that it produces the confidence interval with the shortest length, which can be clearly seen from both Tables 3 and 4.

Table 3. In Section 5.2, with sample size $N = 500$, sample bias (Bias), sample standard deviation (SD), estimated standard error (SE), coverage probability (CP), and length (Length) of 95% confidence interval of the estimator of FullData (using all simulated data), CC (using only completely observed subjects) and of the proposed estimator studied in Section 4.

Parameter		Method	Bias	SD	SE	CP	Length
	$\tilde{\gamma}$	FullData	0.0001	0.0120	0.0132	0.9480	0.0515
		CC	−0.0729	0.0180	0.0183	0.0370	0.0716
		Proposed	−0.0423	0.0500	0.0498	0.8200	0.1926
	β_1	FullData	0.0021	0.1686	0.1649	0.9400	0.6415
		CC	−0.6547	0.2207	0.2114	0.1460	0.8233
		Proposed	0.0354	0.4698	0.4746	0.9320	1.8513
True Nonzero	β_2	Full Data	−0.0275	0.1692	0.1791	0.9440	0.6952
		CC	−0.3501	0.2227	0.2174	0.6180	0.8471
		Proposed	−0.2654	0.5843	0.5609	0.8940	1.9237
	β_5	Full Data	−0.0172	0.1576	0.1756	0.9650	0.6826
		CC	−0.4478	0.2172	0.2161	0.4370	0.8418
		Proposed	−0.1251	0.4037	0.4611	0.9330	1.8063
	β_3	FullData	0.0085	0.1567	0.1890	0.9960	0.7184
		CC	0.0063	0.2067	0.2304	0.9890	0.8890
		Proposed	0.0109	0.0988	0.1690	1.0000	0.4398
	β_4	Full Data	−0.0019	0.1581	0.1900	0.9940	0.7206
		CC	−0.0017	0.2097	0.2307	0.9900	0.8914
		Proposed	0.0126	0.1112	0.1447	1.0000	0.3668
True Zero	β_6	Full Data	0.0045	0.1212	0.1606	0.9980	0.6146
		CC	−0.0053	0.1749	0.1953	0.9900	0.7560
		Proposed	0.0034	0.0664	0.1160	1.0000	0.2555
	β_7	Full Data	0.0014	0.1351	0.1839	0.9980	0.7063
		CC	−0.0055	0.1870	0.2245	0.9950	0.8717
		Proposed	0.0024	0.0386	0.1115	1.0000	0.2538
	β_8	Full Data	−0.0072	0.1295	0.1748	0.9990	0.6653
		CC	−0.0062	0.1795	0.2125	0.9940	0.8251
		Proposed	0.0016	0.0741	0.1066	1.0000	0.2284

Table 4. In Section 5.2, with sample size $N = 1000$, sample bias (Bias), sample standard derivation (SD), estimated standard error (SE), coverage probability (CP), and length (Length) of 95% confidence interval of the estimator of FullData (using all simulated data), CC (using only completely observed subjects) and of the proposed estimator studied in Section 4.

Parameter		Method	Bias	SD	SE	CP	Length
	$\tilde{\gamma}$	FullData	−0.0005	0.0073	0.0088	0.9690	0.0344
		CC	−0.0730	0.0126	0.0130	0.0000	0.0507
		Proposed	−0.0213	0.0311	0.0334	0.8700	0.1293
	β_1	FullData	−0.0005	0.1186	0.1170	0.9300	0.4547
		CC	−0.6655	0.1568	0.1507	0.0090	0.5864
		Proposed	0.0211	0.2911	0.2969	0.9300	1.1631
True Nonzero	β_2	Full Data	−0.0321	0.1175	0.1249	0.9550	0.4861
		CC	−0.3387	0.1477	0.1534	0.3960	0.5972
		Proposed	−0.0979	0.2907	0.3383	0.9230	1.3115
	β_5	Full Data	−0.0225	0.1051	0.1206	0.9590	0.4698
		CC	−0.4485	0.1478	0.1534	0.1770	0.5964
		Proposed	−0.0621	0.2351	0.2526	0.9290	0.9871

Table 4. *Cont.*

Parameter		Method	Bias	SD	SE	CP	Length
True Zero	β_3	FullData	−0.0007	0.0621	0.1162	1.0000	0.4253
		CC	0.0023	0.1414	0.1614	0.9920	0.6180
		Proposed	0.0044	0.0581	0.0910	1.0000	0.2091
	β_4	Full Data	0.0020	0.0632	0.1170	1.0000	0.4271
		CC	−0.0005	0.1333	0.1608	0.9930	0.6207
		Proposed	0.0063	0.0584	0.0887	1.0000	0.2107
	β_6	Full Data	0.0013	0.0571	0.1010	1.0000	0.3670
		CC	−0.0034	0.1159	0.1378	0.9950	0.5313
		Proposed	0.0012	0.0281	0.0688	1.0000	0.1430
	β_7	Full Data	−0.0028	0.0599	0.1144	1.0000	0.4231
		CC	−0.0033	0.1243	0.1584	0.9970	0.6131
		Proposed	0.0016	0.0288	0.0698	1.0000	0.1421
	β_8	Full Data	0.0039	0.0589	0.1080	1.0000	0.3970
		CC	0.0028	0.1256	0.1497	0.9940	0.5752
		Proposed	0.0000	0.0333	0.0644	1.0000	0.1314

6. Real Data Application

The Medical Information Mart for Intensive Care III (MIMIC-III) is an openly available electronic health records (EHR) database, developed by the MIT Lab for Computational Physiology [13], comprising de-identified health-related data associated with intensive care unit patients with rich information including demographics, vital signs, laboratory test, medications, and more.

Our initial motivation for this data analysis is to understand the missingness mechanism for some laboratory test biomarkers in this EHR system. As for the EHR database, since the data are collected in a non-prescheduled fashion, i.e., only available when the patient seeks care or the physician orders care, the visiting process could be potentially informative about the patients' risk categories. Therefore, it is very plausible that the data are missing not at random, or a mix of missing not at random and missing at random [27,28]. When we first conducted the data cleaning process briefly, an interesting phenomenon we observe is that, compared to most biomarkers which usually have <3% missing values, the albumin level in the blood sample, a very indicative biomarker associated with different types of diseases [29], has around 30% missingness.

To further understand this phenomenon, we concentrate on a subset of the data with sample size $N = 1359$ in which 421 samples have missing values in the albumin level but all other variables are complete. We aim to apply the proposed method to the study of the albumin level (Y). The calcium level in the blood sample, free of missing data, has been shown in the biomedical literature that it has high correlation with the albumin level [30–32]; therefore, we adopt the calcium level as the shadow variable Z. Seventeen other variables comprise the vector $\mathbf{U}$, which are either demographics (age and gender), chart events (respiratory rate, glucose, heart rate, systolic blood pressure, diastolic blood pressure, and temperature), other laboratory tests (urea nitrogen, platelets, magnesium, hematocrit, red blood cell, white blood cell, and peripheral capillary oxygen saturation (SpO2)), or aggregated metrics (simplified acute physiology score (SAPS-II) and sequential organ failure assessment score (SOFA)).

We implement the proposed estimator studied in Section 4 to achieve both variable selection and post-selection inference. We also compare it with the CC method which naively fits the regularized linear regression with the ALASSO penalty. For each of the methods, we apply the data perturbation scheme presented in Section 4 with $M = 500$ for standard error estimation. The results are summarized in Table 5. The solution path of the Proposed method, as the tuning parameter λ varies, is also provided in Figure 2.

Table 5. In Section 6, the parameter estimate (Estimate), standard error (SE), and confidence interval (CI) of the estimator of CC (using only completely observed subjects) and of the proposed estimator studied in Section 4 in the MIMIC-III study.

Effect	CC			Proposed		
	Estimate	SE	CI	Estimate	SE	CI
Calcium(shadow)	0.7707	0.0691	[0.6532, 0.9153]	1.5271	0.1796	[1.1815, 1.8835]
Red Blood Cell	0.6491	0.0514	[0.5337, 0.7257]	0.7545	0.1631	[0.3594, 1.0109]
Magnesium	0.0000	0.0686	[−0.2073, 0.0000]	0.2731	0.2452	[0.0000, 0.6609]
SOFA	−0.2720	0.0268	[−0.3135, −0.2099]	−0.1852	0.1040	[−0.3467, 0.0000]
Temperature	−0.0360	0.0351	[−0.0883, 0.0659]	0.0000	0.0964	[0.0000, 0.3132]
White Blood Cell	−0.0245	0.0123	[−0.0416, 0.0000]	0.0000	0.0025	[0.0000, 0.0000]
Age	0.0000	0.0008	[0.0000, 0.0000]	0.0000	0.0017	[0.0000, 0.0000]
Gender	0.0000	0.0240	[−0.0477, 0.0662]	0.0000	0.1320	[−0.4025, 0.0000]
Respiratory Rate	0.0000	0.0034	[−0.0141, 0.0000]	0.0000	0.0008	[0.0000, 0.0000]
Glucose	0.0000	0.0000	[0.0000, 0.0000]	0.0000	0.0005	[0.0000, 0.0000]
Heart Rate	0.0000	0.0025	[−0.0091, 0.0000]	0.0000	0.0004	[0.0000, 0.0000]
Systolic BP	0.0000	0.0045	[−0.0139, 0.0000]	0.0000	0.0000	[0.0000, 0.0000]
Diastolic BP	0.0000	0.0072	[0.0000, 0.0223]	0.0000	0.0000	[0.0000, 0.0000]
Urea Nitrogen	0.0000	0.0004	[0.0000, 0.0000]	0.0000	0.0000	[0.0000, 0.0000]
Platelets	0.0000	0.0000	[0.0000, 0.0000]	0.0000	0.0000	[0.0000, 0.0000]
Hematocrit	0.0000	0.0027	[0.0000, 0.0000]	0.0000	0.0000	[0.0000, 0.0000]
SpO2	0.0000	0.0145	[−0.0479, 0.0000]	0.0000	0.0162	[0.0000, 0.0000]
SAPS-II	0.0000	0.0106	[−0.0051, 0.0269]	0.0000	0.0000	[0.0000, 0.0000]

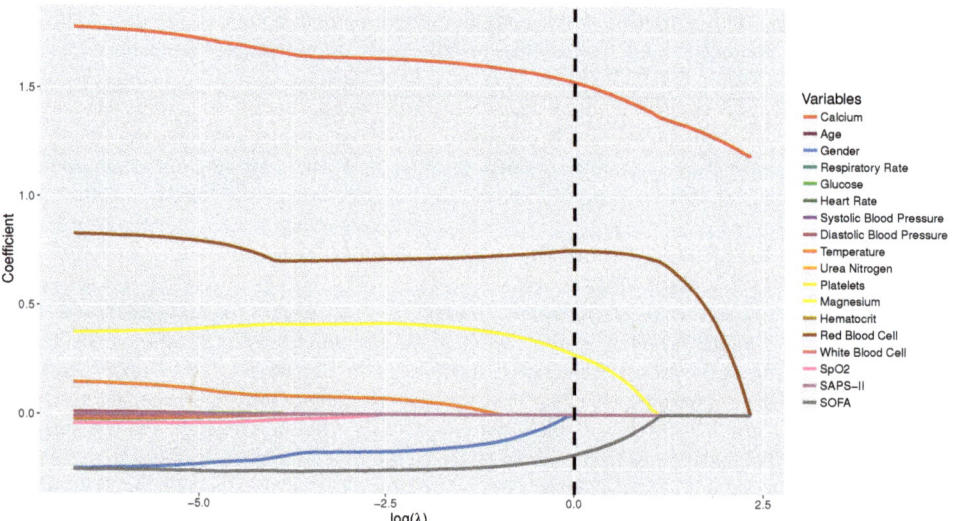

Figure 2. In Section 6, as tuning parameter λ varies, the solution path of the proposed estimator in the MIMIC-III study. The optimal λ, λ^*, equals 1.0030 and $\log \lambda^* = 0.0030$.

In general, both methods achieve the goal of variable selection and post-selection inference by leveraging the regularization technique coupled with the data perturbation strategy, and identify many variables as noise with zero coefficients. In particular, the Proposed method provides larger effects for the calcium level (the shadow variable) and the red blood cell count, whereas a smaller effect for the aggregated SOFA score. The Proposed method simplifies the body temperature and the white blood cell count as nonsignificant variables, which are identified as nonzero but with a very small effect using the CC method. It is also worthwhile to mention that the Proposed method signifies the magnesium level with a quite significant coefficient, which was extensively investigated in the scientific literature [33–35].

7. Discussion

In this paper, we provide a systematic approach for parameter estimation and statistical inference in both traditional linear model where the regularization is not needed and the modern regularized regression setting, when the outcome variable is prone to missing values and the missingness mechanism can be arbitrarily flexible. A pivotal condition rooted in our procedure is the shadow variable Z, which overcomes the model identifiability problem and enables the nuisance-free conditional likelihood process.

Certainly any method would have its own limitations and could be potentially improved. One needs a model $p(z \mid \mathbf{u})$ to implement the proposed estimator in Sections 3 and 4. As its modeling does not involve any missing data, we simply use the parametric maximum likelihood estimation in our algorithm as well as in the theoretical justification. Indeed, any statistical or machine learning method can be used for modeling $p(z \mid \mathbf{u})$. For instance, if one would like to consider a semiparametric model [36], e.g.,

$$p(z \mid \mathbf{u}; \boldsymbol{\eta}, F) = \frac{\exp(\boldsymbol{\eta}^T \mathbf{u} z) f(z)}{\int \exp(\boldsymbol{\eta}^T \mathbf{u} z) dF(z)},$$

where $\boldsymbol{\eta} = (\eta_1, ..., \eta_p)^T$ is a vector of unknown parameters and $f(z)$ is the density of an unknown baseline distribution function F with respect to some dominating measure ν. With this model fitted, W_{ij} can be simplified to $W_{ij} = \exp(-z_{i \setminus j} \boldsymbol{\eta}^T \mathbf{u}_{i \setminus j})$. Therefore, a similar conditional likelihood approach can be used to estimate $\boldsymbol{\eta}$ without estimating the nonparametric component $f(z)$.

Author Contributions: Conceptualization, J.Z.; Experiment, J.Z. and C.C.; Writing, J.Z. and C.C.; Supervision, J.Z. All authors have read and agreed to the published version of the manuscript

Funding: Jiwei Zhao is supported by the National Science Foundation award 1953526.

Conflicts of Interest: The authors declare no conflict of interest.

Appendix A. Proof of Theorem 1

Proof. Note that $\hat{\boldsymbol{\theta}}$ is obtained by setting estimating equation $\frac{\partial L(\hat{\boldsymbol{\theta}}, \hat{\boldsymbol{\eta}})}{\partial \boldsymbol{\theta}} = 0$, which is equivalent to

$$\left\{ \frac{\partial L(\hat{\boldsymbol{\theta}}, \hat{\boldsymbol{\eta}})}{\partial \boldsymbol{\theta}} - \frac{\partial L(\boldsymbol{\theta}_0, \hat{\boldsymbol{\eta}})}{\partial \boldsymbol{\theta}} \right\} + \left\{ \frac{\partial L(\boldsymbol{\theta}_0, \hat{\boldsymbol{\eta}})}{\partial \boldsymbol{\theta}} - \frac{\partial L(\boldsymbol{\theta}_0, \boldsymbol{\eta}_0)}{\partial \boldsymbol{\theta}} \right\} + \frac{\partial L(\boldsymbol{\theta}_0, \boldsymbol{\eta}_0)}{\partial \boldsymbol{\theta}} = 0. \tag{A1}$$

Specifically,

$$\frac{\partial L(\hat{\boldsymbol{\theta}}, \hat{\boldsymbol{\eta}})}{\partial \boldsymbol{\theta}} - \frac{\partial L(\boldsymbol{\theta}_0, \hat{\boldsymbol{\eta}})}{\partial \boldsymbol{\theta}} = \frac{\partial^2 L(\boldsymbol{\theta}_0, \hat{\boldsymbol{\eta}})}{\partial \boldsymbol{\theta} \partial \boldsymbol{\theta}^T} \left(\hat{\boldsymbol{\theta}} - \boldsymbol{\theta}_0 \right) + o_p \left(N^{-\frac{1}{2}} \right), \tag{A2}$$

by Taylor expansion. Similarly,

$$\frac{\partial L(\boldsymbol{\theta}_0, \hat{\boldsymbol{\eta}})}{\partial \boldsymbol{\theta}} - \frac{\partial L(\boldsymbol{\theta}_0, \boldsymbol{\eta}_0)}{\partial \boldsymbol{\theta}} = \frac{\partial^2 L(\boldsymbol{\theta}_0, \boldsymbol{\eta}_0)}{\partial \boldsymbol{\theta} \partial \boldsymbol{\eta}^T} (\hat{\boldsymbol{\eta}} - \boldsymbol{\eta}_0) + o_p \left(N^{-\frac{1}{2}} \right). \tag{A3}$$

With (A2) and (A3) plugging into (A1), we can obtain the following equation,

$$\sqrt{N} \frac{\partial^2 L(\boldsymbol{\theta}_0, \hat{\boldsymbol{\eta}})}{\partial \boldsymbol{\theta} \partial \boldsymbol{\theta}^T} \left(\hat{\boldsymbol{\theta}} - \boldsymbol{\theta}_0 \right) + \sqrt{N} \frac{\partial^2 L(\boldsymbol{\theta}_0, \boldsymbol{\eta}_0)}{\partial \boldsymbol{\theta} \partial \boldsymbol{\eta}^T} (\hat{\boldsymbol{\eta}} - \boldsymbol{\eta}_0) + \sqrt{N} \frac{\partial L(\boldsymbol{\theta}_0, \boldsymbol{\eta}_0)}{\partial \boldsymbol{\theta}} + o_p(1) = 0. \tag{A4}$$

As $\sqrt{N} (\hat{\boldsymbol{\eta}} - \boldsymbol{\eta}_0) = -\mathbf{G}^{-1} \sqrt{N} \frac{1}{N} \sum_{i=1}^{N} \frac{\partial}{\partial \boldsymbol{\eta}} \log \{p(z_i \mid \mathbf{u}_i; \boldsymbol{\eta}_0)\} + o_p(1)$ from the asymptotic property of $\hat{\boldsymbol{\eta}}$, (A4) is equivalent to

$$\sqrt{N}\frac{\partial^2 L(\theta_0,\widehat{\eta})}{\partial\theta\partial\theta^T}\left(\widehat{\theta}-\theta_0\right) + \frac{\partial^2 L(\theta_0,\eta_0)}{\partial\theta\partial\eta^T}\left[-G^{-1}\sqrt{N}\frac{1}{N}\sum_{i=1}^{N}\frac{\partial}{\partial\eta}\log\{p(z_i\mid u_i;\eta_0)\}\right]$$
$$+ \sqrt{N}\frac{\partial L(\theta_0,\eta_0)}{\partial\theta} + o_p(1) = 0.$$

Thus,

$$\sqrt{N}\left(\widehat{\theta}-\theta_0\right)$$
$$= -\left\{\frac{\partial^2 L(\theta_0,\widehat{\eta})}{\partial\theta\partial\theta^T}\right\}^{-1} \times \left\{\frac{\partial^2 L(\theta_0,\eta_0)}{\partial\theta\partial\eta^T}\left[-G^{-1}\sqrt{N}\frac{1}{N}\sum_{i=1}^{N}\frac{\partial}{\partial\eta}\log\{p(z_i\mid u_i;\eta_0)\}\right] + \sqrt{N}\frac{\partial L(\theta_0,\eta_0)}{\partial\theta}\right\}$$
$$+ o_p(1)$$
$$= -A^{-1}\left\{B\left[-G^{-1}\sqrt{N}\frac{1}{N}\sum_{i=1}^{N}\frac{\partial}{\partial\eta}\log\{p(z_i\mid u_i;\eta_0)\}\right] + \sqrt{N}\frac{\partial L(\theta_0,\eta_0)}{\partial\theta}\right\} + o_p(1), \quad (A5)$$

where $\frac{\partial^2 L(\theta_0,\eta_0)}{\partial\theta\partial\theta^T} \xrightarrow{P} A = E\left\{\frac{\partial^2 \phi_{ij}(\theta_0,\eta_0)}{\partial\theta\partial\theta^T}\right\}$, and $\frac{\partial^2 L(\theta_0,\eta_0)}{\partial\theta\partial\eta^T} \xrightarrow{P} B = E\left\{\frac{\partial^2 \phi_{ij}(\theta_0,\eta_0)}{\partial\theta\partial\eta^T}\right\}$. In addition, we need to form a projection of $\frac{1}{N}\sum_{i=1}^{N}\frac{\partial}{\partial\eta}\log\{p(z_i\mid u_i;\eta_0)\}$ in (A5) through

$$\frac{1}{N}\sum_{i=1}^{N}\frac{\partial}{\partial\eta}\log\{p(z_i\mid u_i;\eta_0)\} = \binom{N}{2}^{-1}\sum_{1\leq i<j\leq N}\frac{1}{2}\left[\frac{\partial}{\partial\eta}\log\{p(z_i\mid u_i;\eta_0)\} + \frac{\partial}{\partial\eta}\log\{p(z_j\mid u_j;\eta_0)\}\right],$$

and

$$\frac{\partial L(\theta_0,\eta_0)}{\partial\theta} = \binom{N}{2}^{-1}\sum_{1\leq i<j\leq N}\frac{\partial\phi_{ij}(\theta_0,\eta_0)}{\partial\theta}.$$

To sum up, (A5) can be formed as

$$\sqrt{N}\left(\widehat{\theta}-\theta_0\right) = A^{-1}\sqrt{N}\binom{N}{2}^{-1}\sum_{1\leq i<j\leq N}\left\{BG^{-1}M_{ij}(\eta_0) - N_{ij}(\theta_0,\eta_0)\right\} + o_p(1),$$

where $M_{ij}(\eta_0) = \frac{1}{2}\left[\frac{\partial}{\partial\eta}\log\{p(z_i\mid u_i;\eta_0)\} + \frac{\partial}{\partial\eta}\log\{p(z_j\mid u_j;\eta_0)\}\right]$ and $N_{ij}(\theta_0,\eta_0) = \frac{\partial\phi_{ij}(\theta_0,\eta_0)}{\partial\theta}$. □

Appendix B. Proof of Theorem 2

Proof. Define function

$$q_{ij}(\theta) = \phi_{ij}\left(\theta_0 + \frac{\theta}{\sqrt{N}},\widehat{\eta}\right) - \phi_{ij}(\theta_0,\widehat{\eta}) - \left(\frac{\theta}{\sqrt{N}}\right)^T\frac{\partial\phi_{ij}(\theta_0,\widehat{\eta})}{\partial\theta} = O_p\left(\frac{1}{N}\right), \quad (A6)$$

and we can form a U-statistic based on $q_{ij}(\theta)$ as

$$Q_N(\theta) = \frac{2}{N(N-1)}\sum_{1\leq i<j\leq N}q_{ij}(\theta)$$
$$= L\left(\theta_0 + \frac{\theta}{\sqrt{N}}\right) - L(\theta_0) - \frac{1}{\sqrt{N}}\cdot\frac{2}{N(N-1)}\theta^T\sum_{1\leq i<j\leq N}\frac{\partial\phi_{ij}(\theta_0,\widehat{\eta})}{\partial\theta}.$$

The variance of $Q_N(\theta)$ is bounded by $\mathrm{var}\{Q_N(\theta)\} \leq \frac{2}{N}\mathrm{var}\{q_{ij}(\theta)\}$, from Corollary 3.2 of [37]. Meanwhile, $\frac{2}{N}\mathrm{var}\{q_{ij}(\theta)\} = \frac{2}{N}\left[E\{q_{ij}(\theta)^2\} - E\{q_{ij}(\theta)\}^2\right] \leq \frac{2}{N}E\{q_{ij}(\theta)^2\}$, as $E\{q_{ij}(\theta)\}^2 \geq 0$. As $\phi_{ij}(\theta,\widehat{\eta})$ is convex, that is, differentiable at θ_0, we can conclude

$$\phi_{ij}\left(\theta_0 + \frac{\theta}{\sqrt{N}}, \widehat{\eta}\right) - \phi_{ij}(\theta_0, \widehat{\eta}) \geq \left(\frac{\theta}{\sqrt{N}}\right)^T \frac{\partial \phi_{ij}(\theta_0, \widehat{\eta})}{\partial \theta}, \tag{A7}$$

from which we can obtain $q_{ij}(\theta) \geq 0$. Similarly,

$$\phi_{ij}\left(\theta_0 + \frac{\theta}{\sqrt{N}}, \widehat{\eta}\right) - \phi_{ij}(\theta_0, \widehat{\eta}) \leq \left(\frac{\theta}{\sqrt{N}}\right)^T \frac{\partial \phi_{ij}\left(\theta_0 + \frac{\theta}{\sqrt{N}}, \widehat{\eta}\right)}{\partial \theta}. \tag{A8}$$

From (A6)–(A8), we can conclude

$$0 \leq q_{ij}(\theta) \leq \left(\frac{\theta}{\sqrt{N}}\right)^T \left\{ \frac{\partial \phi_{ij}\left(\theta_0 + \frac{\theta}{\sqrt{N}}, \widehat{\eta}\right)}{\partial \theta} - \frac{\partial \phi_{ij}(\theta_0, \widehat{\eta})}{\partial \theta} \right\}.$$

Therefore, we can bound

$$\frac{2}{N} E\left\{ q_{ij}(\theta)^2 \right\} \leq \frac{2}{N} \left(\frac{1}{\sqrt{N}}\right)^2 E\left[\theta^T \left\{ \frac{\partial}{\partial \theta} \phi_{ij}\left(\theta_0 + \frac{\theta}{\sqrt{N}}, \widehat{\eta}\right) - \frac{\partial \phi_{ij}(\theta_0, \widehat{\eta})}{\partial \theta} \right\} \right]^2.$$

The term $\theta^T \left\{ \frac{\partial}{\partial \theta} \phi_{ij}\left(\theta_0 + \frac{\theta}{\sqrt{N}}, \widehat{\eta}\right) - \frac{\partial \phi_{ij}(\theta_0, \widehat{\eta})}{\partial \theta} \right\} \xrightarrow{p} 0$ as $N \to \infty$. Thus, $\text{var}\{N \cdot Q_N(\theta)\} \xrightarrow{p} 0$ and consequently

$$N \cdot Q_N(\theta) - N \cdot E\{Q_N(\theta)\} \xrightarrow{p} 0. \tag{A9}$$

Meanwhile, $E\{Q_N(\theta)\} = E\left\{ \phi_{ij}\left(\theta_0 + \frac{\theta}{\sqrt{N}}, \widehat{\eta}\right) \right\} - E\{\phi_{ij}(\theta_0, \widehat{\eta})\}$. Eventually from (A9) we have

$$N\left\{ L\left(\theta_0 + \frac{\theta}{\sqrt{N}}\right) - L(\theta_0) \right\} - \theta^T \sqrt{N} \frac{2}{N(N-1)} \sum_{1 \leq i < j \leq N} \frac{\partial \phi_{ij}(\theta_0, \widehat{\eta})}{\partial \theta}$$
$$- N\left[E\left\{ \phi_{ij}\left(\theta_0 + \frac{\theta}{\sqrt{N}}, \widehat{\eta}\right) \right\} - E\{\phi_{ij}(\theta_0, \widehat{\eta})\} \right] \xrightarrow{p} 0. \tag{A10}$$

The third term on the left side of (A10) has convergence properties

$$N\left[E\left\{ \phi_{ij}\left(\theta_0 + \frac{\theta}{\sqrt{N}}, \widehat{\eta}\right) \right\} - E\{\phi_{ij}(\theta_0, \widehat{\eta})\} \right]$$
$$= N\left[E\left\{ \phi_{ij}(\theta_0, \widehat{\eta}) + \left(\frac{\theta}{\sqrt{N}}\right)^T \frac{\partial \phi_{ij}(\theta_0, \widehat{\eta})}{\partial \theta} + \frac{1}{2}\left(\frac{\theta}{\sqrt{N}}\right)^T \frac{\partial^2 \phi_{ij}(\theta_0, \widehat{\eta})}{\partial \theta \partial \theta^T} \frac{\theta}{\sqrt{N}} + o_p\left(\frac{1}{N}\right) \right\} \right.$$
$$\left. - E\{\phi_{ij}(\theta_0, \widehat{\eta})\} \right]$$
$$\xrightarrow{p} \frac{1}{2} \theta^T A \theta.$$

By CLT for U-statistics,

$$\sqrt{N}\left[\frac{2}{N(N-1)} \sum_{1 \leq i < j \leq N} \frac{\partial \phi_{ij}(\theta_0, \widehat{\eta})}{\partial \theta} \right] \xrightarrow{d} N(0, \Sigma).$$

Using Slutsky's theorem, we can simplify (A10) as

$$N\left\{ L\left(\theta_0 + \frac{\theta}{\sqrt{N}}\right) - L(\theta_0) \right\} \xrightarrow{d} \frac{1}{2} \theta^T A \theta + \theta^T W,$$

where $W \sim N(0, \Sigma)$. Based on convexity [38], for every compact set $K \subset \mathbb{R}^{p+1}$, we have

$$\left[N\left\{L\left(\theta_0 + \frac{\theta}{\sqrt{N}}, \hat{\eta}\right) - L(\theta_0, \hat{\eta})\right\} : \theta \in \mathbf{K}\right] \xrightarrow{d} \left\{\frac{1}{2}\theta^{\mathrm{T}}\mathbf{A}\theta + \theta^{\mathrm{T}}\mathbf{W} : \theta \in \mathbf{K}\right\}. \tag{A11}$$

Now we develop large sample properties on the penalty term in objective function with adaptive LASSO penalty. We modify the penalty term as

$$N\sum_{j=1}^{p} \frac{\lambda}{\left|\widehat{\widetilde{\beta}}_j\right|} \left|\widetilde{\beta}_{j,0} + \frac{\widetilde{\beta}_j}{\sqrt{N}}\right| - N\sum_{j=1}^{p} \frac{\lambda}{\left|\widehat{\widetilde{\beta}}_j\right|} \left|\widetilde{\beta}_{j,0}\right|.$$

From Theorem 1, we have already obtained $\sqrt{N}\left(\widehat{\widetilde{\beta}}_j - \widetilde{\beta}_{j,0}\right) = O_p(1)$. Meanwhile, $N\lambda \to \infty$ and $\sqrt{N}\lambda \to 0$. If $\widetilde{\beta}_{j,0} \neq 0$, then $\sqrt{N}\lambda / \left|\widehat{\widetilde{\beta}}_j\right| \xrightarrow{p} 0$ and $\left|\sqrt{N}\widetilde{\beta}_{j,0} + \widetilde{\beta}_j\right| - \left|\sqrt{N}\widetilde{\beta}_{j,0}\right| \to \operatorname{sign}(\widetilde{\beta}_{j,0})\widetilde{\beta}_j$. Eventually

$$N\frac{\lambda}{\left|\widehat{\widetilde{\beta}}_j\right|}\left(\left|\widetilde{\beta}_{j,0} + \frac{\widetilde{\beta}_j}{\sqrt{N}}\right| - \left|\widetilde{\beta}_{j,0}\right|\right) = \sqrt{N}\frac{\lambda}{\left|\widehat{\widetilde{\beta}}_j\right|}\left(\left|\sqrt{N}\widetilde{\beta}_{j,0} + \widetilde{\beta}_j\right| - \left|\sqrt{N}\widetilde{\beta}_{j,0}\right|\right) \xrightarrow{p} 0.$$

If $\widetilde{\beta}_{j,0} = 0$, then $\sqrt{N}\lambda / \left|\widehat{\widetilde{\beta}}_j\right| = N\lambda / \left(\sqrt{N}\left|\widehat{\widetilde{\beta}}_j\right|\right) \xrightarrow{p} \infty$, consequently

$$N\frac{\lambda}{\left|\widehat{\widetilde{\beta}}_j\right|}\left(\left|\widetilde{\beta}_{j,0} + \frac{\widetilde{\beta}_j}{\sqrt{N}}\right| - \left|\widetilde{\beta}_{j,0}\right|\right) = \sqrt{N}\frac{\lambda}{\left|\widehat{\widetilde{\beta}}_j\right|}\left|\widetilde{\beta}_j\right| \xrightarrow{p} \begin{cases} 0, & \text{if } \widetilde{\beta}_j = 0, \\ \infty, & \text{if } \widetilde{\beta}_j \neq 0. \end{cases}$$

Therefore, we can summarize

$$N\sum_{j=1}^{p}\frac{\lambda}{\left|\widehat{\widetilde{\beta}}_j\right|}\left(\left|\widetilde{\beta}_{j,0} + \frac{\widetilde{\beta}_j}{\sqrt{N}}\right| - \left|\widetilde{\beta}_{j,0}\right|\right) \xrightarrow{p} \begin{cases} 0, & \text{if } \widetilde{\beta} = (\widetilde{\beta}_1, \ldots, \widetilde{\beta}_{p_0}, 0, \ldots, 0), \\ \infty, & \text{otherwise.} \end{cases}$$

We have infinity in the limit function, so we cannot use standard argumentation relating to uniform convergence in probability on compacts [39]. However, we can apply slightly more complicated epi-convergence. Thus, based on the works in [23,40,41], we have

$$N\left\{L\left(\theta_0 + \frac{\theta}{\sqrt{N}}\right) - L(\theta_0)\right\} + N\sum_{j=1}^{p}\frac{\lambda}{\left|\widehat{\widetilde{\beta}}_j\right|}\left(\left|\widetilde{\beta}_{j,0} + \frac{\widetilde{\beta}_j}{\sqrt{N}}\right| - \left|\widetilde{\beta}_{j,0}\right|\right) \xrightarrow{e-d} V(\theta), \tag{A12}$$

and

$$V(\theta) = \begin{cases} \frac{1}{2}\theta_T^{\mathrm{T}}\mathbf{A}_1\theta_T + \theta_T^{\mathrm{T}}\mathbf{W}_T, & \text{if } \theta = (\widetilde{\gamma}, \widetilde{\beta}_1, \ldots, \widetilde{\beta}_{p_0}, 0, \ldots, 0), \\ \infty, & \text{otherwise.} \end{cases}$$

and $\mathbf{W}_T \sim N(\mathbf{0}, \Sigma_1)$. Specifically, the left side of (A12) is minimized if $\theta = \sqrt{N}\left(\widehat{\theta}_\lambda - \theta_0\right)$ and $V(\theta)$ has a unique minimizer $\left(-(\mathbf{A}_1^{-1}\mathbf{W}_T)^{\mathrm{T}}, \mathbf{0}^{\mathrm{T}}\right)^{\mathrm{T}}$ by setting $\frac{\partial V(\theta)}{\partial \theta} = 0$. Therefore, convergence of minimizers [40] can be concluded from (A12):

$$\sqrt{N}\left(\widehat{\theta}_{\lambda,T} - \theta_{0,T}\right) \xrightarrow{d} -\mathbf{A}_1^{-1}\mathbf{W}_T \text{ and } \sqrt{N}\left(\widehat{\theta}_{\lambda,T^c} - \theta_{0,T^c}\right) \xrightarrow{d} 0. \tag{A13}$$

For $j \in T$,

$$\operatorname{pr}(j \notin T_N) = \operatorname{pr}\left(\widehat{\widetilde{\beta}}_{j,\lambda} = 0\right) \to 0.$$

Thus, $\operatorname{pr}(T \subset T_N) \to 1$. In addition, $\widehat{\theta}_\lambda$ minimizes the convex objective function $L_\lambda(\theta)$ so that $0 \in \partial L_\lambda(\widehat{\theta}_\lambda)$. As $L_\lambda(\theta)$ might be nondifferentiable and gradient of $L_\lambda(\theta)$ does not exist for some θ, we use $\partial L_\lambda(\theta)$ to represent an arbitrary selection of the subgradient of $L_\lambda(\theta)$. By taking the subgradient of the objective function with adaptive LASSO penalty, we can obtain

$$\partial L_\lambda(\widehat{\theta}_\lambda) = \partial L(\widehat{\theta}_\lambda) + \partial \left(\sum_{j=1}^{p} \frac{\lambda}{\left|\widehat{\widetilde{\beta}}_j\right|} \left|\widehat{\widetilde{\beta}}_{j,\lambda}\right| \right).$$

For $j \notin T$, $\mathrm{pr}\,(j \in T_N)$ can be upper bounded by

$$\mathrm{pr}\left(\partial_j L(\widehat{\theta}_\lambda) + \frac{\lambda}{\left|\widehat{\widetilde{\beta}}_j\right|} \mathrm{sign}\left(\widehat{\widetilde{\beta}}_{j,\lambda}\right) = 0 \right) \leq \mathrm{pr}\left(\sqrt{N}\left|\partial_j L(\widehat{\theta}_\lambda)\right| = \sqrt{N}\frac{\lambda}{\left|\widehat{\widetilde{\beta}}_j\right|} \right), \qquad (\mathrm{A}14)$$

where ∂_j is the j-th coordinate of subgradient and $\sqrt{N}\lambda/\left|\widehat{\widetilde{\beta}}_j\right| \xrightarrow{p} \infty$ as $j \notin T$.

We can expand the subgradient $\sqrt{N}\partial L(\widehat{\theta}_\lambda)$ as

$$\sqrt{N}\partial L(\widehat{\theta}_\lambda) = \sqrt{N}\left\{ \partial L(\widehat{\theta}_\lambda) - \partial L(\theta_0) - \mathbf{A}\left(\widehat{\theta}_\lambda - \theta_0\right) \right\} + \sqrt{N}\partial L(\theta_0) + \sqrt{N}\mathbf{A}\left(\widehat{\theta}_\lambda - \theta_0\right), \qquad (\mathrm{A}15)$$

where $\sqrt{N}\partial L(\theta_0)$ is bounded in probability, $\sqrt{N}\mathbf{A}\left(\widehat{\theta}_\lambda - \theta_0\right) \xrightarrow{D} \sqrt{N}\mathbf{W}$ which is bounded in probability as well. By Theorem 1 of the work in [42],

$$\sup_{|\widehat{\theta}_\lambda - \theta_0| \leq M/\sqrt{N}} \left|\partial L(\widehat{\theta}_\lambda) - \partial L(\theta_0) - \mathbf{A}\left(\widehat{\theta}_\lambda - \theta_0\right)\right| = o_p\left(\frac{1}{\sqrt{N}}\right).$$

Therefore, $\sqrt{N}\left\{ \partial L(\widehat{\theta}_\lambda) - \partial L(\theta_0) - \mathbf{A}\left(\widehat{\theta}_\lambda - \theta_0\right) \right\} \xrightarrow{p} 0$. Finally, $\sqrt{N}\left|\partial_j L(\widehat{\theta}_\lambda)\right|$ is bounded and the right side of (A14) converges to 0, which proves $\mathrm{pr}(j \in T_N) \to 0$ for $j \notin T$. □

References

1. Little, R.J.; Rubin, D.B. *Statistical Analysis with Missing Data*, 2nd ed.; Wiley: Hoboken, NJ, USA, 2002.
2. Shao, J.; Zhao, J. Estimation in longitudinal studies with nonignorable dropout. *Stat. Its Interface* **2013**, *6*, 303–313. [CrossRef]
3. Wang, S.; Shao, J.; Kim, J.K. An instrumental variable approach for identification and estimation with nonignorable nonresponse. *Stat. Sin.* **2014**, *24*, 1097–1116. [CrossRef]
4. Zhao, J.; Shao, J. Semiparametric pseudo-likelihoods in generalized linear models with nonignorable missing data. *J. Am. Stat. Assoc.* **2015**, *110*, 1577–1590. [CrossRef]
5. Miao, W.; Tchetgen Tchetgen, E.J. On varieties of doubly robust estimators under missingness not at random with a shadow variable. *Biometrika* **2016**, *103*, 475–482. [CrossRef]
6. Zhao, J.; Ma, Y. Optimal pseudolikelihood estimation in the analysis of multivariate missing data with nonignorable nonresponse. *Biometrika* **2018**, *105*, 479–486. [CrossRef]
7. Miao, W.; Liu, L.; Tchetgen Tchetgen, E.; Geng, Z. Identification, Doubly Robust Estimation, and Semiparametric Efficiency Theory of Nonignorable Missing Data With a Shadow Variable. *arXiv* **2019**, arXiv:1509.02556.
8. Tchetgen Tchetgen, E.J.; Wirth, K.E. A general instrumental variable framework for regression analysis with outcome missing not at random. *Biometrics* **2017**, *73*, 1123–1131. [CrossRef]
9. Sun, B.; Liu, L.; Miao, W.; Wirth, K.; Robins, J.; Tchetgen Tchetgen, E.J. Semiparametric estimation with data missing not at random using an instrumental variable. *Stat. Sin.* **2018**, *28*, 1965–1983.
10. Zhao, J.; Yang, Y.; Ning, Y. Penalized pairwise pseudo likelihood for variable selection with nonignorable missing data. *Stat. Sin.* **2018**, *28*, 2125–2148. [CrossRef]
11. Jiang, W.; Bogdan, M.; Josse, J.; Miasojedow, B.; Rockova, V.; Group, T. Adaptive Bayesian SLOPE–High-dimensional Model Selection with Missing Values. *arXiv* **2019**, arXiv:1909.06631.

12. Jiang, W.; Josse, J.; Lavielle, M.; Group, T. Logistic regression with missing covariates—Parameter estimation, model selection and prediction within a joint-modeling framework. *Comput. Stat. Data Anal.* **2020**, *145*, 106907. [CrossRef]
13. Johnson, A.E.; Pollard, T.J.; Shen, L.; Li-wei, H.L.; Feng, M.; Ghassemi, M.; Moody, B.; Szolovits, P.; Celi, L.A.; Mark, R.G. MIMIC-III, a freely accessible critical care database. *Sci. Data* **2016**, *3*, 160035. [CrossRef] [PubMed]
14. Zhao, J.; Ma, Y. A versatile estimation procedure without estimating the nonignorable missingness mechanism. *arXiv* **2019**, arXiv:1907.03682.
15. Liang, K.Y.; Qin, J. Regression analysis under non-standard situations: A pairwise pseudolikelihood approach. *J. R. Stat. Soc. Ser. B* **2000**, *62*, 773–786. [CrossRef]
16. Zhao, J.; Shao, J. Approximate conditional likelihood for generalized linear models with general missing data mechanism. *J. Syst. Sci. Complex.* **2017**, *30*, 139–153. [CrossRef]
17. Zhao, J. Reducing bias for maximum approximate conditional likelihood estimator with general missing data mechanism. *J. Nonparametr. Stat.* **2017**, *29*, 577–593. [CrossRef]
18. Yang, Y.; Zhao, J.; Wilding, G.; Kluczynski, M.; Bisson, L. Stability enhanced variable selection for a semiparametric model with flexible missingness mechanism and its application to the ChAMP study. *J. Appl. Stat.* **2020**, *47*, 827–843. [CrossRef]
19. Zhao, J.; Chen, C. Estimators based on unconventional likelihoods with nonignorable missing data and its application to a children's mental health study. *J. Nonparametric Stat.* **2019**, *31*, 911–931. [CrossRef]
20. Tibshirani, R. Regression shrinkage and selection via the lasso. *J. R. Stat. Soc. Ser. B* **1996**, *58*, 267–288. [CrossRef]
21. Fan, J.; Li, R. Variable selection via nonconcave penalized likelihood and its oracle properties. *J. Am. Stat. Assoc.* **2001**, *96*, 1348–1360. [CrossRef]
22. Zhang, C.H. Nearly unbiased variable selection under minimax concave penalty. *Ann. Stat.* **2010**, *38*, 894–942. [CrossRef]
23. Zou, H. The adaptive lasso and its oracle properties. *J. Am. Stat. Assoc.* **2006**, *101*, 1418–1429. [CrossRef]
24. Cai, T.; Tian, L.; Wei, L. Semiparametric Box–Cox power transformation models for censored survival observations. *Biometrika* **2005**, *92*, 619–632. [CrossRef]
25. Kosorok, M.R. *Introduction to Empirical Processes and Semiparametric Inference*; Springer Science & Business Media: Berlin/Heidelberg, Germany, 2007.
26. Minnier, J.; Tian, L.; Cai, T. A perturbation method for inference on regularized regression estimates. *J. Am. Stat. Assoc.* **2011**, *106*, 1371–1382. [CrossRef]
27. Hu, Z.; Melton, G.B.; Arsoniadis, E.G.; Wang, Y.; Kwaan, M.R.; Simon, G.J. Strategies for handling missing clinical data for automated surgical site infection detection from the electronic health record. *J. Biomed. Inform.* **2017**, *68*, 112–120. [CrossRef]
28. Li, J.; Wang, M.; Steinbach, M.S.; Kumar, V.; Simon, G.J. Don't Do Imputation: Dealing with Informative Missing Values in EHR Data Analysis. In Proceedings of the 2018 IEEE International Conference on Big Knowledge (ICBK), Singapore, 17–18 November 2018; pp. 415–422.
29. Phillips, A.; Shaper, A.G.; Whincup, P. Association between serum albumin and mortality from cardiovascular disease, cancer, and other causes. *Lancet* **1989**, *334*, 1434–1436. [CrossRef]
30. Katz, S.; Klotz, I.M. Interactions of calcium with serum albumin. *Arch. Biochem. Biophys.* **1953**, *44*, 351–361. [CrossRef]
31. Butler, S.; Payne, R.; Gunn, I.; Burns, J.; Paterson, C. Correlation between serum ionised calcium and serum albumin concentrations in two hospital populations. *Br. Med. J.* **1984**, *289*, 948–950. [CrossRef]
32. Hossain, A.; Mostafa, G.; Mannan, K.; Prosad Deb, K.; Hossain, M. Correlation Between Serum Albumin Level and Ionized Calcium in Idiopathic Nephrotic Syndrome in Children. *Urol. Nephrol. Open Access. J.* **2015**, *3*, 70–71. [CrossRef]
33. Kroll, M.; Elin, R. Relationships between magnesium and protein concentrations in serum. *Clin. Chem.* **1985**, *31*, 244–246. [CrossRef]
34. Huijgen, H.J.; Soesan, M.; Sanders, R.; Mairuhu, W.M.; Kesecioglu, J.; Sanders, G.T. Magnesium levels in critically ill patients: What should we measure? *Am. J. Clin. Pathol.* **2000**, *114*, 688–695. [CrossRef]

35. Djagbletey, R.; Phillips, B.; Boni, F.; Owoo, C.; Owusu-Darkwa, E.; deGraft Johnson, P.K.G.; Yawson, A.E. Relationship between serum total magnesium and serum potassium in emergency surgical patients in a tertiary hospital in Ghana. *Ghana Med. J.* **2016**, *50*, 78–83. [CrossRef]
36. Luo, X.; Tsai, W.Y. A proportional likelihood ratio model. *Biometrika* **2011**, *99*, 211–222. [CrossRef]
37. Shao, J. *Mathematical Statistics*; Springer Texts in Statistics; Springer: Berlin/Heidelberg, Germany, 2003.
38. Arcones, M.A. Weak convergence of convex stochastic processes. *Stat. Probab. Lett.* **1998**, *37*, 171–182. [CrossRef]
39. Rejchel, W. Model selection consistency of U-statistics with convex loss and weighted lasso penalty. *J. Nonparametric Stat.* **2017**, *29*, 768–791. [CrossRef]
40. Geyer, C.J. On the asymptotics of constrained M-estimation. *Ann. Stat.* **1994**, *22*, 1993–2010. [CrossRef]
41. Pflug, G.C. Asymptotic stochastic programs. *Math. Oper. Res.* **1995**, *20*, 769–789. [CrossRef]
42. Niemiro, W. Least empirical risk procedures in statistical inference. *Appl. Math.* **1993**, *22*, 55–67. [CrossRef]

Publisher's Note: MDPI stays neutral with regard to jurisdictional claims in published maps and institutional affiliations.

© 2020 by the authors. Licensee MDPI, Basel, Switzerland. This article is an open access article distributed under the terms and conditions of the Creative Commons Attribution (CC BY) license (http://creativecommons.org/licenses/by/4.0/).

Article

Consistent Estimation of Generalized Linear Models with High Dimensional Predictors via Stepwise Regression

Alex Pijyan [1], Qi Zheng [2], Hyokyoung G. Hong [1,*] and Yi Li [3]

1. Department of Statistics and Probability, Michigan State University, East Lansing, MI 48824, USA; pijyanal@msu.edu
2. Department of Bioinformatics and Biostatistics, University of Louisville, Louisville, KY 40202, USA; qi.zheng@louisville.edu
3. Department of Biostatistics, University of Michigan, Ann Arbor, MI 48109, USA; yili@umich.edu
* Correspondence: hhong@msu.edu

Received: 1 August 2020; Accepted: 28 August 2020; Published: 31 August 2020

Abstract: Predictive models play a central role in decision making. Penalized regression approaches, such as least absolute shrinkage and selection operator (LASSO), have been widely used to construct predictive models and explain the impacts of the selected predictors, but the estimates are typically biased. Moreover, when data are ultrahigh-dimensional, penalized regression is usable only after applying variable screening methods to downsize variables. We propose a stepwise procedure for fitting generalized linear models with ultrahigh dimensional predictors. Our procedure can provide a final model; control both false negatives and false positives; and yield consistent estimates, which are useful to gauge the actual effect size of risk factors. Simulations and applications to two clinical studies verify the utility of the method.

Keywords: estimation consistency; generalized linear models; high dimensional predictors; model selection; stepwise regression

1. Introduction

In the era of precision medicine, constructing interpretable and accurate predictive models, based on patients' demographic characteristics, clinical conditions and molecular biomarkers, has been crucial for disease prevention, early diagnosis and targeted therapy [1]. When the number of predictors is moderate, penalized regression approaches such as least absolute shrinkage and selection operator (LASSO) by [2] have been used to construct predictive models and explain the impacts of the selected predictors. However, in ultrahigh dimensional settings where p is in the exponential order of n, penalized methods may incur computational challenges [3], may not reach globally optimal solutions and often generate biased estimates [4]. Sure independence screening (SIS) proposed by [5] has emerged as a powerful tool for modeling ultrahigh dimensional data. However, the method relies on a partial faithfulness assumption, which stipulates that jointly important variables must be marginally important, an assumption that may not be always realistic. To relieve this condition, some iterative procedures, such as ISIS [5], have been adopted to repeatedly screen variables based on the residuals from the previous iterations, but with heavy computation and unclear theoretical properties. Conditional screening approaches [see, e.g., [6]] have, to some extent, addressed the challenge. However, screening methods do not directly generate a final model, and post-screening regularization methods, such as LASSO, are recommended by [5] to produce a final model.

For generating a final predictive model in ultrahigh dimensional settings, recent years have seen a surging interest of performing forward regression, an old technique for model selection; see [7–9], among many others. Under some regularity conditions and with some proper stopping criteria, forward regression can achieve screening consistency and sequentially select variables according to metrics such as AIC, BIC or R^2. Closely related to forward selection also, is least angle regression (LARS) [10], a widely used model selection algorithm for high-dimensional models. In the generalized linear model setting [11,12], proposed differential geometrical LARS (dgLARS) based on a differential geometrical extension of LARS.

However, these methods have drawbacks. First, once a variable is identified by the forward selection, it is not removable from the list of selected variables. Hence, false positives are unavoidable without a systematic elimination procedure. Second, most of the existing works focus on variable selection and are silent with respect to estimation accuracy.

To address the first issue, some works have been proposed to add backward elimination steps once forward selection is accomplished, as backward elimination may further eliminate false positives from the variables selected by forward selection. For example, ref. [13,14] proposed a stepwise selection for linear regression models in high-dimensional settings and proved model selection consistency. However, it is unclear whether the results hold for high-dimensional generalized linear models (GLMs); Ref. [15] proposed a similar stepwise algorithm in high-dimensional GLM settings, but with no theoretical properties on model selection. Moreover, none of the relevant works have touched upon the accuracy of estimation.

We extend a stepwise regression method to accommodate GLMs with high-dimensional predictors. Our method embraces both model selection and estimation. It starts with an empty model or pre-specified predictors, scans all features and sequentially selects features, and conducts backward elimination once forward selection is completed. Our proposal controls both false negatives and false positives in high dimensional settings: the forward selection steps recruit variables in an inclusive way by allowing some false positives for the sake of avoiding false negatives, while the backward selection steps eliminate the potential false positives from the recruited variables. We use different stopping criteria in the forward and backward selection steps, to control the numbers of false positives and false negatives. Moreover, we prove that, under a sparsity assumption of the true model, the proposed approach can discover all of the relevant predictors within a finite number of steps, and the estimated coefficients are consistent, a property still unknown to the literature. Finally, our GLM framework enables our work to accommodate a wide range of data types, such as binary, categorical and count data.

To recap, our proposed method distinguishes from the existing stepwise approaches in high dimensional settings. For example, it improves [13,14] by extending the work to a more broad GLM setting and [15] by establishing the theoretical properties.

Compared with the other variable selection and screening works, our method produces a final model in ultrahigh dimensional settings, without applying a pre-screening step which may produce unintended false negatives. Under some regularity conditions, the method identifies or includes the true model with probability going to 1. Moreover, unlike the penalized approaches such as LASSO, the coefficients estimated by our stepwise selection procedure in the final model will be consistent, which are useful for gauging the real effect sizes of risk factors.

2. Method

Let $(\mathbf{X}_i, Y_i)$, $i = 1, \ldots, n$, denote n independently and identically distributed (i.i.d.) copies of $(\mathbf{X}, Y)$. Here, $\mathbf{X} = (1, X_1, \ldots, X_p)^T$ is a $(p+1)$-dimensional predictor vector with $X_0 = 1$ corresponding to the intercept term, and Y is an outcome. Suppose that the conditional density of Y, given $\mathbf{X}$, belongs to a linear exponential family:

$$\pi(Y \mid \mathbf{X}) = \exp\{Y\mathbf{X}^T\beta - b(\mathbf{X}^T\beta) + \mathcal{A}(Y)\}, \tag{1}$$

where $\boldsymbol{\beta} = (\beta_0, \beta_1, \ldots, \beta_p)^{\mathrm{T}}$ is the vector of coefficients; β_0 is the intercept; and $\mathcal{A}(\cdot)$ and $b(\cdot)$ are known functions. Model (1), with a canonical link function and a unit dispersion parameter, belongs to a larger exponential family [16]. Further, $b(\cdot)$ is assumed twice continuously differentiable with a non-negative second derivative $b''(\cdot)$. We use $\mu(\cdot)$ and $\sigma(\cdot)$ to denote $b'(\cdot)$ and $b''(\cdot)$, i.e., the mean and variance functions, respectively. For example, $b(\theta) = \log(1 + \exp(\theta))$ in a logistic distribution and $b(\theta) = \exp(\theta)$ in a Poisson distribution.

Let $L(u,v) = uv - b(u)$ and $\mathbb{E}_n\{f(\xi)\} = n^{-1}\sum_{i=1}^n f(\xi_i)$ denote the mean of $\{f(\xi_i)\}_{i=1}^n$ for a sequence of i.i.d. random variables ξ_i ($i = 1, \ldots, n$) and a non-random function $f(\cdot)$. Based on the i.i.d. observations, the log-likelihood function is

$$\ell(\boldsymbol{\beta}) = n^{-1}\sum_{i=1}^n L(\mathbf{X}_i^{\mathrm{T}}\boldsymbol{\beta}, Y_i) = \mathbb{E}_n\{L(\mathbf{X}^{\mathrm{T}}\boldsymbol{\beta}, Y)\}. \tag{2}$$

We use $\boldsymbol{\beta}_* = (\beta_{*0}, \beta_{*1}, \ldots, \beta_{*p})^{\mathrm{T}}$ to denote the true values of $\boldsymbol{\beta}$. Then the true model is $\mathcal{M} = \{j : \beta_{*j} \neq 0, j \geq 1\} \cup \{0\}$, which consists of the intercept and all variables with nonzero effects. Overarching goals of ultra-high dimensional data analysis are to identify $\mathcal{M}$ and estimate β_{*j} for $j \in \mathcal{M}$. While most of the relevant literature [8,9] is on estimating $\mathcal{M}$, this work is to accomplish both identification of $\mathcal{M}$ and estimation of β_{*j}.

When p is in the exponential order of n, we aim to generate a predictive model that contains the true model with high probability, and provide consistent estimates of regression coefficients. We further introduce the following notation. For a generic index set $S \subset \{0, 1, \ldots, p\}$ and a $(p+1)$-dimensional vector $\mathbf{A}$, we use S^c to denote the complement of a set S and $\mathbf{A}_S = \{A_j : j \in S\}$ to denote the subvector of $\mathbf{A}$ corresponding to S. For instance, if $S = \{2, 3, 4\}$, then $\mathbf{X}_{iS} = (X_{i2}, X_{i3}, X_{i4})^{\mathrm{T}}$. Moreover, denote by $\ell_S(\boldsymbol{\beta}_S) = \mathbb{E}_n\{L(\mathbf{X}_S^{\mathrm{T}}\boldsymbol{\beta}_S, Y)\}$ the log-likelihood of the regression model of Y on $\mathbf{X}_S$ and denote by $\hat{\boldsymbol{\beta}}_S$ the maximizer of $\ell_S(\boldsymbol{\beta}_S)$. Under model (1), we elaborate on the idea of stepwise (details in the supplementary materials) selection, consisting of the forward and backward stages.

Forward stage: We start with F_0, a set of variables that need to be included according to some *a priori* knowledge, such as clinically important factors and conditions. If no such information is available, F_0 is set to be $\{0\}$, corresponding to a null model. We sequentially add covariates as follows:

$$F_0 \subset F_1 \subset F_2 \subset \cdots \subset F_k,$$

where $F_k \subset \{0, 1, \ldots, p\}$ is the index set of the selected covariates upon completion of the kth step, with $k \geq 0$. At the $(k+1)$th step, we append new variables to F_k one at a time and refit GLMs: for every $j \in F_k^c$, we let $F_{k,j} = F_k \cup \{j\}$, obtain $\hat{\boldsymbol{\beta}}_{F_{k,j}}$ by maximizing $\ell_{F_{k,j}}(\boldsymbol{\beta}_{F_{k,j}})$, and compute the increment of log-likelihood,

$$\ell_{F_{k,j}}(\hat{\boldsymbol{\beta}}_{F_{k,j}}) - \ell_{F_k}(\hat{\boldsymbol{\beta}}_{F_k}).$$

Then the index of a new candidate variable is determined to be

$$j_{k+1} = \arg\max_{j \in F_k^c} \ell_{F_{k,j}}(\hat{\boldsymbol{\beta}}_{F_{k,j}}) - \ell_{F_k}(\hat{\boldsymbol{\beta}}_{F_k}).$$

Additionally, we update $F_{k+1} = F_k \cup \{j_{k+1}\}$. We then need to decide whether to stop at the kth step or move on to the $(k+1)$th step with F_{k+1}. To do so, we use the following EBIC criterion:

$$\mathrm{EBIC}(F_{k+1}) = -2\ell_{F_{k+1}}(\hat{\boldsymbol{\beta}}_{F_{k+1}}) + |F_{k+1}|n^{-1}(\log n + 2\eta_1 \log p), \tag{3}$$

where the second term is motivated by [17] and $|F|$ denotes the cardinality of a set F.

The forward selection stops if $\mathrm{EBIC}(F_{k+1}) > \mathrm{EBIC}(F_k)$. We denote the stopping step by k^* and the set of variables selected so far by F_{k^*}.

Backward stage: Upon the completion of forward stage, backward elimination, starting with $B_0 = F_{k^*}$, sequentially drops covariates as follows:

$$B_0 \supset B_1 \supset B_2 \supset \cdots \supset B_k,$$

where B_k is the index set of the remaining covariates upon the completion of the kth step of the backward stage, with $k \geq 0$. At the $(k+1)$th backward step and for every $j \in B_k$, we let $B_{k/j} = B_k \setminus \{j\}$, obtain $\hat{\boldsymbol{\beta}}_{B_{k/j}}$ by maximizing $\ell(\boldsymbol{\beta}_{B_{k/j}})$, and calculating the difference of the log-likelihoods between these two nested models:

$$\ell_{B_k}(\hat{\boldsymbol{\beta}}_{B_k}) - \ell_{B_{k/j}}(\hat{\boldsymbol{\beta}}_{B_{k/j}}).$$

The variable that can be removed from the current set of variables is indexed by

$$j_{k+1} = \arg\min_{j \in B_k} \ell_{B_k}(\hat{\boldsymbol{\beta}}_{B_k}) - \ell_{B_{k/j}}(\hat{\boldsymbol{\beta}}_{B_{k/j}}).$$

Let $B_{k+1} = B_k \setminus \{j_{k+1}\}$. We determine whether to stop at the kth step or move on to the $(k+1)$th step of the backward stage according to the following BIC criterion:

$$\text{BIC}(B_{k+1}) = -2\ell_{B_{k+1}}(\hat{\boldsymbol{\beta}}_{B_{k+1}}) + \eta_2 n^{-1}|B_{k+1}|\log n. \tag{4}$$

If $\text{BIC}(B_{k+1}) > \text{BIC}(B_k)$, we end the backward stage at the kth step. Let k^{**} denote the stopping step and we declare the selected model $B_{k^{**}}$ to be the final model. Thus, $\hat{\mathcal{M}} = B_{k^{**}}$ is the estimate of $\mathcal{M}$. As the backward stage starts with the k^* variables selected by forward selection, k^{**} cannot exceed k^*.

A strength of our algorithm, termed STEPWISE hereafter, is the added flexibility with η_1 and η_2 in the stopping criteria for controlling the false negatives and positives. For example, a smaller value of η_1 close to zero in the forward selection step will likely include more variables, and thus incur more false positives and less false negatives, whereas a larger value of η_1 will recruit too few variables and cause too many false negatives. Similarly, in the backward selection step, a large η_2 would eliminate more variables and therefore further reduce more false positives, and vice versa for a small η_2. While finding optimal η_1 and η_2 is not trivial, our numerical experiences suggest a small η_1 and a large η_2 may well balance the false negatives and positives. When $\eta_2 = 0$, no variables can be dropped after forward selection; hence, our proposal includes forward selection as a special case.

Moreover, [8] proposed a sequentially conditioning approach based on offset terms that absorb the prior information. However, our numerical experiments indicate that the offset approach may be suboptimal compared to our full stepwise optimization approach, which will be demonstrated in the simulation studies.

3. Theoretical Properties

With a column vector $\mathbf{v}$, let $\|\mathbf{v}\|_q$ denote the L_q-norm for any $q \geq 1$. For simplicity, we denote the L_2-norm of $\mathbf{v}$ by $\|\mathbf{v}\|$, and denote $\mathbf{v}\mathbf{v}^T$ by $\mathbf{v}^{\otimes 2}$. We use $C_1, C_2, \ldots$, to denote some generic constants that do not depend on n and may change from line to line. The following regularity conditions are set.

1. There exist a positive integer q satisfying $|\mathcal{M}| \leq q$ and $q\log p = o(n^{1/3})$ and a constant $K > 0$ such that $\sup_{|S| \leq q} \|\boldsymbol{\beta}_S^*\|_1 \leq K$, where $\boldsymbol{\beta}_S^* = \arg\max_{\boldsymbol{\beta}_S} E[\ell_S(\boldsymbol{\beta}_S)]$ is termed the least false value of model S.
2. $\|\mathbf{X}\|_\infty \leq K$. In addition, $E(X_j) = 0$ and $E(X_j^2) = 1$ for $j \geq 1$.
3. Let $\epsilon_i = Y_i - \mu(\boldsymbol{\beta}_*^T \mathbf{X}_i)$. There exists a positive constant M such that the Cramer condition holds, i.e., $E[|\epsilon_i|^m] \leq m! M^m$ for all $m \geq 1$.
4. $|\sigma(a) - \sigma(b)| \leq K|a - b|$ and $\sigma_{\min} := \inf_{|t| \leq K^3} |b''(t)|$ is bounded below.

5. There exist two positive constants, $\kappa_{\min}$ and $\kappa_{\max}$ such that $0 < \kappa_{\min} < \Lambda\left(E\left(\mathbf{X}_S^{\otimes 2}\right)\right) < \kappa_{\max} < \infty$, uniformly in $S \subset \{0, 1, \ldots, p\}$ satisfying $|S| \leq q$, where $\Lambda(\mathbf{A})$ is the collection of all eigenvalues of a square matrix $\mathbf{A}$.
6. $\min_{S:\mathcal{M} \not\subseteq S, |S| \leq q} D_S > C n^{-\alpha}$ for some constants $C > 0$ and $\alpha > 0$ that satisfies $q n^{-1+4\alpha} \log p \to 0$, where $D_S = \max_{j \in S^c \cap \mathcal{M}} \left| E\left[\left(\mu(\boldsymbol{\beta}_*^T \mathbf{X}) - \mu(\boldsymbol{\beta}_S^{*T} \mathbf{X}_S)\right) X_j\right] \right|$.

Condition (1), as assumed in [8,18], is an alternative to the Lipschitz assumption [5,19]. The bound of the model size allowed in the selection procedure or q is often required in model-based screening methods see, e.g., [8,20–22]. The bound should be large enough so that the correct model can be included, but not too large; otherwise, excessive noise variables would be included, leading to unstable and inconsistent estimates. Indeed, Conditions (1) and (6) reveal that the range of q depends on the true model size $|\mathcal{M}|$, the minimum signal strength, $n^{-\alpha}$ and the total number of covariates, p. The upper bound of q is $o((n^{1-4\alpha}/\log p) \wedge (n^{1/3}/\log p))$, ensuring the consistency of EBIC [17]. Condition (1) also implies that the parameter space under consideration can be restricted to $\mathbb{B} := \{\boldsymbol{\beta} \in \mathbb{R}^{p+1} : \|\boldsymbol{\beta}\|_1 \leq K^2\}$, for any model S with $|S| \leq q$. Condition (2), as assumed in [23,24], reflects that data are often standardized at the pre-processing stage. Condition (3) ensures that Y has a light tail, and is satisfied by Gaussian and discrete data, such as binary and count data [25]. Condition (4) is satisfied by common GLM models, such as Gaussian, binomial, Poisson and gamma distributions. Condition (5) represents the sparse Riesz condition [26] and Condition (6) is a strong "irrepresentable" condition, suggesting that $\mathcal{M}$ cannot be represented by a set of variables that does not include the true model. It further implies that adding a signal variable to a mis-specified model will increase the log-likelihood by a certain lower bound [8]. The signal rate is comparable to the conditions required by the other sequential methods, see, e.g., [7,22].

Theorem 1 develops a lower bound of the increment of the log-likelihood if the true model $\mathcal{M}$ is not yet included in a selected model S.

Theorem 1. *Suppose Conditions (1)–(6) hold. There exists some constant C_1 such that with probability at least $1 - 6\exp(-6q \log p)$,*

$$\min_{S:\mathcal{M} \not\subseteq S, |S| < q} \left\{\max_{j \in S^c} \ell_{S \cup \{j\}}(\hat{\boldsymbol{\beta}}_{S \cup \{j\}}) - \ell_S(\hat{\boldsymbol{\beta}}_S)\right\} \geq C_1 n^{-2\alpha}.$$

Theorem 1 shows that, before the true model is included in the selected model, we can append a variable which will increase the log-likelihood by at least $C_1 n^{-2\alpha}$ with probability tending to 1. This ensures that in the forward stage, our proposed STEPWISE approach will keep searching for signal variables until the true model is contained. To see this, suppose at the kth step of the forward stage that F_k satisfies $\mathcal{M} \not\subseteq F_k$ and $|F_k| < q$, and let r be the index selected by STEPWISE. By Theorem 1, we obtain that, for any $\eta_1 > 0$, when n is sufficiently large,

$$\begin{aligned}
\text{EBIC}(F_{k,r}) - \text{EBIC}(F_k) &= -2\ell_{F_{k,r}}(\hat{\boldsymbol{\beta}}_{F_{k,r}}) + (|F_k| + 1)n^{-1}(\log n + 2\eta_1 \log p) \\
&\quad - \left[-2\ell_{F_k}(\hat{\boldsymbol{\beta}}_{F_k}) + |F_k|n^{-1}(\log n + 2\eta_1 \log p)\right] \\
&\leq -2C_1 n^{-2\alpha} + n^{-1}(\log n + 2\eta_1 \log p) < 0,
\end{aligned}$$

with probability at least $1 - 6\exp(-6q \log p)$, where the last inequality is due to Condition (6). Therefore, with high probability the forward stage of STEPWISE continues as long as $\mathcal{M} \not\subseteq F_k$ and $|F_k| < q$. We next establish an upper bound of the number of steps in the forward stage needed to include the true model.

Theorem 2. *Under the same conditions as in Theorem 1 and if*

$$\max_{S:|S|\leq q} \left\{ \max_{j\in S^c \cap \mathcal{M}^c} \left| E\left[\left\{ Y - \mu(\boldsymbol{\beta}_S^{*T}\mathbf{X}_S) \right\} X_j \right] \right| \right\} = o(n^{-\alpha}),$$

then there exists some constant $C_2 > 2$ such that $\mathcal{M} \subset F_k$, for some F_k in the forward stage of STEPWISE and $k \leq C_2|\mathcal{M}|$, with probability at least $1 - 18\exp(-4q\log p)$.

The "max" condition, as assumed in Section 5.3 of [27], relaxes the partial orthogonality assumption that $\mathbf{X}_{\mathcal{M}^c}$ are independent of $\mathbf{X}_\mathcal{M}$, and ensures that with probability tending to 1, appending a signal variable increases log-likelihood more than adding a noise variable does, uniformly over all possible models S satisfying $\mathcal{M} \not\subseteq S, |S| < q$. This entails that the proposed procedure is much more likely to select a signal variable, in lieu of a noise variable, at each step. Since EBIC is a consistent model selection criterion [28,29], the following theorem guarantees termination of the proposed procedure with $\mathcal{M} \subset F_k$ for some k.

Theorem 3. *Under the same conditions as in Theorem 2 and if $\mathcal{M} \not\subset F_{k-1}$ and $\mathcal{M} \subset F_k$, the forward stage stops at the kth step with probability going to $1 - \exp(-3q\log p)$.*

Theorem 3 ensures that the forward stage of STEPWISE will stop within a finite number of steps and will cover the true model with probability at least $1 - q\exp(-3q\log p) \geq 1 - \exp(-2q\log p)$. We next consider the backward stage and provide a probability bound of removing a signal from a set in which the set of true signals $\mathcal{M}$ is contained.

Theorem 4. *Under the same conditions as in Theorem 2, $BIC(S\setminus\{r\}) - BIC(S) > 0$ uniformly over $r \in \mathcal{M}$ and S satisfying $\mathcal{M} \subset S$ and $|S| \leq q$, with probability at least $1 - 6\exp(-6q\log p)$.*

Theorem 4 indicates that with probability at $1 - 6\exp(-6q\log p)$, BIC would decrease when removing a signal variable from a model that contains the true model. That is, with high probability, back elimination is to reduce false positives.

Recall that F_{k^*} denotes the model selected at the end of the forward selection stage. By Theorem 2, $\mathcal{M} \subset F_{k^*}$ with probability at least $1 - 18\exp(-4q\log p)$. Then Theorem 4 implies that at each step of the backward stage, a signal variable will not be removed from the model with probability at least $1 - 6\exp(-6q\log p)$. By Theorem 2, $|F_{k^*}| \leq C_2|\mathcal{M}|$. Thus, the backward elimination will carry out at most $(C_2 - 1)|\mathcal{M}|$ steps. Combining results from Theorems 2 and 3 yields that $\mathcal{M} \subset \hat{\mathcal{M}}$ with probability at least $1 - 18\exp(-4q\log p) - 6(C_2 - 1)|\mathcal{M}|\exp(-6q\log p)$. Let $\hat{\boldsymbol{\beta}}$ be the estimate of $\boldsymbol{\beta}_*$ in model (1) at the termination of STEPWISE. By convention, the estimates of the coefficients of the unselected covariates are 0.

Theorem 5. *Under the same conditions as in Theorem 2, we have that $\mathcal{M} \subseteq \hat{\mathcal{M}}$ and*

$$\|\hat{\boldsymbol{\beta}} - \boldsymbol{\beta}_*\| \to 0$$

in probability.

The theorem warrants that the proposed STEPWISE yields consistent estimates, a property not shared by many regularized methods, including LASSO. Our later simulations verified this. Proof of main theorems and lemmas are provided in Appendix A.

4. Simulation Studies

We compared the proposal with the other competing methods, including the penalized methods, such as least absolute shrinkage and selection operator (LASSO); the differential geometric least angle regression (dgLARS) [11,12]; the forward regression (FR) approach [7]; the sequentially conditioning (SC) approach [8]; and the screening methods, such as sure independence screening (SIS) [5], which is popular in practice. As SIS does not directly generate a predictive model, we applied LASSO for the top $[n/\log(n)]$ variables chosen by SIS and denoted the procedure by SIS+LASSO. As the FR, SC and STEPWISE approaches involve forward searching and to make them comparable, we applied the same stopping rule, for example, Equation (3) with the same γ, to their forward steps. In particular, the STEPWISE approach, with $\eta_1 = \gamma$ and $\eta_2 = 0$, is equivalent to FR and asymptotically equivalent to SC. By varying γ in FR and SC between γ_L and γ_H, we explored the impact of γ on inducing false positives and negatives. In our numerical studies, we fixed $\gamma_H = 10$ and set $\gamma_L = \eta_1$. To choose η_1 and η_2 in (3) and (4) in STEPWISE, we performed 5-fold cross-validation to minimize the mean squared prediction error (MSPE), and reported the results in Table 1. Since the proposed STEPWISE algorithm uses the (E)BIC criterion, for a fair comparison we chose the tuning parameter in dgLARS by using the BIC criterion as well, and coined the corresponding approach as dgLARS(BIC). The regularization parameter in LASSO was chosen via the following two approaches: (1) giving the smallest BIC for the models on the LASSO path, denoted by LASSO(BIC); (2) using the one-standard-error rule, denoted by LASSO(1SE), which chooses the most parsimonious model whose error is no more than one standard error above the error of the best model in cross-validation [30].

Table 1. The values of η_1 and η_2 used in the simulation studies.

	Normal Model	Binomial Model	Poisson Model
Example 1	(0.5, 3)	(0.5, 3)	(1, 3)
Example 2	(0.5, 3)	(1, 3)	(1, 3)
Example 3	(1, 3)	(0.5, 3)	(0.5, 1)
Example 4	(1, 3.5)	(0, 1)	(1, 3)
Example 5	(0.5, 3)	(0.5, 2)	(0.5, 3)

Note: values for η_1 and η_2 were searched on the grid $\{0, 0.25, 0.5, 1\}$ and $\{1, 2, 3, 3.5, 4, 4.5, 5\}$, respectively.

Denote by $\mathbf{X}_i = (X_{i1}, \ldots, X_{ip})^T$ and $\boldsymbol{\beta} = (\beta_1, \ldots, \beta_p)^T$, the covariate vector for subject i $(1, \ldots, n)$ and the true coefficient vector. The following five examples generated $\mathbf{X}_i^T \boldsymbol{\beta}$, the inner product of the coefficient and covariate vectors for each individual, which were used to generate outcomes from the normal, binomial and Poisson models.

Example 1. For each i,

$$c\mathbf{X}_i^T \boldsymbol{\beta} = c \times \left(\sum_{j=1}^{p_0} \beta_j X_{ij} + \sum_{j=p_0+1}^{p} \beta_j X_{ij} \right), \quad i = 1, \ldots, n,$$

where $\beta_j = (-1)^{B_j}(4\log n/\sqrt{n} + |Z_j|)$, for $j = 1, \ldots, p_0$ and $\beta_j = 0$ otherwise B_j was a binary random variable with $P(B_j = 1) = 0.4$ and Z_j was generated by a standard normal distribution; $p_0 = 8$; X_{ij}s were independently generated from a standardized exponential distribution, that is, $\exp(1) - 1$. Here and also in the other examples, c (specified later) controls the signal strengths.

Example 2. This scenario is the same as **Example 1** except that X_{ij} was independently generated from a standard normal distribution.

Example 3. For each i,

$$c\mathbf{X}_i^T \boldsymbol{\beta} = c \times \left(\sum_{j=1}^{p_0} \beta_j X_{ij} + \sum_{j=p_0+1}^{p} \beta_j X_{ij}^* \right), \quad i = 1, \ldots, n,$$

where $\beta_j = 2j$ for $1 \leq j \leq p_0$ and $p_0 = 5$. We simulated every component of $\mathbf{Z}_i = (Z_{ij}) \in R^p$ and $\mathbf{W}_i = (W_{ij}) \in R^p$ independently from a standard normal distribution. Next, we generated $\mathbf{X}_i$ according to $X_{ij} = (Z_{ij} + W_{ij})/\sqrt{2}$ for $1 \leq j \leq p_0$ and $X_{ij}^* = (Z_{ij} + \sum_{j'=1}^{p_0} Z_{ij'})/2$ for $p_0 < j \leq p$.

Example 4. For each i,

$$c\mathbf{X}_i^T \boldsymbol{\beta} = c \times \left(\sum_{j=1}^{500} \beta_j X_{ij} + \sum_{j=501}^{p} \beta_j X_{ij} \right), i = 1, \ldots, n,$$

where the first 500 X_{ij}s were generated from the multivariate normal distribution with mean $\mathbf{0}$ and a covariance matrix with all of the diagonal entries being 1 and $cov(X_{ij}, X_{ij'}) = 0.5^{|j-j'|}$ for $1 \leq j, j' \leq p$. The remaining $p - 500$ X_{ij}s were generated through the autoregressive processes with $X_{i,501} \sim$ Unif(-2, 2), $X_{ij} = 0.5 X_{i,j-1} + 0.5 X_{ij}^*$, for $j = 502, \ldots, p$, where $X_{ij}^* \sim$ Unif(-2, 2) were generated independently. The coefficients β_j for $j = 1, \ldots, 7, 501, \ldots, 507$ were generated from $(-1)^{B_j}(4\log n/\sqrt{n} + |Z_j|)$, where B_j was a binary random variable with $P(B_j = 1) = 0.4$ and Z_j was from a standard normal distribution. The remaining β_j were zeros.

Example 5. For each i,

$$c\mathbf{X}_i^T \boldsymbol{\beta} = c \times \left(-0.5 X_{i1} + X_{i2} + 0.5 X_{i,100} \right), i = 1, \ldots, n,$$

where $\mathbf{X}_i$ were generated from a multivariate normal distribution with mean $\mathbf{0}$ and a covariance matrix with all of the diagonal entries being 1 and $cov(X_{ij}, X_{ij'}) = 0.9^{|j-j'|}$ for $1 \leq j, j' \leq p$. All of the coefficients were zero except for X_{i1}, X_{i2} and $X_{i,100}$.

Examples 1 and 3 were adopted from [7], while **Examples 2 and 4** were borrowed from [5,15], respectively. We then generated the responses from the following three models.

Normal model: $Y_i = c\mathbf{X}_i^T \boldsymbol{\beta} + \epsilon_i$ with $\epsilon_i \sim N(0, 1)$.
Binomial model: $Y_i \sim$ Bernoulli($\exp(c\mathbf{X}_i^T \boldsymbol{\beta})/\{1 + \exp(c\mathbf{X}_i^T \boldsymbol{\beta})\}$).
Poisson model: $Y_i \sim$ Poisson($\exp(c\mathbf{X}_i^T \boldsymbol{\beta})$).

We considered $n = 400$ and $p = 1000$ throughout all of the examples. We specified the magnitude of the coefficients in the GLMs with a constant multiplier, c. For Examples 1–5, this constant was set, respectively for the normal, binomial and Poisson models, to be: (1, 1, 0.3), (1, 1.5, 0.3), (1, 1, 0.1), (1, 1.5, 0.3) and (1, 3, 2). For each parameter configuration, we simulated 500 independent data sets. We evaluated the performances of the methods by the criteria of true positives (TP), false positives (FP), the estimated probability of including the true models (PIT), the mean squared error (MSE) of $\hat{\boldsymbol{\beta}}$ and the mean squared prediction error (MSPE). To compute the MSPE, we randomly partitioned the samples into the training (75%) and testing (25%) sets. The models obtained from the training datasets were used to predict the responses in the testing datasets. Tables 2–4 report the average TP, FP, PIT, MSE and MSPE over 500 datasets along with the standard deviations. The findings are summarized below.

First, the proposed STEPWISE method was able to detect all the true signals with nearly zero FPs. Specifically, in all of the Examples, STEPWISE outperformed the other methods by detecting more TPs with fewer FPs, whereas LASSO, SIS+LASSO and dgLARS included much more FPs.

Second, though a smaller γ in FR and SC led to the inclusion of all TPs with a PIT close to 1, it incurred more FPs. On the other hand, a larger γ may eliminate some TPs, resulting in a smaller PIT and a larger MSPE.

Third, for the normal model, the STEPWISE method yielded an MSE close to 0, the smallest among all the competing methods. The binary and Poisson data challenged all of the methods, and the MSEs for all the methods were non-negligible. However, the STEPWISE method still produced the lowest MSE. The results seemed to verify the consistency of $\hat{\beta}$, which distinguished the proposed STEPWISE method from the other regularized methods and highlighted its ability to provide a more accurate means to characterize the effects of high dimensional predictors.

Table 2. Normal model.

Example	Method	TP	FP	PIT	MSE ($\times 10^{-4}$)	MSPE
1 ($p_0 = 8$)	LASSO(1SE)	8.00 (0.00)	5.48 (6.61)	1.00 (0.00)	2.45	1.148
	LASSO(BIC)	8.00 (0.00)	2.55 (2.48)	1.00 (0.00)	2.58	1.172
	SIS+LASSO(1SE)	8.00 (0.00)	6.59 (4.22)	1.00 (0.00)	1.49	1.042
	SIS+LASSO(BIC)	8.00 (0.00)	6.04 (3.33)	1.00 (0.00)	1.37	1.025
	dgLARS(BIC)	8.00 (0.00)	3.52(2.53)	1.00 (0.00)	2.25	1.130
	SC (γ_L)	8.00 (0.00)	3.01 (1.85)	1.00 (0.00)	1.09	0.895
	SC (γ_H)	7.60 (1.59)	0.00 (0.00)	0.94 (0.24)	14.56	5.081
	FR (γ_L)	8.00 (0.00)	2.96 (2.04)	1.00 (0.00)	1.08	0.896
	FR (γ_H)	7.88 (0.84)	0.00 (0.00)	0.98 (0.14)	3.74	2.040
	STEPWISE	8.00 (0.00)	0.00 (0.00)	1.00 (0.00)	0.21	0.972
2 ($p_0 = 8$)	LASSO(1SE)	8.00 (0.00)	4.74 (4.24)	1.00 (0.00)	2.46	1.154
	LASSO(BIC)	8.00 (0.00)	2.12 (2.02)	1.00 (0.00)	2.62	1.182
	SIS+LASSO	7.99 (0.10)	6.84 (4.57)	0.99 (0.10)	1.65	1.058
	SIS+LASSO(BIC)	7.99 (0.10)	6.11 (3.85)	0.99 (0.10)	1.56	1.041
	dgLARS(BIC)	8.00 (0.00)	3.26(2.62)	1.00 (0.00)	2.28	1.138
	SC (γ_L)	8.00 (0.00)	2.73 (1.53)	1.00 (0.00)	0.98	0.901
	SC (γ_H)	7.30 (2.11)	0.00 (0.00)	0.90 (0.30)	23.70	6.397
	FR (γ_L)	8.00 (0.00)	2.45 (1.65)	1.00 (0.00)	0.92	0.907
	FR (γ_H)	7.94 (0.60)	0.00 (0.00)	0.99 (0.00)	2.69	2.062
	STEPWISE	8.00 (0.00)	0.01 (0.10)	1.00 (0.00)	0.21	0.972
3 ($p_0 = 5$)	LASSO(1SE)	5.00 (0.00)	8.24 (2.63)	1.00 (0.00)	3.07	1.084
	LASSO(BIC)	5.00 (0.00)	12.33 (3.28)	1.00 (0.00)	27.97	2.398
	SIS+LASSO(1SE)	0.97 (0.26)	15.94 (2.93)	0.00 (0.00)	1406.22	76.024
	SIS+LASSO(BIC)	0.97 (0.26)	16.20 (2.81)	0.00 (0.00)	1354.54	71.017
	dgLARS(BIC)	5.00 (0.00)	53.91 (14.44)	1.00 (0.00)	6.63	0.979
	SC (γ_L)	4.48 (0.50)	0.25 (0.44)	0.48 (0.50)	21.74	3.086
	SC (γ_H)	4.48 (0.50)	0.14 (0.35)	0.48 (0.50)	21.70	2.065
	FR (γ_L)	5.00 (0.00)	0.23 (0.66)	1.00 (0.00)	0.27	0.973
	FR (γ_H)	5.00 (0.00)	0.14 (0.35)	1.00 (0.00)	0.15	0.074
	STEPWISE	5.00 (0.00)	0.03 (0.22)	1.00 (0.00)	0.18	0.976
4 ($p_0 = 14$)	LASSO(1SE)	14.00 (0.00)	29.84 (15.25)	1.00 (0.00)	13.97	1.148
	LASSO(BIC)	13.94 (0.24)	4.92 (5.54)	0.94 (0.24)	38.69	1.995
	SIS+LASSO(1SE)	11.44 (1.45)	15.19 (7.29)	0.05 (0.21)	133.38	4.714
	SIS+LASSO(BIC)	11.35 (1.51)	10.98 (7.19)	0.05 (0.21)	137.06	4.940
	dgLARS(BIC)	14.00 (0.00)	13.93 (6.68)	1.00 (0.00)	18.08	1.329
	SC (γ_L)	13.68 (0.60)	0.86 (0.62)	0.75 (0.44)	11.80	1.148
	SC (γ_H)	4.20 (2.80)	0.03 (0.17)	0.03 (0.17)	407.66	6.567
	FR (γ_L)	14.00 (0.00)	0.50 (0.76)	1.00 (0.00)	1.23	0.940
	FR (γ_H)	4.99 (3.07)	0.00 (0.00)	0.03 (0.17)	360.65	6.640
	STEPWISE	14.00 (0.00)	0.00 (0.00)	1.00 (0.00)	0.91	0.958
5 ($p_0 = 3$)	LASSO(1SE)	3.00 (0.00)	22.76 (9.05)	1.00 (0.00)	1.01	0.044
	LASSO(BIC)	3.00 (0.00)	8.29 (3.23)	1.00 (0.00)	1.75	0.054
	SIS+LASSO(1SE)	3.00 (0.00)	8.40 (3.10)	1.00 (0.00)	0.44	0.041
	SIS+LASSO(BIC)	3.00 (0.00)	9.58 (3.36)	1.00 (0.00)	0.29	0.040
	dgLARS(BIC)	3.00 (0.00)	13.39 (4.94)	1.00 (0.00)	1.28	0.048
	SC (γ_L)	3.00 (0.00)	1.47 (0.67)	1.00 (0.00)	0.03	0.038
	SC (γ_H)	2.01 (0.10)	0.01 (0.10)	0.01 (0.10)	4.51	0.008
	FR (γ_L)	3.00 (0.00)	1.21 (1.01)	1.00 (0.00)	0.03	0.038
	FR (γ_H)	3.00 (0.00)	0.00 (0.00)	1.00 (0.00)	0.01	0.003
	STEPWISE	3.00 (0.00)	0.00 (0.00)	1.00 (0.00)	0.01	0.039

Note: TP, true positives; FP, false positives; PIT, probability of including all true predictors in the selected predictors; MSE, mean squared error of $\hat{\beta}$; MSPE, mean squared prediction error; numbers in the parentheses are standard deviations; LASSO(BIC), LASSO with the tuning parameter chosen to give the smallest BIC for the models on the LASSO path; LASSO(1SE), LASSO with the tuning parameter chosen by the one-standard-error rule; SIS+LASSO(BIC), sure independence screening by [5] followed by LASSO(BIC); SIS+LASSO(1SE), sure independence screening followed by LASSO(1SE); dgLARS(BIC), differential geometric least angle regression by [11,12] with the tuning parameter chosen to give the smallest BIC on the dgLARS path; SC(γ), sequentially conditioning approach by [8]; FR(γ), forward regression by [7]; STEPWISE, the proposed method; in FR and SC, the smaller and large values of γ are presented as γ_L and γ_H, respectively; p_0 denotes the number of true signals; LASSO(1SE), LASSO(BIC), SIS and dgLARS were conducted via R packages glmnet [31], ncvreg [32], screening [33] and dglars [34], respectively

Table 3. Binomial model.

Example	Method	TP	FP	PIT	MSE	MSPE
1 ($p_0 = 8$)	LASSO(1SE)	7.99 (0.10)	4.77 (5.56)	0.99 (0.10)	0.021	0.104
	LASSO(BIC)	7.99 (0.10)	3.19 (2.34)	0.99 (0.10)	0.021	0.104
	SIS+LASSO(1SE)	7.94 (0.24)	35.42 (6.77)	0.94 (0.24)	0.119	0.048
	SIS+LASSO(BIC)	7.94 (0.24)	16.83 (21.60)	0.94 (0.24)	0.119	0.073
	dgLARS(BIC)	8.00 (0.00)	3.27 (2.29)	1.00 (0.00)	0.019	0.102
	SC (γ_L)	8.00 (0.00)	2.81 (1.47)	1.00 (0.00)	0.009	0.073
	SC (γ_H)	1.02 (0.14)	0.00 (0.00)	0.00 (0.00)	0.030	0.028
	FR (γ_L)	8.00 (0.00)	3.90 (2.36)	1.00 (0.00)	0.032	0.066
	FR (γ_H)	2.00 (0.00)	0.00 (0.00)	0.00 (0.00)	0.025	0.027
	STEPWISE	7.98 (0.14)	0.08 (0.53)	0.98 (0.14)	0.002	0.094
2 ($p_0 = 8$)	LASSO(1SE)	7.98 (0.14)	3.29 (2.76)	0.98 (0.14)	0.054	0.073
	LASSO(BIC)	7.99 (0.10)	3.84 (2.72)	0.99 (0.10)	0.052	0.067
	SIS+LASSO(1SE)	7.92 (0.27)	28.20 (7.31)	0.92 (0.27)	0.038	0.030
	SIS+LASSO(BIC)	7.92 (0.27)	9.60 (12.92)	0.92 (0.27)	0.051	0.058
	dgLARS(BIC)	7.99 (0.10)	3.94 (2.65)	0.99 (0.10)	0.050	0.067
	SC (γ_L)	7.72 (0.45)	0.39 (0.49)	0.72 (0.45)	0.005	0.063
	SC (γ_H)	1.13 (0.37)	0.00 (0.00)	0.00 (0.00)	0.069	0.044
	FR (γ_L)	7.99 (0.10)	0.66 (0.76)	0.99 (0.10)	0.014	0.051
	FR (γ_H)	2.10 (0.30)	0.00 (0.00)	0.00 (0.00)	0.061	0.033
	STEPWISE	7.99 (0.10)	0.02 (0.14)	0.99 (0.10)	0.004	0.056
3 ($p_0 = 5$)	LASSO(1SE)	4.51 (0.52)	7.36 (2.57)	0.52 (0.50)	0.155	0.051
	LASSO(BIC)	4.98 (0.14)	5.97 (2.25)	0.98 (0.14)	0.118	0.037
	SIS+LASSO(1SE)	0.85 (0.46)	10.66 (3.01)	0.00 (0.00)	0.206	0.186
	SIS+LASSO(BIC)	0.85 (0.46)	12.10 (3.13)	0.00 (0.00)	0.197	0.185
	dgLARS(BIC)	4.92 (0.27)	16.21 (6.21)	0.92 (0.27)	0.112	0.035
	SC (γ_L)	4.32 (0.49)	0.47 (0.50)	0.33 (0.47)	0.016	0.048
	SC (γ_H)	2.62 (1.34)	0.42 (0.50)	0.00 (0.00)	0.104	0.066
	FR (γ_L)	4.98 (0.14)	0.67 (0.79)	0.98 (0.14)	0.020	0.033
	FR (γ_H)	2.98 (0.95)	0.40 (0.49)	0.00 (0.00)	0.087	0.043
	STEPWISE	4.97 (0.17)	0.04 (0.28)	0.97 (0.17)	0.014	0.034
4 ($p_0 = 14$)	LASSO(1SE)	9.96 (1.89)	6.78 (7.92)	0.01 (0.01)	0.112	0.107
	LASSO(BIC)	9.33 (1.86)	2.79 (2.87)	0.00 (0.00)	0.112	0.118
	SIS+LASSO(1SE)	10.03 (1.62)	28.01 (9.54)	0.03 (0.17)	0.098	0.070
	SIS+LASSO(BIC)	8.90 (1.99)	5.42 (10.64)	0.01 (0.10)	0.114	0.120
	dgLARS(BIC)	9.31 (1.85)	2.84 (2.86)	0.00 (0.00)	0.110	0.117
	SC (γ_L)	9.48 (1.40)	2.35 (2.14)	0.00 (0.00)	0.043	0.070
	SC (γ_H)	1.17 (0.40)	0.00 (0.00)	0.00 (0.00)	0.125	0.049
	FR (γ_L)	11.83 (1.39)	1.58 (1.60)	0.09 (0.29)	0.026	0.048
	FR (γ_H)	2.06 (0.24)	0.00 (0.00)	0.00 (0.00)	0.119	0.032
	STEPWISE	11.81 (1.42)	1.52 (1.58)	0.09 (0.29)	0.026	0.048
5 ($p_0 = 3$)	LASSO(1SE)	2.00 (0.00)	1.55 (1.76)	0.00 (0.00)	0.008	0.215
	LASSO(BIC)	2.00 (0.00)	1.86 (1.57)	0.00 (0.00)	0.008	0.213
	SIS+LASSO(1SE)	2.23 (0.42)	10.81 (6.45)	0.23 (0.42)	0.007	0.192
	SIS+LASSO(BIC)	2.10 (0.30)	3.60 (4.65)	0.10 (0.30)	0.007	0.206
	dgLARS(BIC)	2.00 (0.00)	1.64 (1.49)	0.00 (0.00)	0.008	0.213
	SC (γ_L)	2.27 (0.49)	7.16 (3.20)	0.29 (0.46)	0.060	0.166
	SC (γ_H)	1.87 (0.34)	0.03 (0.17)	0.00 (0.00)	0.005	0.030
	FR (γ_L)	2.96 (0.20)	8.88 (5.39)	0.96 (0.20)	0.013	0.147
	FR (γ_H)	1.97 (0.17)	0.03 (0.17)	0.00 (0.00)	0.005	0.019
	STEPWISE	2.89 (0.31)	0.76 (1.70)	0.89 (0.31)	0.001	0.194

Note: abbreviations are explained in the footnote of Table 2.

Table 4. Poisson model.

Example	Method	TP	FP	PIT	MSE	MSPE
1 ($p_0 = 8$)	LASSO(1SE)	7.93 (0.43)	4.64 (4.82)	0.96 (0.19)	0.001	4.236
	LASSO(BIC)	7.99 (0.10)	14.37 (14.54)	0.99 (0.10)	0.001	3.133
	SIS+LASSO(1SE)	7.89 (0.37)	25.37 (8.39)	0.91 (0.29)	0.001	3.247
	SIS+LASSO(BIC)	7.89 (0.37)	17.77 (11.70)	0.91 (0.29)	0.001	3.078
	dgLARS(BIC)	8.00 (0.00)	13.28 (14.31)	1.00 (0.00)	0.001	3.183
	SC (γ_L)	7.96 (0.20)	4.94 (3.46)	0.96 (0.20)	0.001	2.874
	SC (γ_H)	5.05 (1.70)	0.04 (0.24)	0.07 (0.26)	0.001	3.902
	FR (γ_L)	7.93 (0.26)	4.86 (3.73)	0.93 (0.26)	0.001	2.837
	FR (γ_H)	5.13 (1.61)	0.06 (0.31)	0.07 (0.26)	0.001	3.833
	STEPWISE	7.91 (0.29)	2.77 (2.91)	0.91 (0.29)	0.001	3.410
2 ($p_0 = 8$)	LASSO(1SE)	8.00 (0.00)	2.23 (3.52)	1.00 (0.00)	0.001	3.981
	LASSO(BIC)	8.00 (0.00)	8.98 (8.92)	1.00 (0.00)	0.001	3.107
	SIS+LASSO(1SE)	7.98 (0.14)	22.85 (7.08)	0.98 (0.14)	0.001	2.824
	SIS+LASSO(BIC)	7.98 (0.14)	13.55 (8.24)	0.98 (0.14)	0.001	2.937
	dgLARS(BIC)	8.00 (0.00)	8.91 (9.10)	1.00 (0.00)	0.001	3.099
	SC (γ_L)	8.00 (0.00)	3.89 (2.89)	1.00 (0.00)	0.000	2.979
	SC (γ_H)	5.68 (1.45)	0.00 (0.00)	0.12 (0.33)	0.001	3.971
	FR (γ_L)	8.00 (0.00)	3.60 (2.80)	1.00 (0.00)	0.000	3.032
	FR (γ_H)	5.71 (1.42)	0.00 (0.00)	0.10 (0.30)	0.001	3.911
	STEPWISE	7.98 (0.14)	2.00 (2.23)	0.98 (0.14)	0.000	3.589
3 ($p_0 = 5$)	LASSO(1SE)	4.37 (0.51)	6.88 (2.61)	0.38 (0.48)	0.001	1.959
	LASSO(BIC)	4.79 (0.41)	5.62 (2.17)	0.79 (0.41)	0.000	2.044
	SIS+LASSO(1SE)	0.86 (0.47)	10.11 (2.55)	0.00 (0.00)	0.002	3.266
	SIS+LASSO(BIC)	0.86 (0.47)	11.86 (2.99)	0.00 (0.00)	0.002	3.160
	dgLARS(BIC)	4.55 (0.51)	18.29 (6.13)	0.56 (0.49)	0.001	1.877
	SC (γ_L)	4.73 (0.45)	0.53 (0.66)	0.73 (0.45)	0.000	2.479
	SC (γ_H)	2.84 (0.63)	0.40 (0.49)	0.00 (0.00)	0.001	0.664
	FR (γ_L)	4.54 (0.52)	1.98 (2.19)	0.55 (0.50)	0.000	2.128
	FR (γ_H)	2.71 (0.70)	0.43 (0.50)	0.00 (0.00)	0.001	0.605
	STEPWISE	4.54 (0.52)	1.77 (2.01)	0.55 (0.50)	0.000	2.132
4 ($p_0 = 14$)	LASSO(1SE)	10.01 (1.73)	3.91 (6.03)	0.01 (0.10)	0.003	15.582
	LASSO(BIC)	12.11 (1.46)	36.56 (22.43)	0.19 (0.39)	0.002	5.688
	SIS+LASSO(1SE)	10.42 (1.66)	21.41 (8.87)	0.03 (0.17)	0.003	11.316
	SIS+LASSO(BIC)	10.73 (1.66)	32.67 (8.92)	0.03 (0.17)	0.003	8.545
	dgLARS(BIC)	12.05 (1.52)	38.70 (28.97)	0.18 (0.38)	0.002	5.111
	SC (γ_L)	10.33 (1.63)	10.48 (6.66)	0.02 (0.14)	0.002	4.499
	SC (γ_H)	5.32 (1.92)	0.52 (1.37)	0.00 (0.00)	0.003	14.005
	FR (γ_L)	12.00 (1.71)	8.93 (6.36)	0.23 (0.42)	0.001	4.503
	FR (γ_H)	5.65 (2.13)	0.38 (1.15)	0.00 (0.00)	0.003	13.802
	STEPWISE	11.80 (1.72)	5.97 (5.37)	0.19 (0.39)	0.001	5.809
5 ($p_0 = 3$)	LASSO(1SE)	2.00 (0.00)	1.13 (2.85)	0.00 (0.00)	0.003	2.674
	LASSO(BIC)	2.01 (0.10)	2.82 (2.52)	0.01 (0.10)	0.003	2.583
	SIS+LASSO(1SE)	2.87 (0.34)	9.28 (3.85)	0.87 (0.34)	0.002	2.455
	SIS+LASSO(BIC)	2.87 (0.34)	9.88 (4.29)	0.87 (0.34)	0.002	2.355
	dgLARS(BIC)	2.00 (0.00)	2.88 (2.38)	0.00 (0.00)	0.003	2.562
	SC (γ_L)	2.75 (0.44)	3.27 (1.75)	0.75 (0.44)	0.001	2.339
	SC (γ_H)	2.00 (0.00)	0.00 (0.00)	0.00 (0.00)	0.003	1.086
	FR (γ_L)	3.00 (0.00)	2.80 (1.73)	1.00 (0.00)	0.001	2.326
	FR (γ_H)	2.40 (0.49)	0.00 (0.00)	0.40 (0.49)	0.002	0.981
	STEPWISE	3.00 (0.00)	0.35 (0.59)	1.00 (0.00)	0.001	2.977

Note: abbreviations are explained in the footnote of Table 2.

5. Real Data Analysis

5.1. A Study of Gene Regulation in the Mammalian Eye

To demonstrate the utility of our proposed method, we analyzed a microarray dataset from [35] with 120 twelve-week male rats selected for eye tissue harvesting. The dataset contained more than 31,042 different probe sets (Affymetric GeneChip Rat Genome 230 2.0 Array); see [35] for a more detailed description of the data.

Although our method was applicable to the original 31,042 probe sets, many probes turned out to have very small variances and were unlikely to be informative for correlative analyses. Therefore, using variance as the screening criterion, we selected 5000 genes with the largest variances in expressions and correlated them with gene *TRIM32* that has been found to cause Bardet–Biedl syndrome, a genetically heterogeneous disease of multiple organ systems including the retina [36].

We applied the proposed STEPWISE method to the dataset with $n = 120$ and $p = 5000$, and treated the *TRIM32* gene expression as the response variable and the expressions of 5000 genes as the predictors. With no prior biological information available, we started with the empty set. To choose η_1 and η_2, we carried out 5-fold cross-validation to minimize the mean squared prediction error (MSPE) by using the following grid search: $\eta_1 = \{0, 0.25, 0.5, 1\}$ and $\eta_2 = \{1, 2, 3, 4, 5\}$, and set $\eta_1 = 1$ and $\eta_2 = 4$. We also performed the same procedure to choose the γ for FR and SC. The regularization parameters in LASSO and dgLARS were selected to minimize BIC values.

In the forward step, STEPWISE selected the probes of *1376747_at*, *1381902_at*, *1382673_at* and *1375577_at*, and the backward step eliminated probe *1375577_at*. The STEPWISE procedure produced the following final predictive model:

$TRIM32 = 4.6208 + 0.2310 \times (1376747\_at) + 0.1914 \times (1381902\_at) + 0.1263 \times (1382673\_at)$. Table A1 in Appendix B presents the numbers of overlapping genes among competing methods. It shows that the two out of three probes, *1381902_at* and *1376747_at*, selected from our method are also discovered by the other methods, except for dgLARS.

Next, we performed Leave-One-Out Cross-Validation (LOOCV) to obtain the distribution of the model size (MS) and MSPE for the competing methods.

As reported in Table 5 and Figure 1, LASSO, SIS+LASSO and dgLARS tended to select more variables than the forward approaches and STEPWISE. Among all of the methods, STEPWISE selected the fewest variables but with almost the same MSPE as the other methods.

Table 5. Comparisons of MSPE among competing methods using the mammalian eye data set.

	STEPWISE	FR	LASSO	SIS+LASSO	SC	dgLARS
Training set	0.005	0.005	0.005	0.006	0.005	0.014
Testing set	0.011	0.012	0.010	0.009	0.014	0.020

Note: The mean squared prediction error (MSPE) was averaged over 120 splits. LASSO, least absolute shrinkage and selection operator with regularization parameter that gives the smallest BIC; SIS+LASSO, sure independence screening by [5] followed by LASSO; dgLARS, differential geometric least angle regression by [11,12] that gives the smallest BIC; SC(γ), sequentially conditioning approach by [8]; FR(γ), forward regression by [7]; STEPWISE, the proposed method. STEPWISE was performed with $\eta_1 = 1$ and $\eta_2 = 4$; FR and SC were performed with $\gamma = 1$.

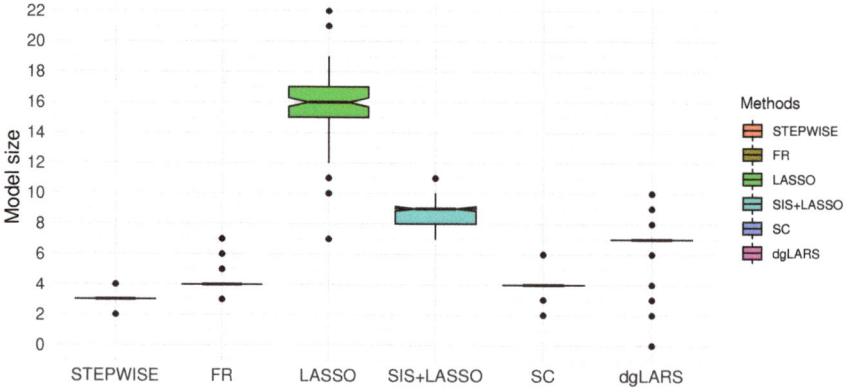

Figure 1. Box plot of model sizes for each method over 120 different training samples from the mammalian eye data set. STEPWISE was performed with $\eta_1 = 1$ and $\eta_2 = 4$, and FR and SC were conducted with $\gamma = 1$.

5.2. An Esophageal Squamous Cell Carcinoma Study

Esophageal squamous cell carcinoma (ESCC), the most common histological type of esophageal cancer, is known to be associated with poor overall survival, making early diagnosis crucial for treatment and disease management [37]. Several studies have investigated the roles of circulating microRNAs (miRNAs) in diagnosis of ESCC [38].

Using a clinical study that investigated the roles of miRNAs on the ESCC [39], we aimed to use miRNAs to predict ESCC risks and estimate their impacts on the development of ESCC. Specifically, with a dataset of serum profiling of 2565 miRNAs from 566 ESCC patients and 4965 controls without cancer, we demonstrated the utility of the proposed STEPWISE method in predicting ESCC with miRNAs.

To proceed, we used a balance sampling scheme (283 cases and 283 controls) in the training dataset. The design of yielding an equal number of cases and controls in the training set has proved to be useful [39] for handling imbalanced outcomes as we encountered here. To validate our findings, samples were randomly divided into a training ($n_1 = 566$, $p = 2565$) and testing set ($n_2 = 4965$, $p = 2565$).

The training set consisted of 283 patients with ESCC (median age of 65 years, 79% male) and 283 control patients (median age of 68 years, 46.3% male), and the testing set consisted of 283 patients with ESCC (median age of 67 years, 85.7% male) and 4682 control patients (median age of 67.5 years, 44.5% male). Control patients without ESCC came from three sources: 323 individuals from National Cancer Center Biobank (NCCB); 2670 individuals from the Biobank of the National Center for Geriatrics and Gerontology (NCGG); and 1972 individuals from Minoru Clinic (MC). More detailed characteristics of cases and controls in the training and testing sets are given in Table 6.

Table 6. Clinicopathological characteristics of study participants of the ESCC data set.

Covariates	Training Set n_1 (%)	Testing set n_2 (%)
Esophageal squamous cell carcinoma (ESCC) patients		
Total number of patients	283	283
Age, median (range)	65 [40, 86]	67 [37, 90]
Gender:		
Male	224 (79.0%)	247 (87.3%)
Female	59 (21.0%)	36 (12.7%)
Stage:		
0	24 (8.5%)	27 (9.5%)
1	127 (44.9%)	128 (45.2%)
2	58 (20.5%)	57 (20.1%)
3	67 (23.7%)	61 (21.6%)
4	7 (2.4%)	10 (3.6%)
Non-ESCC Controls		
Total number of patients	283	4,682
Age, median (range)	68 [27, 92]	67.5 [20, 100]
Gender:		
Male	131 (46.3%)	2,086 (44.5%)
Female	152 (53.7%)	2,596 (55.5%)
Data sources of the controls:		
National Cancer Center Biobank (NCCB)	17 (6.0%)	306 (6.5%)
National Center for Geriatrics and Gerontology (NCGG)	158 (55.8%)	2,512 (53.7%)
Minoru clinic (MC)	108 (38.2%)	1,864 (39.8%)

We defined the binary outcome variable to be 1 if the subject was a case and 0 otherwise. As age and gender (0 = female, 1 = male) are important risk factors for ESCC [40,41] and it is common to adjust for them in clinical models, we set the initial set in STEPWISE to be $F_0 = \{$age, gender$\}$. With $\eta_1 = 0$ and $\eta_2 = 3.5$ that were also chosen from 5-fold CV, our procedure recruited three miRNAs. More specifically, *miR-4783-3p*, *miR-320b*, *miR-1225-3p* and *miR-6789-5p* were selected among 2565 miRNAs by the forward stage from the training set, and then the backward stage eliminated *miR-6789-5p*.

In comparison, with $\gamma = 0$, both FR and SC selected four miRNAs, *miR-4783-3p*, *miR-320b*, *miR-1225-3p* and *miR-6789-5p*. The list of selected miRNAs by different methods are given in Table A2 in Appendix B.

Our findings were biologically meaningful, as the selected miRNAs had been identified by other cancer studies as well. Specifically, *miR-320b* was found to promote colorectal cancer proliferation and invasion by competing with its homologous *miR-320a* [42]. In addition, serum levels of *miR-320* family members were associated with clinical parameters and diagnosis in prostate cancer patients [43]. Reference [44] showed that *miR-4783-3p* was one of the miRNAs that could increase the risk of colorectal cancer death among rectal cancer cases. Finally, *miR-1225-5p* inhibited proliferation and metastasis of gastric carcinoma through repressing insulin receptor substrate-1 and activation of β-catenin signaling [45].

Aiming to identify a final model without resorting to a pre-screening procedure that may miss out on important biomarkers, we applied STEPWISE to reach the following predictive model for ESCC based on patients' demographics and miRNAs:

$\text{logit}^{-1}(-35.70 + 1.41 \times \textit{miR-4783-3p} + 0.98 \times \textit{miR-320b} + 1.91 \times \textit{miR-1225-3p} + 0.10 \times \textit{Age} - 2.02 \times \textit{Gender})$, where $\text{logit}^{-1}(x) = \exp(x)/(1+\exp(x))$.

In the testing dataset, the model had an area under the receiver operating curve (AUC) of 0.99 and achieved a high accuracy of 0.96, with a sensitivity and specificity of 0.97 and 0.95, respectively. Additionally, using the testing cohort, we evaluated the performances of the models sequentially selected by STEPWISE. Starting with a model containing age and gender, STEPWISE selected *miR-4783-3p*, *miR-320b* and *miR-1225-3p* in turn. Figure 2, showing the corresponding receiver operating curves (ROC) for these sequential models, revealed the improvement by sequentially adding predictors to the model and justified the importance of these variables in the final model. In addition, Figure 2e illustrated that adding an extra miRNA selected by FR and SC made little improvement of the model's predictive power.

Furthermore, we conducted subgroup analysis within the testing cohort to study how the sensitivity of the final model differed by cancer stage, one of the most important risk factors. The sensitivities for stages 0, i.e., non-invasive cancer, 9 ($n = 27$), 1 ($n = 128$), 2 ($n = 57$), 3 ($n = 61$) and 4 ($n = 10$) were 1.00, 0.98, 0.97, 0.97 and 1.00, respectively. We next evaluated how the specificity varied across controls coming from different data sources. The specificities for the various control groups, namely, NCCB ($n = 306$), NCGG ($n = 2512$) and MC ($n = 1864$), were 0.99, 0.99 and 0.98, respectively. The results indicated the robust performance of the miRNA-based model toward cancer stages and data sources.

Finally, to compare STEPWISE with the other competing methods, we repeatedly applied the aforementioned balance sampling procedure and split the ESCC data into the training and testing sets 100 times. We obtained MSPE and the average of accuracy, sensitivity, specificity, and AUC. Figure 3 reported the model size of each method. Though STEPWISE selected fewer variables compared to the other variable selection methods (for example, LASSO selected 11–31 variables and dgLARS selected 12–51 variables), it achieved comparable prediction accuracy, specificity, sensitivity and AUC (see Table 7), evidencing the utility of STEPWISE for generating parsimonious models while maintaining competitive predictability.

Table 7. Comparisons of competing methods over 100 independent splits of the ESCC data into training and testing sets.

Training Set	MSPE	Accuracy	Sensitivity	Specificity	AUC
STEPWISE	0.02	0.97	0.98	0.97	1.00
SC	0.01	0.99	0.98	0.98	1.00
FR	0.02	0.99	0.97	0.97	1.00
LASSO	0.01	0.98	1.00	0.97	1.00
SIS+LASSO	0.01	0.99	1.00	0.99	1.00
dgLARS	0.04	0.96	0.99	0.94	1.00
Training Set	**MSPE**	**Accuracy**	**Sensitivity**	**Specificity**	**AUC**
STEPWISE	0.04	0.96	0.97	0.95	0.99
SC	0.03	0.96	0.97	0.96	0.99
FR	0.04	0.96	0.97	0.95	0.99
LASSO	0.03	0.96	0.99	0.95	1.00
SIS+LASSO	0.02	0.97	0.99	0.96	1.00
dgLARS	0.05	0.94	0.98	0.94	1.00

Note: Values were averaged over 100 splits. STEPWISE was performed with $\eta_1 = 0$ and $\eta_2 = 1$. SC and FR were performed with $\gamma = 1$. The regularization parameters in LASSO and dgLARS were selected to minimize the BIC.

We used R software [46] to obtain the numerical results in Sections 4 and 5 with following packages: `ggplot2` [47], `ncvreg` [32], `glmnet` [31], `dglars` [34] and `screening` [33].

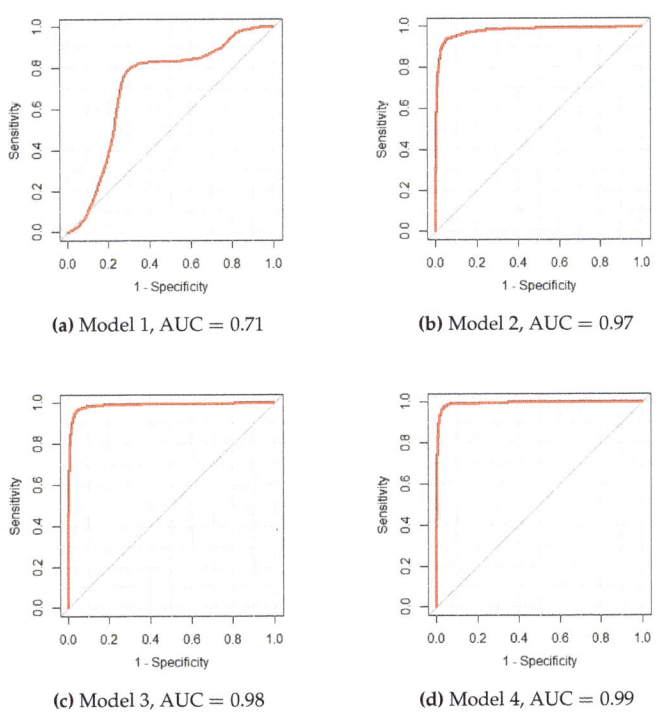

(a) Model 1, AUC = 0.71

(b) Model 2, AUC = 0.97

(c) Model 3, AUC = 0.98

(d) Model 4, AUC = 0.99

Figure 2. *Cont.*

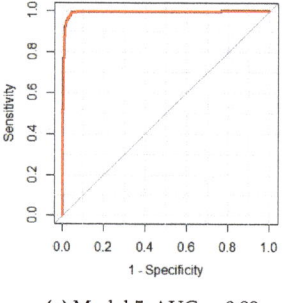

(e) Model 5, AUC = 0.99

Figure 2. Comparisons of ROC curves for the selected models in the ESCC data set by the sequentially selected order: Model 1: $-2.52 + 0.02 \times Age - 1.86 \times Gender$; Model 2: $-20.64 + 0.08 \times Age - 2.12 \times Gender + 2.02 \times miR\text{-}4783\text{-}3p$; Model 3: $-24.21 + 0.09 \times Age - 2.16 \times Gender + 1.44 \times miR\text{-}4783\text{-}3p - 1.31 \times miR\text{-}320b$; Model 4: $-35.70 + 0.10 \times Age - 2.02 \times Gender + 1.40 \times miR\text{-}4783\text{-}3p - 0.98 \times miR\text{-}320b + 1.91 \times miR\text{-}1225\text{-}3p$; Model 5: $-53.10 + 0.10 \times Age - 1.85 \times Gender + 1.43 \times miR\text{-}4783\text{-}3p - 0.92 \times miR\text{-}320b + 1.43 \times miR\text{-}1225\text{-}3p + 2.10 \times miR\text{-}6789\text{-}5p$.

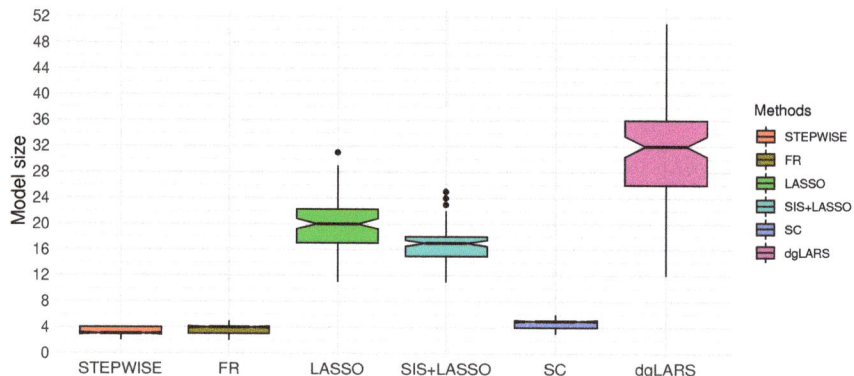

Figure 3. Box plot of model sizes for each method based on 100 ESCC training datasets. Performance of STEPWISE is reported with $\eta_1 = 0$ and $\eta_2 = 3.5$. Performances of SC and FR are reported with $\gamma = 0$.

6. Discussion

We have proposed to apply STEPWISE to produce final models in ultrahigh dimensional settings, without resorting to a pre-screening step. We have shown that the method identifies or includes the true model with probability going to 1, and produces consistent coefficient estimates, which are useful for properly interpreting the actual impacts of risk factors. The theoretical properties of STEPWISE were established under mild conditions, which are worth discussing. As in practice covariates are often standardized for various reasons, Condition (2) is assumed without loss of generality. Conditions (3) and (4) are generally satisfied under common GLM models, including Gaussian, binomial, Poisson and gamma distributions. Condition (5) is also often satisfied in practice. Proposition 2 in [26] may be used as a tool to verify Condition (5) as well. Conditions (1) and (6) are in good faith with the unknown true model size $|\mathcal{M}|$ and minimum signal strength $n^{-\alpha}$ in practice. The "irrepresentable" condition (6) is strong and may not hold in some real datasets, see, e.g., [48,49]. However, the condition holds under some commonly used covariance structures, including AR(1) and compound symmetry structure [48].

As shown in simulation studies and real data analyses, STEPWISE tends to generate models as predictive as the other well-known methods, with fewer variables (Figure 3). Parsimonious models are useful for biomedical studies as they explain data with a small number of important predictors, and offer practitioners a realistic list of biomarkers to investigate. With categorical outcome data frequently observed in biomedical studies (e.g., histology types of cancer), STEPWISE can be extended to accommodate multinomial classification, with more involved notation and computation. We will pursue this elsewhere.

There are several open questions. First, our final model was determined by using (E)BIC, which involves two extra parameters η_1 and η_2. In our numerical experiments, we used cross-validation to choose them, which seemed to work well. However, more in-depth research is needed to find their optimal values to strike a balance between false positives and false negatives. Second, despite our consistent estimates, drawing inferences based on them remains challenging. Statistical inference, which accounts for uncertainty in estimation, is key for properly interpreting analysis results and drawing appropriate conclusions. Our asymptotic results, nevertheless, are a stepping stone toward this important problem.

Supplementary Materials: An R package, STEPWISE, was developed and is available at https://github.com/AlexPijyan/STEPWISE, along with the examples shown in the paper.

Author Contributions: Conceptualization, Q.Z., H.H. and Y.L.; Formal analysis, A.P.; Methodology, A.P, Q.Z., H.H. and Y.L.; Project administration, H.H.; Software, A.P.; Supervision, H.H.; Writing – original draft, Q.Z., H.H. and Y.L.; Writing – review & editing, H.H. and Y.L. All authors have read and agreed to the published version of the manuscript.

Funding: This research was funded partially by grants from NSF (DMS1915099, DMS1952486) and NIH (R01AG056764, U01CA209414, R03AG067611).

Acknowledgments: We are thankful to the Editor, the AE and two referees for insightful suggestions that helped improve the manuscript.

Conflicts of Interest: 'The authors declare no conflict of interest.

Appendix A. Proofs of Main Theorems

Since $b(\cdot)$ is twice continuously differentiable with a nonnegative second derivative $b''(\cdot)$, $b_{\max} := \max_{|t| \leq K^3} |b(t)|$, $\mu_{\max} := \max_{|t| \leq K^3} |b'(t)|$ and $\sigma_{\max} := \sup_{|t| \leq K^3} |b''(t)|$ are bounded above, where L and K are some constants from Conditions (1) and (2), respectively. Let $\mathbb{G}_n\{f(\xi)\} = n^{-1/2} \sum_{i=1}^n (f(\xi_i) - E[f(\xi_i)])$ for a sequence of i.i.d. random variables ξ_i ($i = 1, \ldots, n$) and a non-random function $f(\cdot)$.

Given any β_S, when a variable $X_r, r \in S^c$ is added into the model S, we define the augmented log-likelihood as

$$\ell_{S \cup \{r\}}(\beta_{S+r}) := \mathbb{E}_n \left\{ L \left(\beta_S^T \mathbf{X}_S + \beta_r X_r, Y \right) \right\}. \tag{A1}$$

We use $\hat{\beta}_{S+r}$ to denote the maximizer of (A1). Thus, $\hat{\beta}_{S+r} = \hat{\beta}_{S \cup \{r\}}$. In addition, denote the maximizer of $E[\ell_{S \cup \{r\}}(\beta_{S+r})]$ by β^*_{S+r}. Due to the concavity of the log-likelihood in GLMs with the canonical link, β^*_{S+r} is unique.

Proof of Theorem 1. Given an index set S and $r \in S^c$, let $\mathcal{B}^0_S(d) = \{\beta_S : \|\beta_S - \beta^*_S\| \leq d/(K\sqrt{|S|})\}$ where $d = A_2 \sqrt{q^3 \log p / n}$ with A_2 defined in Lemma A6.

Let Ω be the event that

$$\left\{ \sup_{|S| \leq q, \beta_S \in \mathcal{B}^0_S(d)} \left| \mathbb{G}_n \left[L \left(\beta_S^T \mathbf{X}_S, Y \right) - L \left(\beta_S^{*T} \mathbf{X}_S, Y \right) \right] \right| \leq 20 A_1 d \sqrt{q \log p} \quad \text{and} \right.$$

$$\left. \max_{|S| \leq q} \left| \mathbb{G}_n \left[L(\beta_S^{*T} \mathbf{X}_S, Y) \right] \right| \leq 10 (A_1 K^2 + b_{\max}) \sqrt{q \log p} \right\},$$

where A_1 is some constant defined in Lemma A4. By Lemma A4, $P(\Omega) \geq 1 - 6 \exp(-6q \log p)$. Thus in the rest of the proof, we only consider the sample points in Ω.

In the proof of Lemma A6, we show that $\max_{|S|\leq q} \|\hat{\boldsymbol{\beta}}_S - \boldsymbol{\beta}_S^*\| \leq A_2 K^{-1}(q^2 \log p/n)^{1/2}$ under Ω. Then given an index set S and $\boldsymbol{\beta}_S$ such that $|S| < q$, $\|\boldsymbol{\beta}_S - \boldsymbol{\beta}_S^*\| \leq A_2 K^{-1}(q^2 \log p/n)^{1/2}$, and for any $j \in S^c$,

$$\ell_{S\cup\{j\}}(\boldsymbol{\beta}_{S+j}^*) - \ell_S(\hat{\boldsymbol{\beta}}_S) \geq \inf_{\|\boldsymbol{\beta}_S - \boldsymbol{\beta}_S^*\| \leq A_2 K^{-1}(q^2 \log p/n)^{1/2}} \ell_{S\cup\{j\}}(\boldsymbol{\beta}_{S+j}^*) - \ell_S(\boldsymbol{\beta}_S)$$

$$= n^{-1/2}\mathbb{G}_n\left[L(\boldsymbol{\beta}_{S+j}^{*T}\mathbf{X}_{S\cup\{j\}}, Y)\right] - n^{-1/2}\mathbb{G}_n\left[L(\boldsymbol{\beta}_S^{*T}\mathbf{X}_S, Y)\right]$$

$$- \sup_{\|\boldsymbol{\beta}_S - \boldsymbol{\beta}_S^*\| \leq A_2 K^{-1}(q^2 \log p/n)^{1/2}} \left|n^{-1/2}\mathbb{G}_n\left[L(\boldsymbol{\beta}_S^T \mathbf{X}_S, Y) - L(\boldsymbol{\beta}_S^{*T}\mathbf{X}_S, Y)\right]\right|$$

$$+ E\left[L(\boldsymbol{\beta}_{S+j}^{*T}\mathbf{X}_{S\cup\{j\}}, Y)\right] - E\left[L(\boldsymbol{\beta}_S^{*T}\mathbf{X}_S, Y)\right]$$

$$\geq -20(A_1 K^2 + b_{\max})\sqrt{q \log p/n} - 20 A_1 A_2 q^2 \log p/n + \frac{\sigma_{\min}\kappa_{\min}}{2}\|\boldsymbol{\beta}_{S+j}^* - (\boldsymbol{\beta}_S^{*T}, 0)^T\|^2,$$

where the second inequality follows from the event Ω and Lemma A5.

By Lemma A1, if $\mathcal{M} \not\subseteq S$, there exists $r \in S^c \cap \mathcal{M}$, such that $\|\boldsymbol{\beta}_{S+r}^{*T} - (\boldsymbol{\beta}_S^{*T}, 0)\| \geq C\sigma_{\max}^{-1}\kappa_{\max}^{-1}n^{-\alpha}$. Thus, there exists some constant C_1 that does not depend on n such that

$$\max_{j \in S^c} \ell_{S\cup\{j\}}(\hat{\boldsymbol{\beta}}_{S+j}) - \ell_S(\hat{\boldsymbol{\beta}}_S) \geq \max_{j \in S^c} \ell_{S\cup\{j\}}(\boldsymbol{\beta}_{S+j}^*) - \ell_S(\hat{\boldsymbol{\beta}}_S) \geq \ell_{S\cup\{r\}}(\boldsymbol{\beta}_{S+r}^*) - \ell_S(\hat{\boldsymbol{\beta}}_S)$$

$$\geq -20(A_1 K^2 + b_{\max})\sqrt{q \log p/n} - 20 A_1 A_2 q^2 \log p/n + \frac{C^2 \sigma_{\min}\kappa_{\min} n^{-2\alpha}}{2\sigma_{\max}^2 \kappa_{\max}^2} \geq C_1 n^{-2\alpha}, \quad (A2)$$

where the first inequality follows from $\hat{\boldsymbol{\beta}}_{S+j}$ being the maximizer of (A1) and the second inequality follows from Conditions (1) and (6).

Withdrawing the restriction to Ω, we obtain that

$$P\left(\min_{|S|<q, \mathcal{M} \not\subseteq S} \max_{j \in S^c} \ell_{S\cup\{j\}}(\hat{\boldsymbol{\beta}}_{S\cup\{j\}}) - \ell_S(\hat{\boldsymbol{\beta}}_S) \geq C_1 n^{-2\alpha}\right) \geq 1 - 6\exp(-6q \log p).$$

$\square$

Proof of Theorem 2. We have shown that our forward stage will not stop when $\mathcal{M} \not\subseteq S$ and $|S| < q$ with probability converging to 1.

For any $r \in S^c \cap \mathcal{M}^c$, $\boldsymbol{\beta}_{S+r}^*$ is the unique solution to the equation $E[\{Y - \mu(\boldsymbol{\beta}_{S+r}^T \mathbf{X}_{S\cup\{r\}})\}\mathbf{X}_{S\cup\{r\}}] = \mathbf{0}$. By the mean value theorem,

$$E[\{Y - \mu(\boldsymbol{\beta}_S^{*T}\mathbf{X}_S)\}X_r] = E[\{\mu(\boldsymbol{\beta}_*^T\mathbf{X}) - \mu(\boldsymbol{\beta}_S^{*T}\mathbf{X}_S)\}X_r]$$
$$= E[\{\mu(\boldsymbol{\beta}_*^T\mathbf{X}) - \mu(\boldsymbol{\beta}_S^{*T}\mathbf{X}_S)\}X_r] - E[\{\mu(\boldsymbol{\beta}_*^T\mathbf{X}) - \mu(\boldsymbol{\beta}_{S+r}^{*T}\mathbf{X}_{S\cup\{r\}})\}X_r]$$
$$= (\boldsymbol{\beta}_{S+r}^{*T} - (\boldsymbol{\beta}_S^{*T}, 0))E[\sigma(\tilde{\boldsymbol{\beta}}_{S+r}^T \mathbf{X}_{S\cup\{r\}})\mathbf{X}_{S\cup\{r\}}^{\otimes 2}]\mathbf{e}_r,$$

where $\tilde{\boldsymbol{\beta}}_{S+r}$ is some point between $\boldsymbol{\beta}_{S+r}$ and $(\boldsymbol{\beta}_S^{*T}, 0)^T$ and $\mathbf{e}_r$ is a vector of length $(|S|+1)$ with the rth element being 1.

Since $|\tilde{\boldsymbol{\beta}}_{S+r}^T \mathbf{X}_{S\cup\{r\}}| \leq |\boldsymbol{\beta}_{S+r}^{*T}\mathbf{X}_{S\cup\{r\}}| + |(\boldsymbol{\beta}_S^{*T}, 0)\mathbf{X}_{S\cup\{r\}}| \leq 2K^2$ by Conditions (1) and (2), $|\sigma(\tilde{\boldsymbol{\beta}}_{S+r}^T \mathbf{X}_{S\cup\{r\}})| \geq \sigma_{\min}$ and

$$o(n^{-\alpha}) = \left|E\left[\{Y - \mu(\boldsymbol{\beta}_S^{*T}\mathbf{X}_S)\}X_r\right]\right| \geq \sigma_{\min}\kappa_{\min}\|\boldsymbol{\beta}_{S+r}^{*T} - (\boldsymbol{\beta}_S^{*T}, 0)\|.$$

Therefore, $\max_{S: |S| \leq q, r \in S^c \cap \mathcal{M}^c} \|\boldsymbol{\beta}_{S+r}^{*T} - (\boldsymbol{\beta}_S^{*T}, 0)\| = o(n^{-\alpha})$.

Under Ω that is defined in Theorem 1, $\max_{|S|\leq q} \|\hat{\boldsymbol{\beta}}_S - \boldsymbol{\beta}_S^*\| \leq A_2 K^{-1}(q^2 \log p/n)^{1/2}$. For any $j \in S^c$,

$$\ell_{S\cup\{j\}}(\boldsymbol{\beta}_{S+j}^*) - \ell_S(\hat{\boldsymbol{\beta}}_S) \leq \sup_{\|\boldsymbol{\beta}_S - \boldsymbol{\beta}_S^*\| \leq A_2 K^{-1}(q^2 \log p/n)^{1/2}} \ell_{S\cup\{j\}}(\boldsymbol{\beta}_{S+j}^*) - \ell_S(\boldsymbol{\beta}_S)$$

$$\leq \left| n^{-1/2} \mathbb{G}_n \left[L(\boldsymbol{\beta}_{S+j}^{*T} \mathbf{X}_{S\cup\{j\}}, Y) \right] \right| + \left| n^{-1/2} \mathbb{G}_n \left[L(\boldsymbol{\beta}_S^{*T} \mathbf{X}_S, Y) \right] \right|$$

$$+ \sup_{\|\boldsymbol{\beta}_S - \boldsymbol{\beta}_S^*\| \leq A_2 K^{-1}(q^2 \log p/n)^{1/2}} \left| n^{-1/2} \mathbb{G}_n \left[L(\boldsymbol{\beta}_S^T \mathbf{X}_S, Y) - L(\boldsymbol{\beta}_S^{*T} \mathbf{X}_S, Y) \right] \right|$$

$$+ \left| E \left[L(\boldsymbol{\beta}_{S+j}^{*T} \mathbf{X}_{S\cup\{j\}}, Y) \right] - E \left[L(\boldsymbol{\beta}_S^{*T} \mathbf{X}_S, Y) \right] \right|$$

$$\leq 20(A_1 K^2 + b_{\max})\sqrt{qn^{-1} \log p} + 20 A_1 A_2 q^2 n^{-1} \log p + \sigma_{\max}\kappa_{\max} \|\boldsymbol{\beta}_{S+j}^* - (\boldsymbol{\beta}_S^{*T}, 0)^T\|^2 / 2,$$

where the second inequality follows from the event Ω and Lemma A5. Since $\max_{S:|S|<q, r\in S^c\cap\mathcal{M}^c} \|\boldsymbol{\beta}_{S+r}^* - (\boldsymbol{\beta}_S^{*T}, 0)^T\| = o(n^{-\alpha})$ and $qn^{-1+4\alpha} \log p \to 0$,

$$\max_{S:|S|<q, r\in S^c\cap\mathcal{M}^c} \ell_{S\cup\{r\}}(\boldsymbol{\beta}_{S+r}^*) - \ell_S(\hat{\boldsymbol{\beta}}_S) \leq 20(A_1 K^2 + b_{\max})\sqrt{qn^{-1} \log p} + 20 A_1 A_2 q^2 n^{-1} \log p$$

$$+ \sigma_{\max}\kappa_{\max} \|\boldsymbol{\beta}_{S+j}^* - (\boldsymbol{\beta}_S^{*T}, 0)^T\|^2 / 2 = o(n^{-2\alpha}),$$

with probability at least $1 - 6\exp(-6q \log p)$. Then by Lemma A6,

$$\max_{S:|S|<q, r\in S^c\cap\mathcal{M}^c} \ell_{S\cup\{r\}}(\hat{\boldsymbol{\beta}}_{S+r}) - \ell_S(\hat{\boldsymbol{\beta}}_S)$$

$$\leq \max_{S:|S|<q, r\in S^c\cap\mathcal{M}^c} |\ell_{S\cup\{r\}}(\hat{\boldsymbol{\beta}}_{S+r}) - \ell_{S\cup\{r\}}(\boldsymbol{\beta}_{S+r}^*)| + \max_{S:|S|<q, r\in S^c\cap\mathcal{M}^c} |\ell_{S\cup\{r\}}(\boldsymbol{\beta}_{S+r}^*) - \ell_S(\hat{\boldsymbol{\beta}}_S)|$$

$$\leq A_3 q^2 n^{-1} \log p + o(n^{-2\alpha}) = o(n^{-2\alpha}), \tag{A3}$$

with probability at least $1 - 12\exp(-6q \log p)$.

By Theorem 1, if $\mathcal{M} \not\subseteq S$, the forward stage would select a noise variable with probability less than $18\exp(-6q \log p)$.

For $k > |\mathcal{M}|$, $\mathcal{M} \not\subseteq S_k$ implies that at least $k - |\mathcal{M}|$ noise variables are selected within the k steps. Then for $k = C_2|\mathcal{M}|$ with $C_2 > 2$,

$$P(\mathcal{M} \not\subseteq S_k) \leq \sum_{j=k-|\mathcal{M}|}^{k} \binom{k}{j} \{18\exp(-6q \log p)\}^j \leq |\mathcal{M}| k^{|\mathcal{M}|} \{18\exp(-6q \log p)\}^{k-|\mathcal{M}|}$$

$$\leq 18\exp(-6q \log p + \log|\mathcal{M}| + |\mathcal{M}| \log k) \leq 18\exp(-4q \log p).$$

Therefore, $\mathcal{M} \subset S_{C_2|\mathcal{M}|}$ with probability at least $1 - 18\exp(-4q \log p)$. □

Proof of Theorem 3. By Theorem 2, $\mathcal{M}$ will be included in F_k for some $k < q$ with probability going to 1. Therefore, the forward stage stops at the kth step if $\text{EBIC}(F_{k+1}) > \text{EBIC}(F_k)$.

On the other hand, that $\text{EBIC}(F_{k+1}) < \text{EBIC}(F_k)$ if and only if $2\ell_{F_{k+1}}(\hat{\boldsymbol{\beta}}_{F_{k+1}}) - 2\ell_{F_k}(\hat{\boldsymbol{\beta}}_{F_k}) \geq (\log n + 2\eta_1 \log p)/n$. Thus, to show the forward stage stops at the kth step, we only need to show that with probability tending to 1,

$$2\ell_{F_{k+1}}(\hat{\boldsymbol{\beta}}_{F_{k+1}}) - 2\ell_{F_k}(\hat{\boldsymbol{\beta}}_{F_k}) < (\log n + 2\eta_1 \log p)/n, \tag{A4}$$

for all $\eta_1 > 0$.

To prove (A4), we first verify the conditions (A4) and (A5) in [17]. Given any index S such that $\mathcal{M} \subseteq S$ and $|S| \leq q$, let $\boldsymbol{\beta}_{*S}$ be the subvector of $\boldsymbol{\beta}_*$ corresponding to S. We obtain that

$$E\left[(Y - \mu(\boldsymbol{\beta}_{*S}^T \mathbf{X}_S))\mathbf{X}_S\right] = E\left[E\left[(Y - \mu(\boldsymbol{\beta}_{*\mathcal{M}}^T \mathbf{X}_\mathcal{M}))|\mathbf{X}_S\right]\mathbf{X}_S\right] = 0.$$

This implies $\boldsymbol{\beta}_S^* = \boldsymbol{\beta}_{*S}$.

Given any $\boldsymbol{\pi} \in \mathbb{R}^{|S|}$, let $\mathcal{H}_S := \{h(\boldsymbol{\pi}, \boldsymbol{\beta}_S) = (\sigma_{\max} K^2 |S|)^{-1} \sigma(\boldsymbol{\beta}_S^T \mathbf{X}_S) (\boldsymbol{\pi}^T \mathbf{X}_S)^2, \|\boldsymbol{\pi}\| = 1, \boldsymbol{\beta}_S \in \mathcal{B}_S^0(d)\}$. By Conditions (1) and (2), $h(\boldsymbol{\pi}, \boldsymbol{\beta}_S)$ is bounded between -1 and 1 uniformly over $\|\boldsymbol{\pi}\| = 1$ and $\boldsymbol{\beta}_S \in \mathcal{B}_S^0(d)$.

By Lemma 2.6.15 in [50], the VC indices of $\mathcal{W} := \{(K\sqrt{|S|})^{-1} \boldsymbol{\pi}^T \mathbf{X}_S, \|\boldsymbol{\pi}\| = 1\}$ and $\mathcal{V} := \{\boldsymbol{\beta}_S^T \mathbf{X}_S, \boldsymbol{\beta}_S \in \mathcal{B}_S^0(d)\}$ are bounded by $|S| + 2$. For the definitions of the VC index and covering numbers, we refer to pages 83 and 85 in [50]. The VC index of the class $\mathcal{U} := \{(K^2|S|)^{-1}(\boldsymbol{\pi}^T \mathbf{X}_S)^2, \|\boldsymbol{\pi}\| = 1\}$ is the VC index of the class of sets $\{(\mathbf{X}_S, t) : (K^2|S|)^{-1}(\boldsymbol{\pi}^T \mathbf{X}_S)^2 \leq t, \|\boldsymbol{\pi}\| = 1, t \in \mathbb{R}\}$. Since $\{(\mathbf{X}_S, t) : (K^2|S|)^{-1}(\boldsymbol{\pi}^T \mathbf{X}_S)^2 \leq t\} = \{(\mathbf{X}_S, t) : 0 < (K\sqrt{|S|})^{-1}\boldsymbol{\pi}^T \mathbf{X}_S \leq \sqrt{t}\} \cup \{(\mathbf{X}_S, t) : -\sqrt{t} < (K\sqrt{|S|})^{-1}\boldsymbol{\pi}^T \mathbf{X}_S \leq 0\}$, each set of $\{(\mathbf{X}_S, t) : (K^2|S|)^{-1}(\boldsymbol{\pi}^T \mathbf{X}_S)^2 \leq t, \|\boldsymbol{\pi}\| = 1, t \in \mathbb{R}\}$ is created by taking finite unions, intersections and complements of the basic sets $\{(\mathbf{X}_S, t) : (K\sqrt{|S|})^{-1} \boldsymbol{\pi}^T \mathbf{X}_S < t\}$. Therefore, the VC index of $\{(\mathbf{X}_S, t) : (K^2|S|)^{-1}(\boldsymbol{\pi}^T \mathbf{X}_S)^2 \leq t, \|\boldsymbol{\pi}\| = 1, t \in \mathbb{R}\}$ is of the same order as the VC index of $\{(\mathbf{X}_S, t) : (K\sqrt{|S|})^{-1}\boldsymbol{\pi}^T \mathbf{X}_S < t\}$, by Lemma 2.6.17 in [50].

Then by Theorem 2.6.7 in [50], for any probability measure Q, there exists some universal constant C_3 such that $N(\epsilon, \mathcal{U}, L_2(Q)) \leq (C_3/\epsilon)^{2(|S|+1)}$. Likewise, $N(d\epsilon, \mathcal{V}, L_2(Q)) \leq (C_3/\epsilon)^{2(|S|+1)}$. Given a $\boldsymbol{\beta}_{S,0} \in \mathcal{B}_S^0(d)$, for any $\boldsymbol{\beta}_S$ in the ball $\{\boldsymbol{\beta}_S : \sup_{\mathbf{x}} |\boldsymbol{\beta}_S^T \mathbf{x} - \boldsymbol{\beta}_{S,0}^T \mathbf{x}| < d\epsilon\}$, we have $\sup_{\mathbf{x}} |\sigma(\boldsymbol{\beta}_S^T \mathbf{x}) - \sigma(\boldsymbol{\beta}_{S,0}^T \mathbf{x})| < Kd\epsilon$ by Condition (4). Let $\mathcal{V}' := \{\sigma_{\max}^{-1} \sigma(\boldsymbol{\beta}_S^T \mathbf{X}_S), \boldsymbol{\beta}_S \in \mathcal{B}_S^0(d)\}$. By the definition of covering number, $N(Kd\epsilon, \mathcal{V}', L_2(Q)) \leq (C_3/\epsilon)^{2(|S|+1)}$ Given a $\sigma(\boldsymbol{\beta}_{S,0}^T \mathbf{x})$ and $\boldsymbol{\pi}_0^T \mathbf{x}$, for any $\sigma(\boldsymbol{\beta}_S^T \mathbf{x})$ in the ball $\{\sigma(\boldsymbol{\beta}_S^T \mathbf{x}) : \sup_{\mathbf{x}} |\sigma(\boldsymbol{\beta}_S^T \mathbf{x}) - \sigma(\boldsymbol{\beta}_{S,0}^T \mathbf{x})| \leq Kd\epsilon\}$ and $\boldsymbol{\pi}$ in the ball $\{\boldsymbol{\pi} : \sup_{\mathbf{x}} |(\boldsymbol{\pi}^T \mathbf{x})^2 - (\boldsymbol{\pi}_0^T \mathbf{x})^2| < \epsilon\}$, $(\sigma_{\max} K^2 |S|)^{-1} \sup_{\mathbf{x}} |\sigma(\boldsymbol{\beta}_S^T \mathbf{x})(\boldsymbol{\pi}^T \mathbf{x})^2 - \sigma(\boldsymbol{\beta}_{S,0}^T \mathbf{x})(\boldsymbol{\pi}_0^T \mathbf{x})^2| \leq (\sigma_{\max}^{-1} Kd + (K^2|S|)^{-1})\epsilon$. Thus, $N((\sigma_{\max}^{-1} Kd + (K^2|S|)^{-1})\epsilon, \mathcal{H}_S, L_2(Q)) \leq (C_3/\epsilon)^{4(|S|+1)}$, and consequently $N(\epsilon, \mathcal{H}_S, L_2(Q)) \leq (C_4/\epsilon)^{4(|S|+1)}$ for some constant C_4.

By Theorem 1.1 in [51] and $|S| \leq q$, we can find some constant C_5 such that

$$P\left(\sup_{\|\boldsymbol{\pi}\|=1, \boldsymbol{\beta}_S \in \mathcal{B}_S^0(d)} |\mathbb{G}_n[h(\boldsymbol{\pi}, \boldsymbol{\beta}_S)]| \geq C_5 \sqrt{q \log p}\right)$$

$$\leq \frac{C_4'}{C_5 \sqrt{q \log p}} \left(\frac{C_4' C_5^2 q \log p}{4(|S|+1)}\right)^{4(|S|+1)} \exp(-2C_5^2 q \log p)$$

$$\leq \exp\left(4(|S|+1) \log(C_4' C_5^2 q \log p) - 2C_5^2 q \log p\right) \leq \exp\left(-5q \log p\right),$$

where C_4' is some constant that depends on C_4 only. Thus,

$$P\left(\sup_{|S| \leq q, \|\boldsymbol{\pi}\|=1, \boldsymbol{\beta}_S \in \mathcal{B}_S^0(d)} \left|\mathbb{E}_n\left\{\sigma\left(\mathbf{X}_S^T \boldsymbol{\beta}_S\right)(\boldsymbol{\pi}^T \mathbf{X}_S)^2\right\} - E\left[\sigma\left(\mathbf{X}_S^T \boldsymbol{\beta}_S\right)(\boldsymbol{\pi}^T \mathbf{X}_S)^2\right]\right| \geq C_5 K^2 \sqrt{q^3 \log p/n}\right)$$

$$\leq \sum_{s=|\mathcal{M}|}^{q} \left(\frac{ep}{s}\right)^s \exp(-5q \log p) \leq \exp(-3q \log p). \tag{A5}$$

By Condition (5), $\sigma_{\min} \kappa_{\min} \leq \Lambda\left(E\left[\sigma\left(\mathbf{X}_S^T \boldsymbol{\beta}_S\right) \mathbf{X}_S^{\otimes 2}\right]\right) \leq \sigma_{\max} \kappa_{\max}$, for all $\boldsymbol{\beta}_S \in \mathcal{B}_S^0(d)$ and $S : \mathcal{M} \subseteq S, |S| \leq q$. Then, by (A5),

$$\sigma_{\min} \kappa_{\min}/2 \leq \Lambda\left(\mathbb{E}_n\left\{\sigma\left(\mathbf{X}_S^T \boldsymbol{\beta}_{*S}\right) \mathbf{X}_S^{\otimes 2}\right\}\right) \leq 2\sigma_{\max} \kappa_{\max}$$

uniformly over all S satisfying $\mathcal{M} \subseteq S$ and $|S| \leq q$, with probability at least $1 - \exp(-3q \log p)$. Hence, the condition (A4) in [17] is satisfied with probability at least $1 - \exp(-3q \log p)$.

Additionally, for any $\beta_S \in \mathcal{B}_S^0(d)$,

$$\left| \mathbb{E}_n \left\{ \sigma\left(\mathbf{X}_S^T \beta_S\right) \left(\pi^T \mathbf{X}_S\right)^2 \right\} - \mathbb{E}_n \left\{ \sigma\left(\mathbf{X}_S^T \beta_{*S}\right) \left(\pi^T \mathbf{X}_S\right)^2 \right\} \right|$$

$$\leq \left| n^{-1/2} \mathbb{G}_n \left\{ \sigma\left(\mathbf{X}_S^T \beta_S\right) \left(\pi^T \mathbf{X}_S\right)^2 \right\} \right| + \left| n^{-1/2} \mathbb{G}_n \left\{ \sigma\left(\mathbf{X}_S^T \beta_{*S}\right) \left(\pi^T \mathbf{X}_S\right)^2 \right\} \right|$$

$$+ \left| E\left[\sigma\left(\mathbf{X}_S^T \beta_S\right) \left(\pi^T \mathbf{X}_S\right)^2 \right] - E\left[\sigma\left(\mathbf{X}_S^T \beta_{*S}\right) \left(\pi^T \mathbf{X}_S\right)^2 \right] \right|$$

$$\leq 2C_5 K^2 \sqrt{q^3 \log p / n} + \mu_{\max} \|\beta_S - \beta_{*S}\| \sqrt{|S|} K \lambda_{\max}.$$

Hence, the condition (A5) in [17] is satisfied uniformly over all S such that $\mathcal{M} \subseteq S$ and $|S| \leq q$, with probability at least $1 - \exp(-3q \log p)$.

Then (A4) can be shown by following the proof of Equation (3.2) in [17]. Thus, our forward stage stops at the kth step with probability at least $1 - \exp(-3q \log p)$. □

Proof of Theorem 4. Suppose that a covariate X_r is removed from S. For any $r \in \mathcal{M}$, since $\mathcal{M} \not\subseteq S \setminus \{r\}$ and r is the only element that is in $(S \setminus \{r\})^c \cap \mathcal{M}$, by Lemma A1 and (A2)

$$\ell_S(\hat{\beta}_S) - \ell_{S \setminus \{r\}}(\hat{\beta}_{S \setminus \{r\}}) \geq \ell_S(\beta_S^*) - \ell_{S \setminus \{r\}}(\hat{\beta}_{S \setminus \{r\}})$$
$$= \ell_{S \setminus \{r\} \cup \{r\}}(\beta_{S \setminus \{r\}+r}^*) - \ell_{S \setminus \{r\}}(\hat{\beta}_{S \setminus \{r\}}) \geq C_1 n^{-2\alpha},$$

with probability at least $1 - 6 \exp(-6q \log p)$. From the proof of Theorem 1, we have for any $\eta_2 > 0$, $\text{BIC}(S) - \text{BIC}(S \setminus \{r\}) \leq -2C_1 n^{-2\alpha} + \eta_2 n^{-1} \log n < 0$, uniformly over $r \in \mathcal{M}$ and S satisfying $\mathcal{M} \subset S$ and $|S| \leq q$, with probability at least $1 - 6 \exp(-6q \log p)$. □

Proof of Theorem 5. By Theorems 1–3, we have that the event $\Omega_1 := \{|\hat{\mathcal{M}}| \leq q \text{ and } \mathcal{M} \subseteq \hat{\mathcal{M}}\}$ holds with probability at least $1 - 25 \exp(-2q \log p)$. Thus, in the rest of the proof, we restrict our attention on Ω_1.

As shown in the proof of Theorem 3, we obtain that $\beta_{\hat{\mathcal{M}}}^* = \beta_{*\hat{\mathcal{M}}}$. Then by Lemma A6, we have $\|\hat{\beta}_{\hat{\mathcal{M}}} - \beta_{\hat{\mathcal{M}}}^*\| \leq A_2 K^{-1} \sqrt{q^2 \log p / n}$ with probability at least $1 - 6 \exp(-6q \log p)$. Withdrawing the attention on Ω_1, we obtain that

$$\|\hat{\beta} - \beta_*\| = \|\hat{\beta}_{\hat{\mathcal{M}}} - \beta_{*\hat{\mathcal{M}}}\| = \|\hat{\beta}_{\hat{\mathcal{M}}} - \beta_{\hat{\mathcal{M}}}^*\| \leq A_2 K^{-1} \sqrt{q^2 \log p / n},$$

with probability at least $1 - 31 \exp(-2q \log p)$. □

Additional Lemmas and Proofs

Lemma A1. *Given a model S such that $|S| < q$, $\mathcal{M} \not\subseteq S$, under Condition (6),*
(i): $\exists r \in S^c \cap \mathcal{M}$, such that $\beta_{S+r}^* \neq (\beta_S^{*T}, 0)^T$.
(ii): Suppose Conditions (1), (2) and (6') hold. $\exists r \in S^c \cap \mathcal{M}$, such that $\|\beta_{S+r}^{*T} - (\beta_S^{*T}, 0)\| \geq C \sigma_{\max}^{-1} \kappa_{\max}^{-1} n^{-\alpha}$.

Proof. As β_{S+j}^* is the maximizer of $E\left[\ell_{S \cup \{j\}}(\beta_{S+j})\right]$, by the concavity of $E\left[\ell_{S \cup \{j\}}(\beta_{S+j})\right]$, β_{S+j}^* is the solution to the equation $E\left[\left(Y - \mu(\beta_S^{*T} \mathbf{X}_S + \beta_j X_j)\right) \mathbf{X}_{S \cup \{j\}}\right] = \mathbf{0}$,
(i): Suppose that $\beta_{S+j}^* = (\beta_S^{*T}, 0)^T, \forall j \in S^c \cap \mathcal{M}$. Then,

$$0 = E[(Y - \mu(\beta_S^{*T} \mathbf{X}_S)) X_j] = E[(\mu(\beta_*^T \mathbf{X}) - \mu(\beta_S^{*T} \mathbf{X}_S)) X_j]$$
$$\Rightarrow \max_{j \in S^c \cap \mathcal{M}} |E[(\mu(\beta_*^T \mathbf{X}) - \mu(\beta_S^{*T} \mathbf{X}_S)) X_j]| = 0,$$

which violates the Condition (6). Therefore, we can find a $r \in S^c \cap \mathcal{M}$, such that $\beta^*_{S+r} \neq (\beta^{*T}_S, 0)^T$.
(ii): Let $r \in S^c \cap \mathcal{M}$ satisfy that $|E[(\mu(\beta^T_* \mathbf{X}) - \mu(\beta^{*T}_S \mathbf{X}_S))X_r]| > Cn^{-\alpha}$. Without loss of generality, we assume that X_r is the last element of $\mathbf{X}_{S \cup \{r\}}$. By the mean value theorem,

$$\begin{aligned}
&E[(\mu(\beta^T_* \mathbf{X}) - \mu(\beta^{*T}_S \mathbf{X}_S))X_r] \\
&= E[(\mu(\beta^T_* \mathbf{X}) - \mu(\beta^{*T}_S \mathbf{X}_S))X_r] - E[(\mu(\beta^T_* \mathbf{X}) - \mu(\beta^{*T}_{S+r} \mathbf{X}_{S \cup \{r\}}))X_r] \\
&= E[(\mu(\beta^{*T}_{S+r} \mathbf{X}_{S \cup \{r\}}) - \mu((\beta^{*T}_S, 0)\mathbf{X}_{S \cup \{r\}}))X_r] \\
&= (\beta^{*T}_{S+r} - (\beta^{*T}_S, 0)) E[\sigma(\tilde{\beta}^T_{S+r} \mathbf{X}_{S \cup \{r\}}) \mathbf{X}^{\otimes 2}_{S \cup \{r\}}] \mathbf{e}_r,
\end{aligned} \qquad (A6)$$

where $\tilde{\beta}_{S+r}$ is some point between β^*_{S+r} and $(\beta^{*T}_S, 0)^T$ and $\mathbf{e}_r$ is a vector of length $(|S|+1)$ with the rth element being 1.

As $\tilde{\beta}_{S+r}$ is some point between β^*_{S+r} and $(\beta^{*T}_S, 0)^T$, $|\tilde{\beta}^T_{S+r} \mathbf{X}_{S \cup \{r\}}| \leq |\beta^{*T}_{S+r} \mathbf{X}_{S \cup \{r\}}| + |(\beta^{*T}_S, 0) \mathbf{X}_{S \cup \{r\}}| \leq 2K^2$, by Conditions (1) and (2). Thus, $|\sigma(\tilde{\beta}^T_{S+r} \mathbf{X}_{S \cup \{r\}})| \leq \sigma_{\max}$. By (A6) and Condition (5),

$$\begin{aligned}
Cn^{-\alpha} &\leq \left|E\left[\left(\mu(\beta^T_* \mathbf{X}) - \mu(\beta^{*T}_S \mathbf{X}_S)\right)X_r\right]\right| \\
&\leq \|\beta^{*T}_{S+r} - (\beta^{*T}_S, 0)\| \sigma_{\max} \lambda_{\max}\left(E[\mathbf{X}^{\otimes 2}_{S \cup \{r\}}]\right) \|\mathbf{e}_r\| \leq \sigma_{\max} \kappa_{\max} \|\beta^{*T}_{S+r} - (\beta^{*T}_S, 0)\|.
\end{aligned}$$

Therefore, $\|\beta^{*T}_{S+r} - (\beta^{*T}_S, 0)\| \geq C\sigma^{-1}_{\max} \kappa^{-1}_{\max} n^{-\alpha}$. □

Lemma A2. *Let $\xi_i, i = 1, \ldots, n$ be n i.i.d random variables such that $|\xi_i| \leq B$ for a constant $B > 0$. Under Conditions (1)–(3), we have $E[|Y_i \xi_i - E[Y_i \xi_i]|^m] \leq m!(2B(\sqrt{2}M + \mu_{\max}))^m$, for every $m \geq 1$.*

Proof. By Conditions (1) and (2), $|\beta^T_* \mathbf{X}_i| \leq KL$, $\forall i \geq 1$ and consequently $|\mu(\beta^T_* \mathbf{X}_i)| \leq \mu_{\max}$. Then by Condition (3),

$$\begin{aligned}
E[|Y_i|^m] &= E[|\epsilon_i + \mu(\beta^T_* \mathbf{X}_i)|^m] \leq \sum_{t=0}^{m} \binom{m}{t} E[|\epsilon_i|^t] \mu^{m-t}_{\max} \\
&\leq \sum_{t=0}^{m} t! \binom{m}{t} M^t \mu^{m-t}_{\max} \leq m!(M + \mu_{\max})^m,
\end{aligned}$$

for every $m \geq 1$. By the same arguments, it can be shown that, for every $m \geq 1$, $E[|Y_i \xi_i - E[Y_i \xi_i]|^m] \leq E[(|Y_i \xi_i| + |E[Y_i \xi_i]|)^m] \leq m!(2B(M + \mu_{\max}))^m$. □

Lemma A3. *Under Conditions (1)–(3), when n is sufficiently large such that $28\sqrt{q \log p/n} < 1$, we have $\sup_{\beta \in \mathbb{B}} |\mathbb{E}_n\{L(\beta^T \mathbf{X}, Y)\}| \leq 2(M + \mu_{\max})K^3 + b_{\max}$, with probability $1 - 2\exp(-10q \log p)$.*

Proof. By Conditions (2), $\sup_{\beta \in \mathbb{B}} |\beta^T \mathbf{X}| \leq K^3$. Thus,

$$\begin{aligned}
\sup_{\beta \in \mathbb{B}} \left|\mathbb{E}_n\left\{L(\beta^T \mathbf{X}, Y)\right\}\right| &\leq \sup_{\beta \in \mathbb{B}} \left|\mathbb{E}_n\left\{\left|Y\beta^T \mathbf{X}\right|\right\}\right| + b_{\max} \\
&\leq \left(\left|\mathbb{E}_n\{|Y| - E[|Y|]\}\right| + E[|Y|]\right)K^3 + b_{\max} \\
&\leq \left(\left|\mathbb{E}_n\{|Y| - E[|Y|]\}\right|\right)K^3 + (M + \mu_{\max})K^3 + b_{\max},
\end{aligned}$$

where the last inequality follows from that $E[|Y|] \leq M + \mu_{\max}$ as shown in the proof of Lemma A2.

Let $\xi_i = 1\{Y_i > 0\} - 1\{Y_i < 0\}$. Thus $|\xi_i| \leq 1$. By Lemma A2, we have $E\left[||Y_i| - E[|Y_i|]|^m\right] \leq m!(2(M + \mu_{\max}))^m$. Applying Bernstein's inequality (e.g., Lemma 2.2.11 in [50]) yields that

$$P\left(|\mathbb{E}_n\{|Y| - E[|Y|]\}| > 10(M + \mu_{\max})\sqrt{q\log p/n}\right)$$
$$\leq 2\exp\left(-\frac{1}{2}\frac{196q\log p}{4 + 20\sqrt{q\log p/n}}\right) \leq 2\exp(-10q\log p), \quad (A7)$$

when n is sufficiently large such that $20\sqrt{q\log p/n} < 1$. Since $10(M + \mu_{\max})\sqrt{q\log p/n} = o(1)$, then

$$P\left(\sup_{\beta \in \mathbb{B}}|\mathbb{E}_n\{L(\beta^T\mathbf{X}, Y)\}| \geq 2(M + \mu_{\max})K^3 + b_{\max}\right) \leq 2\exp(-10q\log p).$$

□

Lemma A4. *Given an index set S and $r \in S^c$, let $\mathcal{B}_S^0(d) = \{\beta_S : \|\beta_S - \beta_S^*\| \leq d/(K\sqrt{|S|})\}$ and $A_1 := (M + 2\mu_{\max})$. Under Conditions (1)–(3), when n is sufficiently large such that $10\sqrt{q\log p/n} < 1$, we have*

1. $|\mathbb{G}_n\left[L(\beta_S^T\mathbf{X}_S, Y) - L(\beta_S^{*T}\mathbf{X}_S, Y)\right]| \leq 20A_1 d\sqrt{q\log p}$, *uniformly over $\beta_S \in \mathcal{B}_S^0(d)$ and $|S| \leq q$, with probability at least $1 - 4\exp(-6q\log p)$.*
2. $|\mathbb{G}_n\left[L(\beta_S^{*T}\mathbf{X}_S, Y)\right]| \leq 10(A_1 K^2 + b_{\max})\sqrt{q\log p}$, *uniformly over $|S| \leq q$, with probability at least $1 - 2\exp(-8q\log p)$.*

Proof. : (1): Let $\mathcal{R}_{|S|}(d)$ be a $|S|$-dimensional ball with center at 0 and radius $d/(K\sqrt{|S|})$. Then $\mathcal{B}_S^0(d) = \mathcal{R}_{|S|}(d) + \beta_S^*$. Let $\mathcal{C}_{|S|} := \{\mathcal{C}(\xi_k)\}$ be a collection of cubes that cover the ball $R_{|S|}(d)$, where $\mathcal{C}(\xi_k)$ is a cube containing ξ_k with sides of length $d/(K\sqrt{|S|}n^2)$ and ξ_k is some point in $\mathcal{R}_{|S|}(d)$. As the volume of $\mathcal{C}(\xi_k)$ is $(d/(K\sqrt{|S|}n^2))^{|S|}$ and the volume of $\mathcal{R}_{|S|}(d)$ is less than $(2d/(K\sqrt{|S|}))^{|S|}$, we can select ξ_ks so that no more than $(4n^2)^{|S|}$ cubes are needed to cover $\mathcal{R}_{|S|}(d)$. We thus assume $|\mathcal{C}_{|S|}| \leq (4n^2)^{|S|}$. For any $\xi \in \mathcal{C}(\xi_k)$, $\|\xi - \xi_k\| \leq d/(Kn^2)$. In addition, let $T_{1S}(\xi) := \mathbb{E}_n[Y\xi^T\mathbf{X}_S]$, $T_{2S}(\xi) := \mathbb{E}_n[b((\beta_S^* + \xi)^T\mathbf{X}_S) - b(\beta_S^{*T}\mathbf{X}_S)]$, and $T_S(\xi) := T_{1S}(\xi) - T_{2S}(\xi)$. Given any $\xi \in \mathcal{R}_{|S|}(d)$, there exists $\mathcal{C}(\xi_k) \in \mathcal{C}_{|S|}$ such that $\xi \in \mathcal{C}(\xi_k)$. Then

$$|T_S(\xi) - E[T_S(\xi)]| \leq |T_S(\xi) - T_S(\xi_k)| |T_S(\xi_k) - E[T_S(\xi_k)]| + |E[T_S(\xi)] - E[T_S(\xi_k)]|$$
$$=: I + II + III.$$

We deal with III first. By the mean value theorem, there exists a $\tilde{\xi}$ between ξ and ξ_k such that

$$|E[T_S(\xi_k)] - E[T_S(\xi)]| = \left|E[Y(\xi_k - \xi)^T\mathbf{X}_S] + E[\mu\left((\beta_S^* + \tilde{\xi})^T\mathbf{X}_S\right)(\xi_k - \xi)^T\mathbf{X}_S]\right|$$
$$\leq E[|Y|]\|\xi_k - \xi\|\|\mathbf{X}_S\| + \mu_{\max}\|\xi_k - \xi\|\|\mathbf{X}_S\| \leq (M + 2\mu_{\max})d\sqrt{|S|}n^{-2} = A_1 d\sqrt{|S|}n^{-2}, \quad (A8)$$

where the last inequality follows from Lemma A2 and $A_1 = M + 2\mu_{\max}$.

Next, we evaluate II. By Condition (2), $|\mathbf{X}_{iS}^T\xi| \leq \|\mathbf{X}_{iS}\|\|\xi\| \leq d/(K\sqrt{|S|})\sqrt{|S|}K = d$, for all $\xi \in \mathcal{R}_{|S|}(d)$. Then by Lemma A2,

$$E\left[|Y\xi_k^T\mathbf{X}_S - E[Y\xi_k^T\mathbf{X}_S]|^m\right] \leq m!(2(M + \mu_{\max})d)^m.$$

By Bernstein's inequality, when n is sufficiently large such that $10\sqrt{q\log p/n} \le 1$.

$$P\left(|T_{1S}(\boldsymbol{\xi}_k) - E[T_{1S}(\boldsymbol{\xi}_k)]| > 10(M+\mu_{\max})d\sqrt{qn^{-1}\log p}\right)$$
$$\le 2\exp\left(-\frac{1}{2}\frac{100q\log p}{4+20\sqrt{q\log p/n}}\right) \le 2\exp(-10q\log p). \quad (A9)$$

Since $|b((\boldsymbol{\beta}_S^* + \boldsymbol{\xi}_k)^T \mathbf{X}_S) - b(\boldsymbol{\beta}_S^{*T}\mathbf{X}_S)| \le \mu_{\max}d$, by the same arguments used for (A9), we have

$$P\left(|T_{2S}(\boldsymbol{\xi}_k) - E[T_{2S}(\boldsymbol{\xi}_k)]| > 10\mu_{\max}d\sqrt{qn^{-1}\log p}\right) \le 2\exp(-10q\log p). \quad (A10)$$

Combining (A9) and (A10) yields that uniformly over $\boldsymbol{\xi}_k$

$$|T_S(\boldsymbol{\xi}_k) - E[T_S(\boldsymbol{\xi}_k)]| \le 10A_1 d\sqrt{qn^{-1}\log p}, \quad (A11)$$

with probability at least $1 - 2(4n^2)^{|S|}\exp(-10q\log p)$.

We now assess I. Following the same arguments as in Lemma A3,

$$P\left(\sup_{\boldsymbol{\xi}\in\mathcal{C}(\boldsymbol{\xi}_k)}|T_S(\boldsymbol{\xi}) - T_S(\boldsymbol{\xi}_k)| > (2M+3\mu_{\max})d\sqrt{|S|n^{-2}}\right) \le 2\exp(-8q\log p). \quad (A12)$$

Since $\sqrt{|S|}n^{-2} = o(\sqrt{qn^{-1}\log p})$, combining (A8), (A11) and (A12) together yields that

$$P\left(\sup_{\boldsymbol{\xi}\in\mathcal{R}_{|S|}(d)}|T_S(\boldsymbol{\xi}) - E[T_S(\boldsymbol{\xi})]| \ge 20A_1 d\sqrt{qn^{-1}\log p}\right)$$
$$\le 2(4n^2)^{|S|}\exp(-10q\log p) + 2\exp(-8q\log p) \le 4\exp(-8q\log p).$$

By the combinatoric inequality $\binom{p}{s} \le (ep/s)^s$, we obtain that

$$P\left(\sup_{|S|\le q, \boldsymbol{\beta}_S\in\mathcal{B}_S^0(d_1)}\left|\mathbb{G}_n\left[L\left(\boldsymbol{\beta}_S^T\mathbf{X}_S, Y\right) - L\left(\boldsymbol{\beta}_S^{*T}\mathbf{X}_S, Y\right)\right]\right| \ge 20A_1 d\sqrt{q\log p}\right)$$
$$\le \sum_{s=1}^{q}(ep/s)^s 4\exp(-8q\log p) \le 4\exp(-6q\log p).$$

(2): We evaluate the mth moment of $L(\boldsymbol{\beta}_S^*\mathbf{X}_S, Y)$.

$$E\left[(Y\boldsymbol{\beta}_S^*\mathbf{X}_S - b(\boldsymbol{\beta}_S^*\mathbf{X}_S))^m\right] \le E\left[\sum_{t=0}^{m}\binom{m}{t}|Y|^t K^{2t}b_{\max}^{m-t}\right]$$
$$\le \sum_{t=0}^{m}\binom{m}{t}t!((M+\mu_{\max})K^2)^t b_{\max}^{m-t} \le m!((M+\mu_{\max})K^2 + b_{\max})^m.$$

Then, by Bernstein's inequality,

$$P\left(|\mathbb{G}_n[L(\boldsymbol{\beta}_S^{*T}\mathbf{X}_S, Y)]| > 10(A_1 K^2 + b_{\max})\sqrt{q\log p}\right) \le 2\exp(-10q\log p).$$

By the same arguments used in (i), we obtain that

$$P\left(\sup_{|S|\leq q}\left|\mathbb{G}_n\left[L\left(\boldsymbol{\beta}_S^{*T}\mathbf{X}_S,Y\right)\right]\right|\geq 10(A_1K^2+b_{\max})\sqrt{q\log p}\right)$$
$$\leq \sum_{s=1}^q (ep/s)^s 2\exp(-10q\log p) \leq 2\exp(-8q\log p).$$

□

Lemma A5. *Given a model S and $r \in S^c$, under Conditions (1), (2) and (5), for any $\|\boldsymbol{\beta}_S - \boldsymbol{\beta}_S^*\| \leq K/\sqrt{|S|}$, $\sigma_{\min}\kappa_{\min}\|\boldsymbol{\beta}_S - \boldsymbol{\beta}_S^*\|^2/2 \leq E\left[\ell_S(\boldsymbol{\beta}_S^*)\right] - E\left[\ell_S(\boldsymbol{\beta}_S)\right] \leq \sigma_{\max}\kappa_{\max}\|\boldsymbol{\beta}_S - \boldsymbol{\beta}_S^*\|^2/2.$*

Proof. Due to the concavity of the log-likelihood in GLMs with the canonical link, $E\left[Y\mathbf{X}_S - \mu(\boldsymbol{\beta}_S^{*T}\mathbf{X}_S)\mathbf{X}_S\right] = 0$. Then for any $\|\boldsymbol{\beta}_S - \boldsymbol{\beta}_S^*\| \leq K/\sqrt{|S|}$, $|\boldsymbol{\beta}^T\mathbf{X}_S| \leq |\boldsymbol{\beta}^{*T}\mathbf{X}_S| + |(\boldsymbol{\beta} - \boldsymbol{\beta}^*)^T\mathbf{X}_S| \leq K^2 + K/\sqrt{|S|} \times K\sqrt{|S|} = 2KL$. Thus, by Taylor's expansion,

$$E\left[\ell_S(\boldsymbol{\beta}_S)\right] - E\left[\ell_S(\boldsymbol{\beta}_S^*)\right] = -\frac{1}{2}(\boldsymbol{\beta}_S - \boldsymbol{\beta}_S^*)^T E\left[\sigma\left(\tilde{\boldsymbol{\beta}}_S^T\mathbf{X}_S\right)\mathbf{X}_S^{\otimes 2}\right](\boldsymbol{\beta}_S - \boldsymbol{\beta}_S^*),$$

where $\tilde{\boldsymbol{\beta}}_S$ is between $\boldsymbol{\beta}_S$ and $\boldsymbol{\beta}_S^*$. By Condition (5), $\sigma_{\min}\kappa_{\min}\|\boldsymbol{\beta}_S - \boldsymbol{\beta}_S^*\|^2/2 \leq E\left[\ell_S(\boldsymbol{\beta}_S^*)\right] - E\left[\ell_S(\boldsymbol{\beta}_S)\right] \leq \sigma_{\max}\kappa_{\max}\|\boldsymbol{\beta}_S - \boldsymbol{\beta}_S^*\|^2/2$.
□

Lemma A6. *Under Conditions (1)–(6), there exist some constants A_2 and A_3 that do not depend on n, such that $\|\hat{\boldsymbol{\beta}}_S - \boldsymbol{\beta}_S^*\| \leq A_2K^{-1}\sqrt{q^2\log p/n}$ and $|\ell_S(\hat{\boldsymbol{\beta}}_S) - \ell_S(\boldsymbol{\beta}_S^*)| \leq A_3 q^2\log p/n$ hold uniformly over $S : |S| \leq q$, with probability at least $1 - 6\exp(-6q\log p)$.*

Proof. Define

$$\Omega(d) := \left\{\sup_{|S|\leq q, \boldsymbol{\beta}_S \in \mathcal{B}_S^0(d)}\left|\mathbb{G}_n\left[L\left(\boldsymbol{\beta}_S^T\mathbf{X}_S,Y\right) - L\left(\boldsymbol{\beta}_S^{*T}\mathbf{X}_S,Y\right)\right]\right| < 20A_1d\sqrt{q\log p}\right\}.$$

By Lemma A4, the event $\Omega(d)$ holds with probability at least $1 - 4\exp(-6q\log p)$. Thus, in the proof of Lemma A6, we shall assume $\Omega(d)$ hold with $d = A_2\sqrt{q^3\log p/n}$ for some $A_2 > 20(\sigma_{\min}\kappa_{\min})^{-1}K^2A_1$.

For any $\|\boldsymbol{\beta}_S - \boldsymbol{\beta}_S^*\| = A_2K^{-1}\sqrt{q^2\log p/n}$, since $\sqrt{q^2\log p/n} \leq \sqrt{q^3\log p/n}/\sqrt{|S|}$, $\boldsymbol{\beta}_S \in \mathcal{B}_S^0(d)$. By Lemma A5,

$$\ell_S(\boldsymbol{\beta}_S^*) - \ell_S(\boldsymbol{\beta}_S)$$
$$= \left(\ell_S(\boldsymbol{\beta}_S^*) - E\left[\ell_S(\boldsymbol{\beta}_S^*)\right] - (\ell_S(\boldsymbol{\beta}_S) - E\left[\ell_S(\boldsymbol{\beta}_S)\right])\right) + (E\left[\ell_S(\boldsymbol{\beta}_S^*)\right] - E\left[\ell_S(\boldsymbol{\beta}_S)\right])$$
$$\geq \sigma_{\min}\kappa_{\min}\|\boldsymbol{\beta}_S - \boldsymbol{\beta}_S^*\|^2/2 - 20A_1d\sqrt{q\log p/n}$$
$$= \sigma_{\min}\kappa_{\min}A_2^2 q^2\log p/(K^2n) - 20A_1A_2 q^2\log p/n > 0.$$

Thus,
$$\inf_{|S|\leq q, \|\boldsymbol{\beta}_S - \boldsymbol{\beta}_S^*\| = A_2K^{-1}\sqrt{q^2\log p/n}} \ell_S(\boldsymbol{\beta}_S^*) - \ell_S(\boldsymbol{\beta}_S) > 0.$$

Then by the concavity of $\ell_S(\cdot)$, we obtain that $\max_{|S|\leq q}\left\|\hat{\boldsymbol{\beta}}_S - \boldsymbol{\beta}_S^*\right\| \leq A_2K^{-1}\sqrt{q^2n^{-1}\log p}$.

On the other hand, for any $\|\beta_S - \beta_S^*\| \le A_2 K^{-1}\sqrt{q^2 \log p/n}$,

$$|\ell_S(\beta_S^*) - \ell_S(\beta_S)|$$
$$\le \left|\ell_S(\beta_S^*) - E[\ell_S(\beta_S^*)] - (\ell_S(\beta_S) - E[\ell_S(\beta_S)])\right| + |E[\ell_S(\beta_S^*)] - E[\ell_S(\beta_S)]|$$
$$\le \sigma_{\max}\kappa_{\max}\|\beta_S - \beta_S^*\|^2/2 + 20 A_1 d\sqrt{q\log p/n} \le A_3 q^2 n^{-1}\log p,$$

where $A_3 := 4\sigma_{\max}\lambda_{\max}A_2^2 K^{-2} + 20 A_1 A_2$. □

Appendix B. Additional Results in the Applications

Table A1. Comparison of genes selected by each competing method from the mammalian eye data set.

	STEPWISE	FR	LASSO	SIS+LASSO	SC	dgLARS
STEPWISE	3	3	2	2	2	0
FR		4	2	2	2	0
LASSO			16	5	2	0
SIS+LASSO				9	2	0
SC					4	0
dgLARS						7

Note: Diagonal and off-diagonal elements of the table represent the model sizes for each method and the number of overlapping genes selected by the two methods corresponding to the row and column, respectively.

Table A2. Selected miRNAs for ESCC training dataset.

Methods	Selected miRNAs
STEPWISE	miR-4783-3p; miR-320b; miR-1225-3p
FR	miR-4783-3p; miR-320b; miR-1225-3p; 6789-5p
SC	miR-4783-3p; miR-320b; miR-1225-3p; 6789-5p
LASSO	miR-6789-5p; miR-6781-5p; miR-1225-3p; miR-1238-5p; miR-320b; miR-6794-5p; miR-6877-5p; miR-6785-5p; miR-718; miR-195-5p
SIS+LASSO	miR-6785-5p; miR-1238-5p; miR-1225-3p; miR-6789-5p; miR-320b; miR-6875-5p; miR-6127; miR-1268b; miR-6781-5p; miR-125a-3p
dgLARS	miR-891b; miR-6127; miR-151a-5p; miR-195-5p; ; miR-3688-5p miR-125b-1-3p; miR-1273c; miR-6501-5p; miR-4666a-5p; miR-514a-3p

Note: LASSO, SIS+LASSO, dgLARS selected 20, 17 and 33 miRNAs, respectively, and we only reported top 10 miRNAs.

References

1. Prosperi, M.; Min, J.S.; Bian, J.; Modave, F. Big data hurdles in precision medicine and precision public health. *BMC Med. Inform. Decis. Mak.* **2018**, *18*, 139. [CrossRef] [PubMed]
2. Tibshirani, R. Regression shrinkage and selection via the lasso. *J. R. Stat. Soc. Ser. B-Stat. Methodol.* **1996**, *58*, 267–288. [CrossRef]
3. Flynn, C.J.; Hurvich, C.M.; Simonoff, J.S. On the sensitivity of the lasso to the number of predictor variables. *Stat. Sci.* **2017**, *32*, 88–105. [CrossRef]
4. van de Geer, S.A. On the asymptotic variance of the debiased Lasso. *Electron. J. Stat.* **2019**, *13*, 2970–3008. [CrossRef]
5. Fan, J.; Lv, J. Sure independence screening for ultrahigh dimensional feature space (with discussion). *J. R. Stat. Soc. Ser. B-Stat. Methodol.* **2008**, *70*, 849–911. [CrossRef]
6. Barut, E.; Fan, J.; Verhasselt, A. Conditional sure independence screening. *J. Am. Stat. Assoc.* **2016**, *111*, 1266–1277. [CrossRef] [PubMed]

7. Wang, H. Forward regression for ultra-high dimensional variable screening. *J. Am. Stat. Assoc.* **2009**, *104*, 1512–1524. [CrossRef]
8. Zheng, Q.; Hong, H.G.; Li, Y. Building generalized linear models with ultrahigh dimensional features: A sequentially conditional approach. *Biometrics* **2019**, *76*, 1–14. [CrossRef]
9. Hong, H.G.; Zheng, Q.; Li, Y. Forward regression for Cox models with high-dimensional covariates. *J. Multivar. Anal.* **2019**, *173*, 268–290. [CrossRef]
10. Efron, B.; Hastie, T.; Johnstone, I.; Tibshirani, R. Least angle regression. *Ann. Stat.* **2004**, *32*, 407–499.
11. Augugliaro, L.; Mineo, A.M.; Wit, E.C. Differential geometric least angle regression: a differential geometric approach to sparse generalized linear models. *J. R. Stat. Soc. Ser. B-Stat. Methodol.* **2013**, *75*, 471–498. [CrossRef]
12. Pazira, H.; Augugliaro, L.; Wit, E. Extended differential geometric LARS for high-dimensional GLMs with general dispersion parameter. *Stat. Comput.* **2018**, *28*, 753–774. [CrossRef]
13. An, H.; Huang, D.; Yao, Q.; Zhang, C.H. Stepwise searching for feature variables in high-dimensional linear regression. 2008. Available online: http://eprints.lse.ac.uk/51349/ (accessed on 20 August 2020).
14. Ing, C.K.; Lai, T.L. A stepwise regression method and consistent model selection for high-dimensional sparse linear models. *Stat. Sin.* **2011**, *21*, 1473–1513. [CrossRef]
15. Hwang, J.S.; Hu, T.H. A stepwise regression algorithm for high-dimensional variable selection. *J. Stat. Comput. Simul.* **2015**, *85*, 1793–1806. [CrossRef]
16. McCullagh, P. *Generalized Linear Models*; Routledge: Abingdon-on-Thames, UK, 1989.
17. Chen, J.; Chen, Z. Extended BIC for small-n-large-P sparse GLM. *Stat. Sin.* **2012**, *22*, 555–574. [CrossRef]
18. Bühlmann, P.; Yu, B. Sparse boosting. *J. Mach. Learn. Res.* **2006**, *7*, 1001–1024.
19. van de Geer, S.A. High-dimensional generalized linear models and the lasso. *Ann. Stat.* **2008**, *36*, 614–645. [CrossRef]
20. Chen, J.; Chen, Z. Extended Bayesian information criteria for model selection with large model spaces. *Biometrika* **2008**, *95*, 759–771. [CrossRef]
21. Fan, Y.; Tang, C.Y. Tuning parameter selection in high dimensional penalized likelihood. *J. R. Stat. Soc. Ser. B-Stat. Methodol.* **2013**, *75*, 531–552. [CrossRef]
22. Cheng, M.Y.; Honda, T.; Zhang, J.T. Forward variable selection for sparse ultra-high dimensional varying coefficient models. *J. Am. Stat. Assoc.* **2016**, *111*, 1209–1221. [CrossRef]
23. Zhao, S.D.; Li, Y. Principled sure independence screening for Cox models with ultra-high-dimensional covariates. *J. Multivar. Anal.* **2012**, *105*, 397–411. [CrossRef] [PubMed]
24. Kwemou, M. Non-asymptotic oracle inequalities for the Lasso and group Lasso in high dimensional logistic model. *ESAIM-Prob. Stat.* **2016**, *20*, 309–331. [CrossRef]
25. Jiang, Y.; He, Y.; Zhang, H. Variable selection with prior information for generalized linear models via the prior LASSO method. *J. Am. Stat. Assoc.* **2016**, *111*, 355–376. [CrossRef]
26. Zhang, C.H.; Huang, J. The sparsity and bias of the Lasso selection in high-dimensional linear regression. *Ann. Stat.* **2008**, *36*, 1567–1594. [CrossRef]
27. Fan, J.; Song, R. Sure independence screening in generalized linear models with NP-dimensionality. *Ann. Stat.* **2010**, *38*, 3567–3604. [CrossRef]
28. Luo, S.; Chen, Z. Sequential Lasso cum EBIC for feature selection with ultra-high dimensional feature space. *J. Am. Stat. Assoc.* **2014**, *109*, 1229–1240. [CrossRef]
29. Luo, S.; Xu, J.; Chen, Z. Extended Bayesian information criterion in the Cox model with a high-dimensional feature space. *Ann. Inst. Stat. Math.* **2015**, *67*, 287–311. [CrossRef]
30. Hastie, T.; Tibshirani, R.; Friedman, J. *The Elements of Statistical Learning: Data mining, Inference, and Prediction*; Springer: Berlin/Heidelberger, Germany, 2009.
31. Simon, N.; Friedman, J.; Hastie, T.; Tibshirani, R. Regularization Paths for Cox's Proportional Hazards Model via Coordinate Descent. *J. Stat. Softw.* **2011**, *39*, 1–13. [CrossRef]
32. Breheny, P.; Huang, J. Coordinate descent algorithms for nonconvex penalized regression, with applications to biological feature selection. *Ann. Appl. Stat.* **2011**, *5*, 232–253. [CrossRef]
33. Wang, X.; Leng, C. R package: screening, 2016. Available Online: https://github.com/wwrechard/screening (accessed on 20 August 2020).
34. Augugliaro, L.; Mineo, A.M.; Wit, E.C. dglars: An R Package to Estimate Sparse Generalized Linear Models. *J. Stat. Softw.* **2014**, *59*, 1–40. [CrossRef]

35. Scheetz, T.E.; Kim, K.Y.A.; Swiderski, R.E.; Philp, A.R.; Braun, T.A.; Knudtson, K.L.; Dorrance, A.M.; DiBona, G.F.; Huang, J.; Casavant, T.L.; et al. Regulation of gene expression in the mammalian eye and its relevance to eye disease. *Proc. Natl. Acad. Sci. USA* **2006**, *103*, 14429–14434. [CrossRef] [PubMed]
36. Chiang, A.P.; Beck, J.S.; Yen, H.J.; Tayeh, M.K.; Scheetz, T.E.; Swiderski, R.E.; Nishimura, D.Y.; Braun, T.A.; Kim, K.Y.A.; Huang, J.; et al. Homozygosity mapping with SNP arrays identifies TRIM32, an E3 ubiquitin ligase, as a Bardet–Biedl syndrome gene (BBS11). *Proc. Natl. Acad. Sci. USA* **2006**, *103*, 6287–6292. [CrossRef] [PubMed]
37. He, S.; Peng, J.; Li, L.; Xu, Y.; Wu, X.; Yu, J.; Liu, J.; Zhang, J.; Zhang, R.; Wang, W. High expression of cytokeratin CAM5.2 in esophageal squamous cell carcinoma is associated with poor prognosis. *Medicine* **2019**, *98*, e17104. [CrossRef]
38. Li, B.X.; Yu, Q.; Shi, Z.L.; Li, P.; Fu, S. Circulating microRNAs in esophageal squamous cell carcinoma: association with locoregional staging and survival. *Int. J. Clin. Exp. Med.* **2015**, *8*, 7241–7250.
39. Sudo, K.; Kato, K.; Matsuzaki, J.; Boku, N.; Abe, S.; Saito, Y.; Daiko, H.; Takizawa, S.; Aoki, Y.; Sakamoto, H.; et al. Development and validation of an esophageal squamous cell carcinoma detection model by large-scale microRNA profiling. *JAMA Netw. Open* **2019**, *2*, e194573–e194573. [CrossRef]
40. Zhang, Y. Epidemiology of esophageal cancer. *World J. Gastroenterol* **2013**, *19*, 5598–5606. [CrossRef]
41. Mathieu, L.N.; Kanarek, N.F.; Tsai, H.L.; Rudin, C.M.; Brock, M.V. Age and sex differences in the incidence of esophageal adenocarcinoma: results from the Surveillance, Epidemiology, and End Results (SEER) Registry (1973–2008). *Dis. Esophagus* **2014**, *27*, 757–763. [CrossRef]
42. Zhou, J.; Zhang, M.; Huang, Y.; Feng, L.; Chen, H.; Hu, Y.; Chen, H.; Zhang, K.; Zheng, L.; Zheng, S. MicroRNA-320b promotes colorectal cancer proliferation and invasion by competing with its homologous microRNA-320a. *Cancer Lett.* **2015**, *356*, 669–675. [CrossRef]
43. Lieb, V.; Weigelt, K.; Scheinost, L.; Fischer, K.; Greither, T.; Marcou, M.; Theil, G.; Klocker, H.; Holzhausen, H.J.; Lai, X.; et al. Serum levels of miR-320 family members are associated with clinical parameters and diagnosis in prostate cancer patients. *Oncotarget* **2018**, *9*, 10402–10416. [CrossRef]
44. Mullany, L.E.; Herrick, J.S.; Wolff, R.K.; Stevens, J.R.; Slattery, M.L. Association of cigarette smoking and microRNA expression in rectal cancer: insight into tumor phenotype. *Cancer Epidemiol.* **2016**, *45*, 98–107. [CrossRef]
45. Zheng, H.; Zhang, F.; Lin, X.; Huang, C.; Zhang, Y.; Li, Y.; Lin, J.; Chen, W.; Lin, X. MicroRNA-1225-5p inhibits proliferation and metastasis of gastric carcinoma through repressing insulin receptor substrate-1 and activation of β-catenin signaling. *Oncotarget* **2016**, *7*, 4647–4663. [CrossRef]
46. R Core Team. *R: A Language and Environment for Statistical Computing*; R Foundation for Statistical Computing: Vienna, Austria, 2018.
47. Wickham, H. *ggplot2: Elegant Graphics for Data Analysis*; Springer: Berlin/Heidelberger, Germany, 2016.
48. Zhao, P.; Yu, B. On model selection consistency of Lasso. *J. Mach. Learn. Res.* **2006**, *7*, 2541–2563.
49. Bühlmann, P.; Van De Geer, S. *Statistics for High-dimensional Data: Methods, Theory and Applications*; Springer: Berlin/Heidelberger, Germany, 2011.
50. Vaart, A.W.; Wellner, J.A. *Weak Convergence and Empirical Processes: with Applications to Statistics*; Springer: Berlin/Heidelberger, Germany, 1996.
51. Talagrand, M. Sharper bounds for Gaussian and empirical processes. *Ann. Probab.* **1994**, *22*, 28–76. [CrossRef]

© 2020 by the authors. Licensee MDPI, Basel, Switzerland. This article is an open access article distributed under the terms and conditions of the Creative Commons Attribution (CC BY) license (http://creativecommons.org/licenses/by/4.0/).

Article

Ant Colony System Optimization for Spatiotemporal Modelling of Combined EEG and MEG Data

Eugene A. Opoku [1,*], Syed Ejaz Ahmed [2], Yin Song [1] and Farouk S. Nathoo [1]

1. Department of Mathematics and Statistics, University of Victoria, Victoria, BC V8P 5C2, Canada; yinsong@uvic.ca (Y.S.); nathoo@uvic.ca (F.S.N.)
2. Department of Mathematics and Statistics, Brock University, St. Catharines, ON L2S 3A1, Canada; sahmed5@brocku.ca
* Correspondence: eopoku@uvic.ca

Abstract: Electroencephalography/Magnetoencephalography (EEG/MEG) source localization involves the estimation of neural activity inside the brain volume that underlies the EEG/MEG measures observed at the sensor array. In this paper, we consider a Bayesian finite spatial mixture model for source reconstruction and implement Ant Colony System (ACS) optimization coupled with Iterated Conditional Modes (ICM) for computing estimates of the neural source activity. Our approach is evaluated using simulation studies and a real data application in which we implement a nonparametric bootstrap for interval estimation. We demonstrate improved performance of the ACS-ICM algorithm as compared to existing methodology for the same spatiotemporal model.

Keywords: ant colony system; bayesian spatial mixture model; inverse problem; nonparamteric boostrap; EEG/MEG data

Citation: Opoku, E.A.; Ahmed, S.E.; Song, Y.; Nathoo, F.S. Ant Colony System Optimization for Spatiotemporal Modelling of Combined EEG and MEG Data. *Entropy* **2021**, *23*, 329. https://doi.org/10.3390/e23030329

Academic Editor: David Holcman

Received: 21 January 2021
Accepted: 7 March 2021
Published: 11 March 2021

Publisher's Note: MDPI stays neutral with regard to jurisdictional claims in published maps and institutional affiliations.

Copyright: © 2021 by the authors. Licensee MDPI, Basel, Switzerland. This article is an open access article distributed under the terms and conditions of the Creative Commons Attribution (CC BY) license (https://creativecommons.org/licenses/by/4.0/).

1. Introduction

Electroencephalography (EEG) and Magnetoencephalography (MEG) are two non-invasive approaches for measuring electrical activity of the brain with high temporal resolution. These neuroimaging techniques allow us to study brain dynamics and the complex informational exchange processes in the human brain. They are widely used in many clinical and research applications [1,2], though estimating the evoked-response activity within the brain from electromagnetic fields measured outside of the skull remains a challenging inverse problem with infinitely many different sources within the brain that can produce the same observed data [3].

Proposed solutions to the MEG/EEG inverse problem have been based on distributed source and dipolar methods [4]. In the case of distributed source methods, every location on a fine grid within the brain has associated neural activation source parameters. In this case, the number of unknown current sources exceeds the number of MEG or EEG sensors and estimation thus requires constraints through regularization or priors to obtain a solution. For such methods, various steps have been taken to regularize the solution by choosing minimum-norm solutions or by limiting the spatiotemporal variation of the solution. These approaches impose L_2 or L_1 [5] norm regularization constraints that serve to stabilize and condition the source parameter estimates. However, these methods do not consider the temporal nature of the problem. In [6], the authors propose a dynamic state-space model that accounts for both spatial and temporal correlations within and across candidate intra-cortical sources using Bayesian estimation and Kalman filtering. Dipolar methods, on the other hand, assume that the actual current distribution can be explained by a small set of current dipoles with unknown locations, amplitudes and orientations (see [4] for review). Hence, the resulting inverse problem becomes non-linear and a number of dipoles is to be estimated. Proposed solutions to this problem include algorithms such as simulated annealing [7] to address nonlinear optimization in the localization of neuromagnetic sources.

From the perspective of Bayesian approaches, the ill-posed nature of the inverse problem requires incorporation of prior assumptions when choosing an appropriate solution out of an infinite set of candidates. For instance, the authors of [8] consider Gaussian scale mixture models, with flexible, large covariance components representing spatial patterns of neural activity. The authors of [9] also propose a hierarchical linear model with Gaussian errors in a Parametric Empirical Bayes (PEB) framework whose random terms are drawn from multivariate Gaussian distributions and covariances factor into temporal and spatial components at the sensor and source levels. The authors of [10] propose an application of empirical Bayes to the source reconstruction problem with automatic selection of multiple cortical sources. The authors of [11] developed the Mesostate-Space Model (MSM) based on the assumption that the unknown neural brain activity can be specified in terms of a set of locally distributed and temporally coherent meso-sources for either MEG or EEG data, while the authors of [12] extend this approach to propose a Switching Mesostate-Space Model (SMSM) to allow flexibility by accounting for complex brain processes that cannot be characterized by linear and stationary Gaussian dynamics.

By extending and building on the MSM, the authors of [13] developed a Bayesian spatial finite mixture model incorporating the following two conditions, taken directly from [13]:

1. Relaxing the assumption of independent mixture allocation variables and modeling mixture allocations using the Potts model, which allows for spatial dependence in allocations.
2. Formulate the model for combined MEG and EEG data for joint source localization.

This spatiotemporal model describes a joint model that combines MEG and EEG data, in which brain neural activity is modeled from the Gaussian spatial mixture model. The neural source activity is described in terms of a few hidden states, with each state having its own dynamics and a Potts model used in representing the spatial dependence in the mixture model.

For the Bayesian mixture model formulated, an Iterated Conditional Modes (ICM) algorithm was developed by the authors of [13] for simultaneous point estimation and model selection for the number of mixture components in the latent process. Whilst ICM is a very simple and computationally efficient algorithm, convergence of this algorithm is sensitive to starting values and local optima. This issue was left unresolved in [13]. Here we investigate the potential for finding better solutions, and focus on implementing a population-based optimization algorithm-based Ant Colony System (ACS) [14].

ACS is a metaheuristic optimization algorithm inspired by the biological behavior of ants constructing solutions based on their collective foraging behavior [14]. ACS has been successfully applied in several areas such as clustering, data mining and image segmentation problems [15–17]. ACS is a constructive algorithm that uses an analogue of ant trail pheromones to learn about good features of solutions in combinatorial optimization problems. New solutions are generated using a parameterized probabilistic model, the parameters of which are updated using previously generated solutions so as to direct the search towards promising areas of the solution space. The model used in ACS is known as pheromone, an artificial analogue of the chemical substance used by real ants to mark trails from the nest to food sources. Based on this representation, each artificial ant constructs a part of the solution based on concentration of pheromone information released by other ants. The amount of pheromone deposited by an ant reflects the quality of the good solutions built and the traversed path. The pheromone deposited and volatilized adds solution components to partial solutions. After some time and based on more ants' communications through pheromone information, they tend to follow the same optimal paths yielding the optimal solution, in our context maximization of the posterior density.

As an alternative to the ICM algorithm, we thus implement the ACS algorithm coupled with a local search ICM algorithm to provide a new approach to model estimation and potentially better estimates of the model parameters. This approach is evaluated and found to provide significant improvements. Within the context of a simpler spatial mixture, ACS has been implemented for a Gaussian Potts mixture model in [18] and has been shown to outperform both the Simulated Annealing and ICM algorithms for parameter and mixture component estimation. The theoretical guarantees associated with simulated annealing to reach a global optimum is dependent on the choice of a cooling schedule. The choice of an optimal cooling schedule can be difficult in practice for large spatiotemporal models. ACS has also proved to be competitive with genetic and other optimization algorithms in several tasks, mainly in image classification and the traveling salesman problem [19,20].

Ant Colony Optimization (ACO) algorithms are implemented to solve Constraint Satisfaction Problems (CSP) where ACO solutions to CSP face the challenge of high cost and low solution quality. Motivated by this challenge, the authors of [21] proposed Ant Colony Optimization based on information Entropy (ACOE). The idea is based on incorporating a local search that uses a crossover operation to optimize the best solution according to the feedback of information entropy. This is performed by comparing the difference of the information entropy between the current global best solution and the best solution in the current iteration. Datasets from four classes of binary CSP test cases were generated and then ACOE was implemented for comparison. Results showed that ACOE outperformed Particle Swarm Optimization (PSO), a Differential Evolution (DE) algorithm and Artificial Bee Colony (ABC) in terms of the solution quality, data distribution and convergence performance.

To our knowledge, this is the first attempt at solving the neuroelectromagnetic inverse problem for combined EEG/MEG data using a population-based optimization approach combined with a spatial mixture model. The primary contribution of this paper is the design and implementation of the ACS algorithm to the dynamic spatial model and its evaluation. Importantly, we demonstrate improved results both in the estimation of neural activity and model selection uniformly across all conditions considered.

The rest of the paper proceeds as follows. The posterior distribution of the model and the design and implementation of the ACS algorithm are presented in Section 2. In Section 3 our algorithm is investigated using simulation studies and comparisons made with an existing approach developed in [13] . Section 4 provides an illustration on real data and the development of a nonparametric bootstrap for interval estimation in a study of scrambled face perception. The paper concludes with a conclusion and directions for future work in Section 4.

Related Works

Merging EEG and MEG aims at accounting for information missed by one modality and captured by the other. Fused reconstruction therefore appears promising to reach high temporal and spatial resolutions in brain function imaging. The authors of [22] address the added value of combining EEG and MEG data for distributed source localization, building on the flexibility of parametric empirical Bayes, namely for EEG–MEG data fusion, group level inference and formal hypothesis testing. The proposed approach follows a two-step procedure by first using unimodal or multimodal inference to derive a cortical solution at the group level, and second by using this solution as a prior model for single subject-level inference based on either unimodal or multimodal data. Another popular approach for non-globally optimized solutions of the MEG/EEG inverse problem is based on the use of adaptive Beamformers (BF). However, the BFs are known to fail when dealing with correlated sources acting like poorly tuned spatial filters with a low signal-to-noise ratio (SNR) of the output time series and often meaningless cortical maps of power distribution. To address this limitation, the authors of [23] developed a novel data covariance approach to supply robustness to the beamforming technique when operating in an environment with correlated sources. To reduce the impact of the low spatial resolution of MEG and EEG, the authors of [24] developed a unifying framework for quantifying the spatial fidelity of MEG/EEG source estimates. This method quantifies the spatial fidelity of MEG/EEG estimates from simulated patch activations over the entire neocortex superposed on measured resting-state data. This approach grants more generalizability in the evaluation process that allows for, e.g., comparing linear and non-linear estimates in the whole brain for different Signal-to-Noise Ratios (SNR), number of active sources and activation waveforms. The authors of [25] discuss a solution to the source reconstruction problem and developed a novel hierarchical multiscale Bayesian algorithm for electromagnetic brain imaging using MEG and EEG within the context of sources that vary in spatial extent. In this Bayesian algorithm, the sensor data measurements are defined using a generative probabilistic graphical model that is hierarchical across spatial scales of brain regions and voxels. This algorithm enables robust reconstruction of sources that have different spatial extent, from spatially contiguous clusters of dipoles to isolated dipolar sources.

In [26], the authors propose a methodological framework for inverse-modeling of propagating cortical activity. Within this framework, cortical activity is represented in the spatial frequency domain, which is more natural than the dipole domain when dealing with spatially continuous activity. In dealing with multi-subject MEG/EEG source imaging, he authors of [27] propose a sparse multi-task regression that takes into account inter-subject variabilities known as the Minimum Wasserstein Estimates (MWE). This work jointly localizes sources for a population of subjects by casting the estimation as a multi-task regression problem in three key ideas. First, it proposes to use non-linear registration to obtain subject-specific lead field matrices that are spatially aligned. Second, it copes with the issue of inter-subject spatial variability of functional activations using optimal transport. Finally, it makes use of non-convex sparsity priors and joint inference of source estimates to obtain accurate source amplitudes. Various applications for MEG/EEG source reconstruction have been applied in the clinical setting for detection of epileptic spikes [28], identification of seizure onset zone [29] and presurgical workup of epilepsy patients [30–32].

2. Methods

This section describes the Bayesian spatial mixture model developed in [13] and the ACS-ICM algorithm.

2.1. Model

We provide details and mathematical description of the joint model below. Let $M(t) = (M_1(t), M_2(t), ..., M_{n_M}(t))'$ and $E(t) = (E_1(t), E_2(t), ..., E_{n_E}(t))'$ denote the MEG and EEG, respectively, at time t, $t = 1, \ldots, T$; where n_M and n_E denote the number of MEG and EEG sensors, the model assumes:

$$M(t) = X_M S(t) + \epsilon_M(t), \quad \epsilon_M(t)|\sigma_M^2 \stackrel{iid}{\sim} MVN(0, \sigma_M^2 H_M), \quad t = 1, \ldots, T,$$

$$E(t) = X_E S(t) + \epsilon_E(t), \quad \epsilon_E(t)|\sigma_E^2 \stackrel{iid}{\sim} MVN(0, \sigma_E^2 H_E), \quad t = 1, \ldots, T,$$

where X_M and X_E denote $n_M \times P$ and $n_E \times P$ forward operators, respectively computed based on Maxwell's equations under the quasi-static assumption [33] for EEG and MEG; H_M and H_E are known $n_M \times n_M$ and $n_E \times n_E$ matrices, respectively, which can be obtained from baseline data providing information on the covariance structure of EEG and MEG sensor noise; and $S(t) = (S_1(t), \ldots S_P(t))'$ represents the magnitude and polarity of neural currents sources over a fine grid covering the cortical surface. In this case, P represents a large number of point sources of potential neural activity within the brain covering the cortical surface. It is assumed that the P cortical locations are embedded in a 3D regular grid composed of N_v voxels to allow efficient computational implementation. Given this grid of voxels, a mapping $v : \{1, ..., P\} \to \{1, ..., N_v\}$ is defined such that $v(j)$ is the index of the voxel containing the jth cortical location. We assume a latent Gaussian mixture with allocations at the level of voxels:

$$S_j(t)|\boldsymbol{\mu}(t), \boldsymbol{\alpha}, \mathbf{Z} \stackrel{ind}{\sim} \prod_{l=1}^{K} N(\mu_l(t), \alpha_l)^{Z_{v(j)l}}, \qquad (1)$$

$j = 1, \ldots, P$, $t = 1, \ldots, T$; where $\mathbf{Z} = (\mathbf{Z}_1', \mathbf{Z}_2', \ldots, \mathbf{Z}_{N_v}')'$ is a labeling process defined over the grid of voxels such that for each $v \in \{1, \ldots, N_v\}$, $\mathbf{Z}_v' = (Z_{v1}, Z_{v2}, \ldots, Z_{vK})$ with $Z_{vl} \in \{0,1\}$ and $\sum_{l=1}^{K} Z_{vl} = 1$; $\boldsymbol{\mu}(t) = (\mu_1(t), \mu_2(t), \ldots, \mu_K(t))' = (\mu_1(t), \boldsymbol{\mu}^A(t)')'$, where $\boldsymbol{\mu}^A(t) = (\mu_2(t), \ldots, \mu_K(t))'$ denotes the mean of the "active" states over different components of activity and $\mu_1(t) = 0$ for all t, so that the first component corresponds to an "inactive" state. The variability of the lth mixture component about its mean $\mu_l(t)$ is represented by $\alpha_l, l = 1, \ldots, K$.

The labeling process assigns each voxel to a latent state and is assumed to follow a Potts model:

$$P(\mathbf{Z}|\beta) = \frac{\exp\{\beta \sum_{h \sim j} \delta(\mathbf{Z}_j, \mathbf{Z}_h)\}}{G(\beta)}, \quad \delta(\mathbf{Z}_j, \mathbf{Z}_h) = 2\mathbf{Z}_j'\mathbf{Z}_h - 1,$$

where $G(\beta)$ is the normalizing constant for this probability mass function, $\beta \geq 0$ is a hyper-parameter that governs the strength of spatial cohesion, and $i \sim j$ indicates that voxels i and j are neighbors, with a first-order neighborhood structure over the 3D regular grid. The mean temporal dynamics for active components is assumed to follow a first-order vector autoregressive process:

$$\boldsymbol{\mu}^A(t) = \mathbf{A}\boldsymbol{\mu}^A(t-1) + \mathbf{a}(t), \quad \mathbf{a}(t)|\sigma_a^2 \stackrel{i.i.d}{\sim} MVN(\mathbf{0}, \sigma_a^2 \mathbf{I})$$

$t = 2, \ldots, T$, $\boldsymbol{\mu}^A(1) \sim MVN(\mathbf{0}, \sigma_{\mu_1}^2 \mathbf{I})$, with $\sigma_{\mu_1}^2$ fixed and known, but σ_a^2 unknown and assigned an inverse-Gamma (a_a, b_a) hyper-prior. Although in [13] a pseudo-likelihood approximation is adopted to the normalizing constant of the Potts model and then assigned a uniform prior to the spatial parameter to control the degree of spatial correlation, we fixed the inverse temperature parameter and vary it as part of a sensitivity analysis.

For model selection, the number of mixture components, the value of K, in Equation (1) will not be known prior and so it is estimated simultaneously with model parameters. Thus this approach achieves simultaneous point estimation and model selection. We can obtain a simple estimate for the number of mixture components based on the estimated allocation variables $\hat{\mathbf{Z}}$ when the algorithm is run with a sufficiently large value of K. This is achieved by running the algorithm with a value of K that is larger than the expected number of mixture components. For example, the value of K can be set as $K = 15$ when running the algorithm. The jth location on the cortex is allocated to one of the mixture components based on the estimated value of $\hat{\mathbf{Z}}_{v(j)}$, where $\hat{\mathbf{Z}}_{v(j)} = (\hat{Z}_{v(j)_1}, \hat{Z}_{v(j)_2}, \ldots, \hat{Z}_{v(j)_K})'$ and $\hat{Z}_{v(j)_l} = 1$ if location j is allocated to component $l \in \{1, \ldots, K\}$. When the algorithm is run with a value of K that is large, there will result empty mixture components that have not been assigned any voxel locations under $\hat{\mathbf{Z}}$. In a sense these empty components have been automatically pruned out as redundant. The estimated number of mixture components can be obtained by counting the number of non-empty mixture components as follows:

$$\hat{K} = \sum_{l=1}^{K} I\{\sum_{v=1}^{n_v} \hat{Z}_{v_l} \neq 0\}.$$

This estimator requires us to run our algorithm only once for a single value of K and then the resulting number of mixture components assigned a location in $\hat{\mathbf{Z}}$ is determined and $\hat{K} \leq K$.

2.2. Ant Colony System

Ant Colony System (ACS) is a population-based optimization algorithm introduced in [14]. The basic structure of this algorithm is designed to solve the traveling salesman problem in which the aim is to find the shortest path to cover a given set of cities without revisiting any one of them. The inspiring source and development of this algorithm is the observation of the foraging behavior of real ants in their colony. This behavior is exploited in artificial ant colonies for the search of approximate solutions to discrete optimization problems, for continuous optimization problems, and for important problems in telecommunications, such as routing and load balancing, telecommunication network design, or problems in bioinformatics [34,35]. At the core of this algorithm is the communication between the ants by means of chemical pheromone trails, which enables them to collectively find short paths between their nest and food source. The framework of this algorithm can be categorized into four main parts: (1) construction of an agent ant solution, (2) local pheromone update of the solution, (3) improving solution by local search, and (4) global pheromone update of the best solution.

At each step of this constructive algorithm a decision is made concerning which solution component to add to the sequence of solution components already built. These decisions are dependent on the pheromone information, which represents the learned experience of adding a particular solution component given the current state of the solution under construction. The accumulated amount of pheromone mirrors the quality of the solution constructed based on the value of the objective function. The pheromone update aims to concentrate the search in regions of the search space containing high quality solutions while there is a stochastic component facilitating random exploration of the search space. In particular, the reinforcement of solution components depending on the solution quality is an important ingredient of ACS algorithms. To learn which components contribute to good solutions can help assembling them into better solutions. In general, the ACS approach attempts to solve an optimization problem by iterating the following two steps: (1) candidate solutions are constructed using a pheromone model, that is, a parameterized probability distribution over the solution space; (2) the candidate solutions are used to modify the pheromone values in a way that is deemed to bias future sampling toward high quality solutions.

The posterior distribution of the dynamic model takes the form $P(\Theta|E, M) = P(\Theta, E, M)/P(E, M)$, where:

$$P(\Theta, E, M) = P(E, M|\Theta)P(\Theta) = P(E|\Theta)P(M|\Theta)P(\Theta)$$

$$= \prod_{t=1}^{T} MVN(E(t); X_E S(t), \sigma_E^2 H_E) \times MVN(M(t); X_M S(t), \sigma_M^2 H_M)$$

$$\times IG(\sigma_E^2; a_E, b_E) \times IG(\sigma_M^2; a_M, b_M) \times [\prod_{j=1}^{p}\prod_{t=1}^{T}\prod_{l=1}^{K} N(S_j(t); \mu_l(t), \alpha_l)^{Z_{v(j)l}}]$$

$$\times [\prod_{t=2}^{T} MVN(\mu^A(t); A\mu^A(t-1), \sigma_a^2 I)] \times MVN(\mu^A(1); 0, \sigma_{\mu_1}^2 I) \times \text{Potts}(Z; \beta) \quad (2)$$

$$\times \prod_{l=1}^{K} IG(\alpha_l; a_\alpha, b_\alpha) \times \times [\prod_{i=1}^{K-1}\prod_{j=1}^{K-1} N(A_{ij}; 0, \sigma_A^2)] \times IG(\sigma_a^2; a_a, b_a)$$

where $MVN(x; \mu, \Sigma)$ denotes the density of the $\dim(x)$-dimensional multivariate normal distribution with mean μ and covariance Σ evaluated at x; $IG(x; a, b)$ denotes the density of the inverse gamma distribution with parameters a and b evaluated at x; $N(x; \mu, \sigma^2)$ denotes the density of the normal distribution with mean μ and variance σ^2 evaluated at x; $\text{Potts}(Z; \beta)$ is the joint probability mass function of the Potts model with parameter β evaluated at Z. Equation (2) represents the objective function to be maximized over Θ. The goal is to optimize over $\Theta = \{S(t), Z, \mu(t), \alpha, \sigma_E^2, \sigma_M^2, A, \sigma_a^2\}$ maximizing the posterior (2).

ACS is based on set of agents, each representing artificial ants that construct solutions as sequences of solution components. Agent ant k builds a solution by allocating label ℓ from a set of voxel labels $\Lambda = \{1, \ldots K\}$ to the voxel $s \in \{1, \ldots, N_v\}$ based on a probabilistic transition rule $p^k(s, \ell)$. The transition rule quantifies the probability of ant k, assigning voxel s to label ℓ. This transition rule depends on the pheromone information $\tau(s, \ell)$ of the coupling (s, ℓ) representing the quality of assigning voxel s to label ℓ based on experience gathered by ants in the previous iteration. We let:

$$\ell = \begin{cases} \arg\max_u \tau(s, u) & \text{if } q \leq q_0 \\ p^k(s, \ell) & \text{if } q > q_0 \end{cases}$$

$$p^k(s, \ell) = \frac{\tau(s, \ell)}{\sum_{u \in \Lambda} \tau(s, u)} \qquad (3)$$

where ℓ is a label for voxel s selected according to the transition rule above; $q \sim Uniform(0,1)$; $q_0 \in (0,1)$ is a tuning parameter. An artificial ant chooses, with probability q_0, the solution component that maximizes the pheromone function $\tau(s, \ell)$ or it performs, with probability $1 - q_0$, a probabilistic construction step according to (3). The ACS pheromone system consists of two update rules; one rule is applied whilst constructing solutions (local pheromone update rule) and the other rule is applied after all ants have finished constructing a solution (global pheromone update rule). After assigning a label to a voxel, an ant modifies the amount of pheromone of the chosen couples (s, ℓ) by applying a local pheromone update (4):

$$\tau(s, \ell) \leftarrow (1 - \rho)\tau(s, \ell) + \rho\tau_0 \qquad (4)$$

where $\rho \in (0,1)$ is a tuning parameter that controls evaporation of the pheromone and τ_0 is the initial pheromone value. This operation simulates the natural process of pheromone evaporation preventing the algorithm from converging too quickly (all ants constructing the same solution) and getting trapped into a poor solution. In practice, the effect of this local pheromone update is to decrease the pheromone values via evaporation $(1 - \rho)\tau(s, \ell)$ on the visited solution components, making these components less desirable for the subsequents ants. The value of the evaporation rate indicates the relative importance of the pheromone values from one iteration to the following one. If ρ takes a value near 1, then the pheromone trail will not have a lasting effect, and this mechanism increases the random exploration of the search space within each iteration and helps avoid a too rapid convergence of the algorithm toward a sub-optimal region of the parameter space, whereas a small value will increase the importance of the pheromone, favoring the exploitation of the search space near the current solution.

To improve all solutions constructed and also update the other model parameters, we considered incorporating ICM as a local search method. Here, the ICM algorithm is used for both updating the model parameters and also for a local search over the mixture allocation variables. Thus, the update steps corresponding to ACS are combined with running ICM to convergence at each iteration. Finally, after all solutions have been constructed by combined ACS and ICM steps, the quality of all solutions is evaluated using the objective function where the corresponding best solution is selected. We use a global update rule, where pheromone evaporation is again applied on the best solution chosen. Assuming voxel j is assigned to label v for the best solution, the global update is given as:

$$\tau(j, v) \leftarrow \begin{cases} (1 - \rho)\tau(j, v) + \rho\tau_0, & \\ (1 - \rho)\tau(j, k), & \text{and for all } k \neq v \end{cases}$$

The steps described are performed repeatedly until a change in the objective function becomes negligible and the model parameters from the best solution are returned as the final parameter estimates. The optimal values for the tuning parameters (q_o, τ_o, ρ) used in our ACS-ICM algorithm depend on the data. The strategy we adopt for choosing the tuning parameters is by using an outer level optimization on top of the ACS-ICM algorithm to optimize over tuning parameters (q_o, τ_o, ρ) within updates at the outer level based on the Nelder–Mead algorithm [36] applied to optimize over tuning parameters.

In order to reduce the dimension of parameter space and computing time, we apply clustering to the estimated neural sources. This is achieved by implementing a K-means algorithm to cluster the P locations on the cortex into a smaller number of $J \leq P$ clusters, assuming that $S_j(t) = S_l(t)$ for cortical locations l, j belonging to the same cluster. We investigated different values of $J = 250, 500, 1000$ in our simulation studies. Within the ICM algorithm, the labeling process Z is updated using an efficient chequerboard updating scheme [13]. The update scheme starts with partitioning Z into two blocks $Z = \{Z_W, Z_B\}$ based on a three-dimensional chequerboard arrangement, where Z_W corresponds to "white" voxels and Z_B corresponds to "black" voxels. Under the Markov random field prior with a first-order neighborhood structure, the elements of Z_W are conditionally independent given Z_B, the remaining parameters, and the data E, M. This allows us to update Z_W in a single step, which involves simultaneously updating its elements from their full conditional distributions. The variables Z_B are updated in the same way.

It is well-known that the ICM algorithm is sensitive to initial values and the authors of [13] found this to be the case with the ICM algorithm developed for the spatiotemporal mixture model. The solution obtained, and even the convergence of the algorithm depend rather heavily on the starting values chosen. In the case of ACS-ICM, regardless of the initial values, the algorithm finds a better solution with the optimal tuning parameters and this solution tends to be quite stable. This is because ACS-ICM is a stochastic search procedure in which the pheromone update concentrates the search in regions of the search space containing high quality solutions to reach an optimum. When considering a stochastic optimization algorithm, there are at least two possible types of convergence that can be considered: convergence in value and convergence in solution. With convergence in value, we are interested in evaluating the probability that the algorithm will generate an optimal solution at least once. On the contrary, with convergence in solution we are interested in evaluating the probability that the algorithm reaches a state that keeps generating the same optimal solution. The convergence proofs are presented in [37,38]. The authors of [37] proved convergence with a probability of $1 - \epsilon$ for the optimal solution and more in general for any optimal solution in [38] of the ACS algorithm. This supports the argument that theoretically the application of ACS-ICM to source reconstruction should improve ICM .

The local search ICM algorithm procedure is presented in Algorithm 1 and the ACS-ICM algorithm is presented in Algorithm 2. Convergence of the ICM algorithms is monitored by examining the relative change of the Frobenius norm of the estimated neural sources on consecutive iterations.

Algorithm 1 presents a detailed description of the ICM algorithm. The ICM algorithm requires full conditional distributions of each model parameter where the mode of the distribution is taken as the update step for the parameter. The full conditional distribution are described and presented in [13]. This ICM algorithm is embedded in our ACS-ICM algorithm.

Algorithm 1 Iterated Conditional Modes (ICM) Algorithm.

1: $\Theta = \{S(t), Z, \mu(t), \alpha, \sigma_E^2, \sigma_M^2, A, \sigma_a^2\} \leftarrow$ Initial Value
2: Converged $\leftarrow 0$
3: **while** Converged $= 0$ **do**
4:      $\sigma_M^2 \leftarrow \left[\sum_{t=1}^T \frac{1}{2} (M(t) - X_M S(t))' H_M^{-1} (M(t) - X_M S(t)) + b_M \right] / \left[a_M + \frac{TN_M}{2} + 1 \right]$
5:      $\sigma_E^2 \leftarrow \left[\sum_{t=1}^T \frac{1}{2} (E(t) - X_E S(t))' H_E^{-1} (E(t) - X_E S(t)) + b_E \right] / \left[a_E + \frac{TN_E}{2} + 1 \right]$
6:      $\sigma_a^2 \leftarrow \left[\sum_{t=2}^T \frac{1}{2} (\mu^A(t) - A\mu^A(t-1))'(\mu^A(t) - A\mu^A(t-1)) + b_a \right] / \left[a_a + \frac{(T-1)(K-1)}{2} + 1 \right]$
7:      $vec(A) \leftarrow \left(\frac{1}{\sigma_a^2} \left(\sum_{t=2}^T \mu^A(t)' K r_t \right) \times C_1^{-1} \right)'$, where $C_1 = \frac{1}{\sigma_A^2} I_{(K-1)^2} + \frac{1}{\sigma_a^2} \left(\sum_{t=2}^T K r_t' K r_t \right)$,
     and $Kr_t = \left(\mu^A(t-1)' \otimes I_{K-1} \right)$
8:      **for** $l = 1, \ldots, K$ **do**
9:        $\alpha_l \leftarrow \left[\frac{\sum_{j=1}^P \sum_{t=1}^T Z_{v(j)l} (S_j(t) - \mu_l(t))^2}{2} + b_\alpha \right] / \left[\frac{T \sum_{j=1}^P Z_{v(j)l}}{2} + a_\alpha + 1 \right]$
10:      **end for**
11:      $\mu(1) \leftarrow \left(\left(\sum_{j=1}^P (S_j(1) \vec{I}_{K-1})' D_j + \frac{1}{\sigma_a^2} \mu^A(2)' A \right) \times B_1^{-1} \right)'$, where $B_1 = \sum_{j=1}^P D_j + \frac{1}{\sigma_a^2} A'A + \frac{1}{\sigma_{\mu_1}^2} I_{K-1}$, $D_j = \text{Diag}(\frac{Z_{v(j)l}}{\alpha_l}, l = 2, \ldots, K)$, $\vec{I}_{K-1} = (1, 1, \ldots, 1)'$ with $\dim(\vec{I}_{K-1}) = K - 1$
12:      **for** $t = 2, \ldots, T - 1$ **do**
13:        $\mu(t) \leftarrow \left(\left(\sum_{j=1}^P (S_j(t) \vec{I}_{K-1})' D_j + \frac{1}{\sigma_a^2} (\mu^A(t+1))' A + \frac{1}{\sigma_a^2} (\mu^A(t-1)' A') \right) \times B_2^{-1} \right)'$
     where $B_2 = \sum_{j=1}^P D_j + \frac{1}{\sigma_a^2}(A'A + I_{K-1})$
14:      **end for**
15:      $\mu(T) \leftarrow \left(\left(\sum_{j=1}^P (S_j(T) \vec{I}_{K-1})' D_j + \frac{1}{\sigma_a^2} (\mu^A(T-1)' A') \right) \times B_3^{-1} \right)'$
     where $B_3 = \sum_{j=1}^P D_j + \frac{1}{\sigma_a^2} I_{k-1}$
16:      **for** $j = 1, \ldots, P$ **do**
17:        $S_j \leftarrow -\frac{1}{2} \Sigma_{S_j} W_{2j}$ $\triangleright S_j = (S_j(1), S_j(2), \ldots, S_j(T))'$
     $\Sigma_{S_j}^{-1} = W_{1j} I_T$, $W_{2j}' = (W_{2j}(1), W_{2j}(2), \ldots, W_{2j}(T))$
     where $W_{1j} = \frac{1}{\sigma_M^2} \left(X_M[,j]' H_M^{-1} X_M[,j] \right) + \frac{1}{\sigma_E^2} \left(X_E[,j]' H_E^{-1} X_E[,j] \right) + \sum_{l=1}^K \frac{Z_{v(j)l}}{\alpha_l}$
     $W_{2j}(t) = \frac{1}{\sigma_M^2} \left(-2M(t)' H_M^{-1} X_M[,j] + 2(\sum_{v \neq j} X_M[,v] S_v(t))' H_M^{-1} X_M[,j] \right)$
     $+ \frac{1}{\sigma_E^2} \left(-2E(t)' H_E^{-1} X_E[,j] + 2(\sum_{v \neq j} X_E[,v] S_v(t))' H_E^{-1} X_E[,j] \right) - 2 \sum_{l=1}^K \frac{\mu_l(t)}{\alpha_l}$
     $X_M[,j]$, $X_E[,j]$ denote the jth column of X_E and X_M
18:      **end for**
19:      Let $\mathbb{B}$ denote the indices for "black" voxels and $\mathbb{W}$ denote the indices for "white" voxels.

Algorithm 1 *Cont.*

20: **for** $\kappa \in \mathbb{B}$ *simultaneously* **do**

21: $\quad Z_{\kappa q} \leftarrow 1$ and $Z_{\kappa l} \leftarrow 0, \forall l \neq q$
where $q = \text{argmax}_{h \in \{1,\ldots,K\}} P(h)$, and

22: $\quad P(h) = \dfrac{\alpha_h^{-TN_{j\kappa}/2} \times \exp\left(-\frac{1}{2}\sum_{j|v(j)=\kappa} \alpha_h^{-1} \sum_{t=1}^{T}(S_j(t)-\mu_h(t))^2 + 2\beta \sum_{v \in \delta_\kappa} Z_{vh}\right)}{\sum_{l=1}^{K} \alpha_l^{-TN_{j\kappa}/2} \times \exp\left(-\frac{1}{2}\sum_{j|v(j)=\kappa} \alpha_l^{-1} \sum_{t=1}^{T}(S_j(t)-\mu_l(t))^2 + 2\beta \sum_{v \in \delta_\kappa} Z_{vl}\right)}$

where $N_{j\kappa}$ is the number of cortical locations contained in voxel κ.

23: **end for**

24: **for** $\kappa \in \mathbb{W}$ *simultaneously* **do**

25: $\quad Z_{\kappa q} \leftarrow 1$ and $Z_{\kappa l} \leftarrow 0, \forall l \neq q$
where $q = \text{argmax}_{h \in \{1,\ldots,K\}} P(h)$, and

26: $\quad P(h) = \dfrac{\alpha_h^{-TN_{j\kappa}/2} \times \exp\left(-\frac{1}{2}\sum_{j|v(j)=\kappa} \alpha_h^{-1} \sum_{t=1}^{T}(S_j(t)-\mu_h(t))^2 + 2\beta \sum_{v \in \delta_\kappa} Z_{vh}\right)}{\sum_{l=1}^{K} \alpha_l^{-TN_{j\kappa}/2} \times \exp\left(-\frac{1}{2}\sum_{j|v(j)=\kappa} \alpha_l^{-1} \sum_{t=1}^{T}(S_j(t)-\mu_l(t))^2 + 2\beta \sum_{v \in \delta_\kappa} Z_{vl}\right)}$

where $N_{j\kappa}$ is the number of cortical locations contained in voxel κ.

27: **end for**

28: Check for convergence. Set Converged = 1 if so.

29: **end while**

Algorithm 2 Ant Colony System (ACS)-ICM Algorithm.

1: $\Theta \leftarrow$ Initial Value; set tuning parameters τ_o, q_o, ρ and N_{ants}.

2: Initialize pheromone information $\tau(i, \ell) = \tau_o$, for each $(i, \ell) \in \{1, \ldots N_v\} \times \{1, \ldots K\}$ representing information gathered by ants.

3: Construct candidate solutions for each of N_{ants} ants. For ant j, we find a candidate voxel labeling $\mathbf{Z}^{(j)} = (\mathbf{Z}_1'^{(j)}, \mathbf{Z}_2'^{(j)}, \ldots, \mathbf{Z}_{N_v}'^{(j)})'$. This is done sequentially for each ant j.

- Construct candidate by assigning label l to voxel s using the transition probability rule:

$$\ell = \begin{cases} \arg\max_u \tau(s,u) & \text{if } q \leq q_o \\ p(s,\ell) & \text{if } q > q_o \end{cases}$$

where if $q > q_o$ the label for voxel s is drawn randomly from $\{1, \ldots, K\}$ with probability

$$p(s, \ell) = \frac{\tau(s, \ell)}{\sum_{u \in \Lambda} \tau(s, u)},$$

and where $q \sim \text{uniform}[0, 1]$.

- Assuming voxel s is assigned label ℓ set:
$$\tau(s, \ell) \leftarrow (1-\rho)\tau(s, \ell) + \rho\tau_o$$
and for all $k \neq l$
$$\tau(s, \ell) \leftarrow (1-\rho)\tau(s, k)$$
where ρ is a tuning parameter in (0,1), which represents evaporation of the pheromone trails and $\tau_o > 0$.

Algorithm 2 *Cont.*

4: For all ants, improve candidate solutions by running ICM to convergence (this also allows an update to the other model parameters)
$\Theta = \{\{\boldsymbol{\mu}^A(1), \boldsymbol{\mu}^A(2), ..., \boldsymbol{\mu}^A(T)\}, \{\alpha_1, \alpha_2, ..., \alpha_k\}, \sigma_E^2, \sigma_M^2, \{S_j(t), t=1,2..., T, j=1,2,..., P\}, A, \sigma_a^2\}$.

5: For all N_{ants} solutions, evaluate the quality of each ant's solution using objective function: $P(\Theta, E, M)$. Keep track of the best value. The current solution for each ant serves as the starting value for the next iteration.

6: Apply global updating of the pheromone function. For the best solution, (s, ℓ) update the pheromone as follows:
Assuming voxel s is assigned label ℓ set:
$$\tau(s, \ell) \leftarrow (1-\rho)\tau(s, \ell) + \rho\tau_0$$
and for all $k \neq \ell$:
$$\tau(s, \ell) \leftarrow (1-\rho)\tau(s, k)$$
Check for convergence via increase in $\log P(\Theta, E, M)$. Go back to step 3

7: Return all voxel labeling $\mathbf{Z}$ and model parameters Θ from the best solution.

3. Simulation Studies

In this section, we use a simulation study to evaluate the performance of our algorithm. The simulation study assesses the quality of the source estimates and the optimized objective function values obtained when using our proposed algorithm in comparison to the existing ICM algorithm developed in [13]. We then make comparisons between ACS-ICM and the ICM algorithm applied to combined simulated EEG and MEG data.

3.1. Proposed Approach

The MEG and EEG data were both generated from four scenarios with two, three, four and nine latent states corresponding to regions of neural activity. In each of the four cases, one of the states is inactive, while the remaining states represent different regions of brain activity generated by Gaussian signals. The temporal profile of brain activity at each of the brain locations in the activated regions is depicted in Appendix A, Figures A1 and A2. We projected the source activity at 8196 brain locations from the cortex onto the MEG and EEG sensor arrays using the forward operators X_M and X_E. The simulated data were then obtained by adding Gaussian noise at each sensor, where the variance of the noise at each sensor was set to be 5% of the temporal variance of the signal at that sensor. The number of mixture components K was set to be the true number of latent states (either two, three, four, or nine) in the model. We simulated 500 replicate datasets and both ACS-ICM and ICM were applied to each dataset. For each simulated dataset we applied our algorithm with J = 250, 500, 1000 clusters so as to evaluate how the performance varies as this tuning parameter changes. We initialized both algorithms using the same starting values. For each replicate we computed the correlation between the estimated sources and the true sources $\text{Corr}[(S(1)', S(2)', \ldots, S(T)'), (\hat{S}(1)', \hat{S}(2)', \ldots, \hat{S}(T)')]$ as a measure of agreement. This measure was also averaged over the 500 replicate datasets to compute average correlation. In addition, we estimated the Mean-Squared Error (MSE) of the estimator $\hat{S}_j(t)$ based on the R = 500 simulation replicates for each brain location j and time point t. The Total MSE (TMSE) was computed by adding all the MSE's over brain locations and time points. This was done separately for locations in active and inactive regions.

In our simulation studies, the ACS-ICM algorithm had four tuning parameters. The first denoted as $q_0 \in (0,1)$ controlled the degree of stochasticity, with larger values corresponding to less stochasticity and thus less random exploration of the parameter space. When a solution is chosen, another tuning parameter τ_0 controlled the amount of pheromone reinforcing this solution in the information available to the other ants. A third tuning parameter ρ controlled the evaporation of pheromone, and finally a fourth tuning parameter N_{ants} controlled the number of ants. The number of ants (N_{ants}) was fixed at 10, a value for which we have seen generally good performance. This was chosen based on computing efficiency and similar results (objective function values) from using $N_{ants} \geq 10$.

The remaining optimal tuning parameters (q_o, τ_0, ρ) for all simulations cases were chosen using an outer level optimization using the Nelder–Mead algorithm.

3.2. Simulation Results

3.2.1. Evaluation of Neural Source Estimation

We present the average correlation between the estimated values and the truth for the algorithms considered in our study in Appendix A, Table A1. Inspecting Table A1, we observe that for all cases considered for the true number of latent states (either two, three, four, or nine), the estimates obtained from the ACS-ICM algorithm yielded a higher average correlation than those obtained from ICM. In addition, with respect to the number of clusters, ACS-ICM resulted in a higher average correlation than ICM uniformly for all cluster sizes (250, 500, 1000). In summary, the average correlation was significantly improved when estimates were computed using the ACS-ICM algorithm for both large and small numbers of latent states as well as cluster sizes. In addition, we present in Appendix A, Figure A4, violin plots comparing the correlation values obtained from each of the algorithms for different simulation cases across all replicates. These plots show the entire distribution and provide a better assessment of each algorithm for simulation replicates. Observing Figure A4, we can see that ACS-ICM provides the highest correlation values uniformly in all simulation scenarios.

The TMSE for all simulation scenarios is presented in Appendix A, Table A2. To improve the readability of the results from TMSE values, we computed the relative percentage improvement in TMSE of the neural source estimators from ICM to ACS-ICM. Here, using ICM as the reference algorithm, the relative percentage improvement is defined as the ratio of the difference in TMSE between ICM and ACS-ICM to its ICM TMSE value multiplied by 100. The results of this computation are presented in Table 1. In all simulation scenarios for Table 1, ACS-ICM performed better and showed a significant improvement as compared to ICM. Specifically with respect to the number of clusters, ACS-ICM was roughly 10% better than ICM with respect to TMSE when the cluster size was 250. For both small and large numbers of latent states, ACS-ICM was better than ICM in the active region with significant improvements. This shows that ACS-ICM outperforms ICM in active regions using both small and large numbers of latent states. The total MSEs were decomposed into total variance and total squared bias for the same distinct cases of the simulation depicted in Table 1. From the results, when we considered active regions with different numbers of clusters, and observed that ACS-ICM was better than ICM based on the total squared bias due to the percentage of relative change. Based on the total variance we also noticed a similar positive change from ICM to ACS-ICM uniformly for all values of K. It is also clear that for inactive regions, ACS-ICM was better than ICM for both total variance and squared bias for all simulation cases considered. Overall, these results from the TMSE demonstrate a significant improvement obtained from our algorithm when considering total squared bias and variance for our simulation studies. This improvement was observed uniformly across all conditions.

We present in Figure 1 boxplots comparing the final objective function values obtained from each of the algorithms for the different simulation scenarios across all replicates. Again, a clear pattern emerged showing that ACS-ICM yielded the highest objective function values uniformly in all cases. Overall, ACS-ICM outperformed the ICM algorithm uniformly with respect to both neural source estimates and the values of the objective function. This indicates the superiority of the ACS-ICM algorithm over ICM for computing neural source estimates for the spatiotemporal model.

3.2.2. Evaluation of Mixture Component Estimation

In addition to evaluating point estimation and objective function maximization, we also evaluated model selection, comparing $\hat{K}_{ACS}$ and $\hat{K}_{ICM}$, that is, the estimators obtained from ACS-ICM and ICM, respectively. We focused on estimating the number of mixture components and evaluating the sampling distribution of $\hat{K}_{ACS}$ and $\hat{K}_{ICM}$. The following five scenarios were considered in our experiments:

1. Two latent states with Gaussian source activity in the active regions depicted in Appendix A, Figure A1, panel (a).
2. Three latent states with Gaussian source activity in the active regions depicted in Appendix A, Figure A1, panel (c).
3. Four latent states with Gaussian source activity in the active regions depicted in Appendix A, Figure A1, panel (e).
4. Four latent states with Gaussian and sinusoidal source activity in the active regions depicted in Appendix A, Figure A1, panel (g).
5. Nine latent states with Gaussian source activity in the active regions depicted in Appendix A, Figure A2, panel (a).

Table 1. Simulation study I-Percentage of relative improvement in Total Mean-Squared Error (TMSE) of the neural source estimators decomposed into variance and squared bias from ICM to ACS-ICM. This total was obtained separately for locations in active regions and then for the inactive region.

		Active Region			Inactive Region		
Algorithm	Clusters	TMSE (%) %	(Bias)² %	Variance %	TMSE (%) %	(Bias)² %	Variance %
				$K = 2$			
ICM→ACS-ICM	250	9.78	11.11	8.93	9.93	6.15	13.16
ICM→ACS-ICM	500	6.63	4.39	8.57	9.48	9.71	9.26
ICM→ACS-ICM	1000	2.95	2.04	4.03	3.86	1.57	5.70
				$K = 3$			
ICM→ACS-ICM	250	5.10	7.80	2.76	4.97	4.31	5.60
ICM→ACS-ICM	500	24.85	25.42	24.31	15.89	18.26	13.61
ICM→ACS-ICM	1000	36.57	53.61	20	10.83	10	11.61
				$K = 4$			
ICM→ACS-ICM	250	12.24	11.11	12.88	8.90	14.19	4.94
ICM→ACS-ICM	500	17.94	14.75	20.71	2.86	3.10	2.64
ICM→ACS-ICM	1000	29.28	30.30	30.65	2.76	3.62	2.06
				$K = 9$			
ICM→ACS-ICM	250	31.14	22.77	27.65	15.44	13.07	17.58
ICM→ACS-ICM	500	14.83	11.40	18.20	17.0	20.39	13.71
ICM→ACS-ICM	1000	23.79	25.52	22.08	8.65	7.03	10.14

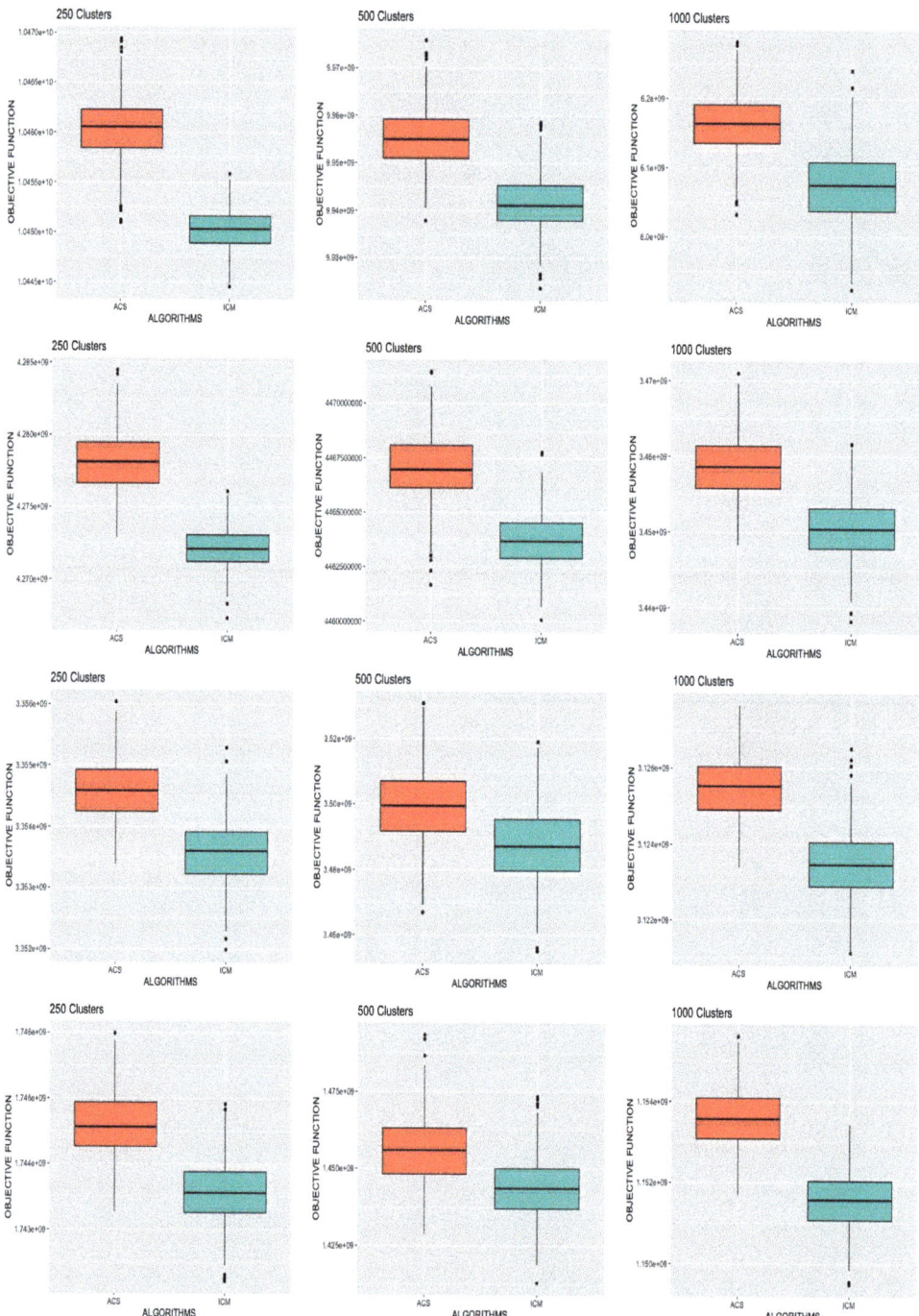

Figure 1. Box-plots comparing the objective function values obtained in the simulation studies for the ICM and ACS-ICM algorithms. The first row corresponds to the case when $K = 2$, second row corresponds to when $K = 3$, third row is when $K = 4$ and the last row is when $K = 9$.

We simulated the data for each of the five scenarios considered, and added 5% Gaussian noise at the sensors with 1000 replicate datasets used in each case. The algorithms were run with an upper bound of $K = 10$ for each of the 5000 simulated datasets. For each dataset, we computed the value of the estimator, and histograms representing the sampling distributions are presented in Figure 2, for each of the five cases above illustrating the sampling distribution of $\hat{K}_{ICM}$ (panels (a)–(e)) and $\hat{K}_{ACS}$ (panels (f)–(j)) corresponding to the first and second row, respectively. Observing Figure 2, where the true signals are well separated in the simulation experiments, in all cases except for the case with a larger number of latent states ($K = 9$), the mode of the sampling distributions corresponds to the true number of latent states for both the ACS-ICM and ICM algorithms. In the case of nine neural sources, ACS-ICM gave better and improved results than ICM. Additionally, Table 2 reports both the bias and mean-squared error of the estimators from ACS-ICM ($\hat{K}_{ACS}$) and ICM ($\hat{K}_{ICM}$). From Table 2, both ACS-ICM and ICM are biased and over-estimated for the small number of latent states but underestimated for the large number of latent states. More importantly, the estimate for the number of mixture components obtained from ACS-ICM exhibited the best performance in terms of both bias and MSE uniformly for all cases considered. This is based on $|\text{Bias}(\hat{K}_{ACS})| < |\text{Bias}(\hat{K}_{ICM})|$ and $\text{MSE}(\hat{K}_{ACS}) < \text{MSE}(\hat{K}_{ICM})$.

We repeated the simulation studies for all five cases but where true signals are less well separated by altering the true signals depicted in Appendix A, Figure A3. We present histograms depicted in Figure 3, for each of the five cases above, illustrating the sampling distribution of $\hat{K}_{ICM}$ (panels (a)–(e)) and $\hat{K}_{ACS}$ (panels (f)–(j)). In this case, the mode of the sampling distribution corresponds to the true number of latent states when $K = 2$ and $K = 3$ but not for the case with four and nine latent states with both algorithms. In Table 2 we compare the bias and mean square error of $\hat{K}_{ICM}$ and $\hat{K}_{ACS}$ under this simulation settings. Similarly, under these settings, ACS-ICM outperformed ICM in terms of the bias and mean square error; thus, $|\text{Bias}(\hat{K}_{ACS})| < |\text{Bias}(\hat{K}_{ICM})|$ and $\text{MSE}(\hat{K}_{ACS}) < \text{MSE}(\hat{K}_{ICM})$. In summary, for model selection, based on the results presented in Table 2, ACS-ICM showed an overall better performance over ICM uniformly for all eight conditions considered.

Whereas the ACS-ICM algorithm showed superiority in terms of quality of source estimates, a drawback is that it is computationally expensive relative to ICM due to its population-based and iterative procedure. Notwithstanding, this might not be a serious challenge for source localization problems, which do not require real-time solutions in most situations. With regards to computation time, on the Niagara cluster running R software on a single core (Intel Skylake 2.4 GHz, AVX512), ICM computed source estimates in approximately 2 min whereas ACS-ICM computed estimates in roughly 6 h and 30 min.

Table 2. Simulation study II—bias and Mean Square Error (MSE) of estimated number of mixture components ($\hat{K}$) from the 1000 simulation replicates when the algorithms were run with $K = 10$.

	$K = 2$		$K = 3$		$K = 4$		$K = 9$	
Algorithm	Bias($\hat{K}$)	MSE($\hat{K}$)	Bias($\hat{K}$)	MSE($\hat{K}$)	Bias($\hat{K}$)	MSE($\hat{K}$)	Bias($\hat{K}$)	MSE($\hat{K}$)
	The case where the true signals were well-separated							
ICM	0.11	0.13	0.06	0.42	0.20	0.44	−2.54	6.19
ACS-ICM	0.04	0.06	0.02	0.38	0.10	0.31	−2.01	4.46
	The case where the true signals were less well-separated							
ICM	0.11	0.13	0.525	0.58	−1.02	1.63	−4.83	16.12
ACS-ICM	0.05	0.07	0.35	0.41	−1.00	1.31	−3.68	10.47

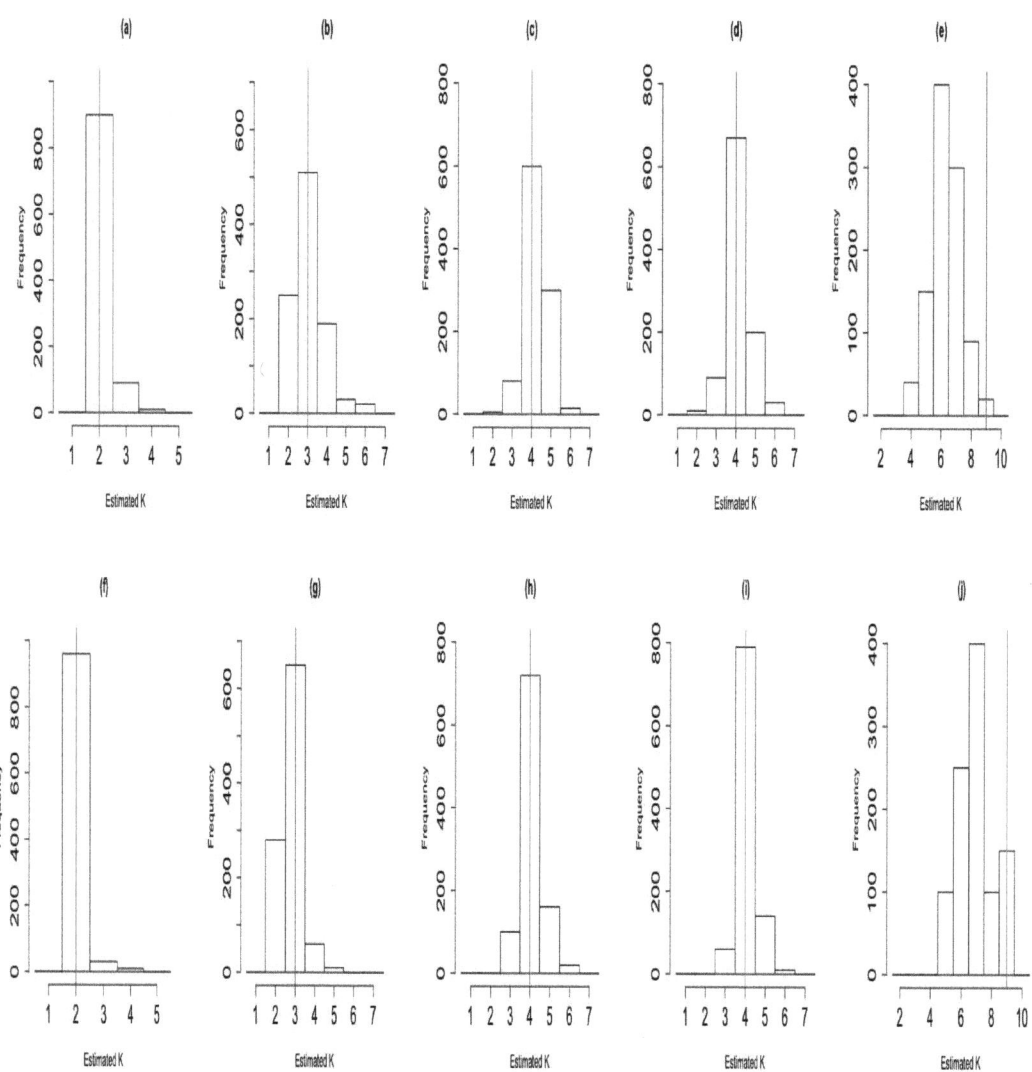

Figure 2. Histograms illustrating the sampling distribution of $\hat{K}$ in the case where the true signals were well separated in the simulation studies. The first row corresponds to the sampling distribution of $\hat{K}_{ICM}$; panel (**a**), $K = 2$; panel (**b**), $K = 3$; panel (**c**), $K = 4$ with three Gaussian signals; panel (**d**), $K = 4$ with two Gaussian signals and one sinusoid; panel (**e**), $K = 9$ with eight Gaussian signals. The second row corresponds to the sampling distribution of $\hat{K}_{ACS}$; panel (**f**), $K = 2$; panel (**g**), $K = 3$; panel (**h**), $K = 4$ with three Gaussian signals; panel (**i**), $K = 4$ with two Gaussian signals and one sinusoid; panel (**j**), $K = 9$ with eight Gaussian signals. In each case the vertical red line indicates the true number of latent states underlying the simulated data.

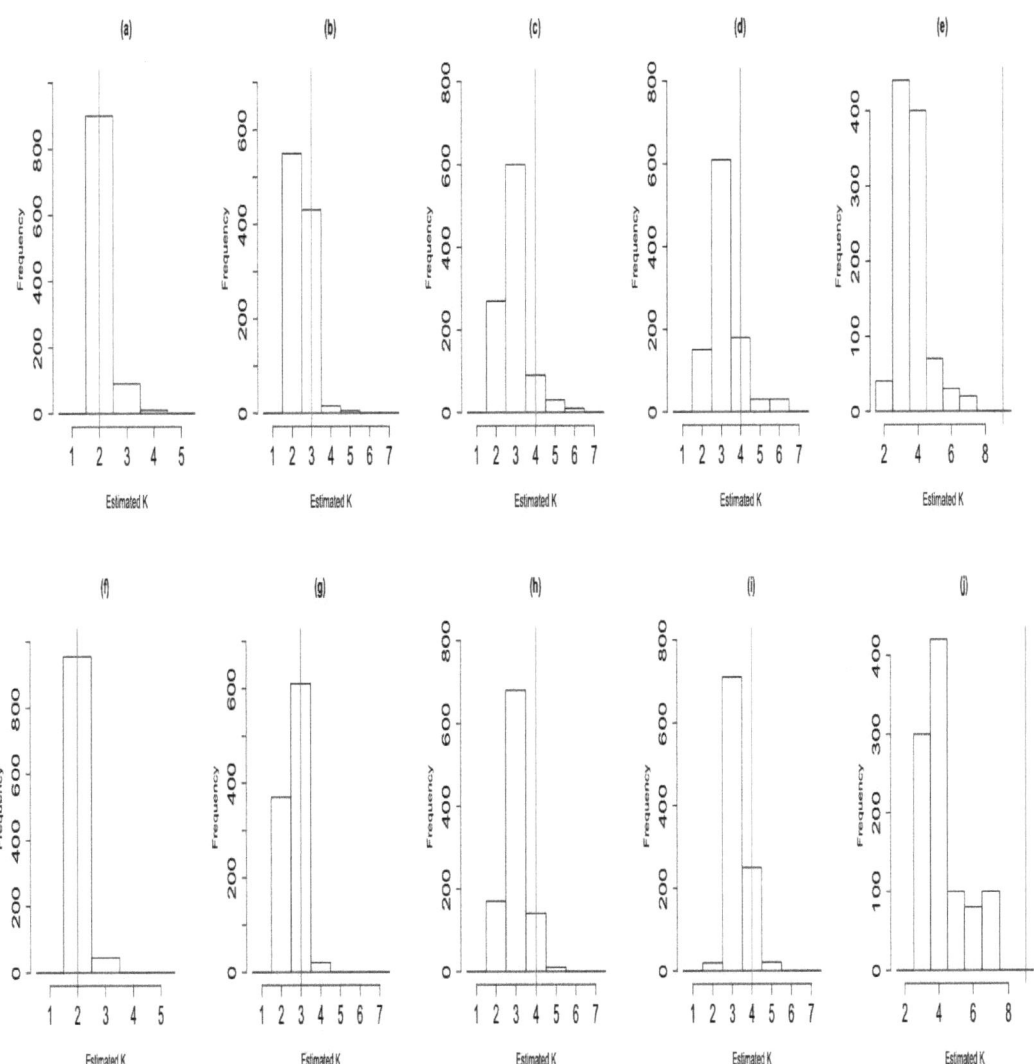

Figure 3. Histograms illustrating the sampling distribution of $\hat{K}$ in the case where the true signals were less well-separated in the simulation studies. The first row corresponds to the sampling distribution of $\hat{K}_{ICM}$; panel (**a**), $K = 2$; panel (**b**), $K = 3$; panel (**c**), $K = 4$ with three Gaussian signals; panel (**d**), $K = 4$ with two Gaussian signals and one sinusoid; panel (**e**), $K = 9$ with eight Gaussian signals. The second row corresponds to the sampling distribution of $\hat{K}_{ACS}$; panel (**f**), $K = 2$; panel (**g**), $K = 3$; panel (**h**), $K = 4$ with three Gaussian signals; panel (**i**), $K = 4$ with two Gaussian signals and one sinusoid; panel (**j**), $K = 9$ with eight Gaussian signals. In each case the vertical red line indicates the true number of latent states underlying the simulated data.

4. Application to Scrambled Face Perception MEG/EEG Data

In this section, we present the application of our methodology for comparison with EEG and MEG data measuring an event-related response to the visual presentation of scrambled faces in a face perception study. In addition, we demonstrate how a nonparametric bootstrap can be used to obtain standard errors, confidence intervals and T-maps. The data from both MEG and EEG were obtained from a single subject in an experimental paradigm that involved repeated random presentation of a picture showing either a face or a scrambled face while the subject was required to make a symmetry judgement. The scrambled faces were created through 2D Fourier transformation, random phase permutation, inverse transformation and outline-masking of each face. The experiment involved a sequence of trials, each lasting 1800 ms, where in each trial the subject was presented with one of the pictures for a period of 600 ms while being required to make a four-way, left–right symmetry judgment while brain activity was recorded over the array. Both scrambled faces and unscrambled faces were presented to the subject; however, our analysis will focus only on trials involving scrambled faces. This produced a multivariate time series for each trial, and the trial-specific time series were then averaged across trials to create a single multivariate time series; the average evoked response is depicted in Figure 4, panel (a), for MEG data, and panel (c), for EEG data. Looking from a spatial perspective, at a given time point, each array recorded a spatial field such as that depicted in Figure 4, panel (b), which shows the MEG spatial field at a particular time point, and Figure 4, panel (d), which shows the EEG spatial field at the same time point. The degree of inter-trial variability was quite low. This experiment was conducted while EEG data were recorded, and then again on the same subject while MEG data were recorded.

The EEG data were acquired on a 128-sensors ActiveTwo system with a high sampling rate of 2048 Hz and down-sampled to 200 Hz. The EEG data were re-referenced to the average over good channels. The resulting EEG data were a trial-specific multivariate time series and contained 128 sensors, 161 time points and 344 trials. For real data analysis, the trial-specific time series were averaged across trials to produce a single average evoked response. The MEG data were acquired on 274 sensors with a CTF/VSM system, with a high sampling rate of 480 Hz and down-sampled to 200 Hz. The MEG data obtained were a trial-specific multivariate time series and contained 274 sensors, 161 time points and 336 trials. We obtained a temporal segment of the data from time point t = 50 to t = 100, resulting in 51 time points for both the EEG and MEG data. The trial-specific time series were averaged across trials to produce a single average evoked response. Detailed description of the data and related analysis can be found in [9,39,40]. In addition, a link to the open access data repository used for analysis can be found here: https://www.fil.ion.ucl.ac.uk/spm/data/mmfaces (accessed on 14 November 2020).

We set the upper bound at $K = 10$ mixture components, voxels as $n_v = 560$, $\beta = 0.3$ (hyperparameter of Potts model) and a cluster size of $J = 250$. For our real data application, the optimal tuning parameters $(q_o, \tau_0, \rho, N_{ants}) = (0.43, 0.05, 0.64, 10)$ were selected similarly using the Nelder–Mead algorithm. First, the ICM algorithm was run to convergence and the estimates obtained were used as the initial values for the ACS-ICM algorithm. Our primary interest lies in the estimated neural sources $\hat{S}(t)$ and we computed the total power of these estimated sources obtained from both algorithms at each brain location, which was then mapped onto the cortex. The cortical maps showing the spatial patterns from the estimated power of the reconstructed sources are displayed in Figure 5. The first and second row depict the corresponding results obtained from the ICM and ACS-ICM algorithms, respectively. As shown in Figure 5, the greatest power occurred on the bilateral ventral occipital cortex for both estimated sources from the ACS-ICM and ICM algorithms. Interestingly, the results from ACS-ICM estimates also differed when compared with the results from ICM in the left ventral frontal and right ventral temporal regions. In particular, the ACS-ICM estimate detected higher power, whereas ICM showed low activation in these regions. The estimated source locations of these region is responsible for high-level visual processing. Therefore, the cortical power map seems to represent regions that would be expected to show scrambled face-related activity. To compare the general quality of the estimates from ACS-ICM versus ICM, we show the plot of the final objective function values obtained from the algorithms in Figure 6. We see clearly that the application of ACS-ICM has led to higher quality estimates with much larger posterior density values.

The ACS-ICM algorithm used to maximize the posterior distribution produces only point estimates of the neural source activity. In addition to the point estimates, we applied a nonparametric bootstrap on the trial-specific multivariate time series to obtain confidence interval estimates and characterize the variability in our source estimates, which is another extension to [13]. The interval estimates were constructed by resampling the trial-specific MEG/EEG time series data with replacement. From each resampled dataset, we obtained the average evoked response and then run the ACS-ICM algorithm for a total of 400 nonparametric bootstrap replicates. This procedure was made feasible using parallel computation on a large number of computing cores. We constructed a cortical map of the bootstrap standard deviations of the total power of the estimated source. To account for uncertainty in our point estimates, we constructed a T-map and this is depicted in Figure 7. A T-map is the ratio of the ACS-ICM point estimate of the source activity to its bootstrap standard deviations. The T-map represents the best depiction of reconstructed power since it accounts for standard errors that a simple map of the point estimates does not. Broadly, the T-map seems to indicate similar results to those obtained from point estimates, in particular with respect to high power activation on the bilateral ventral occipital cortex and right ventral temporal region. An interesting observation from the T-map is the detection of a high signal in the left ventral temporal region but a low activation from the point estimate.

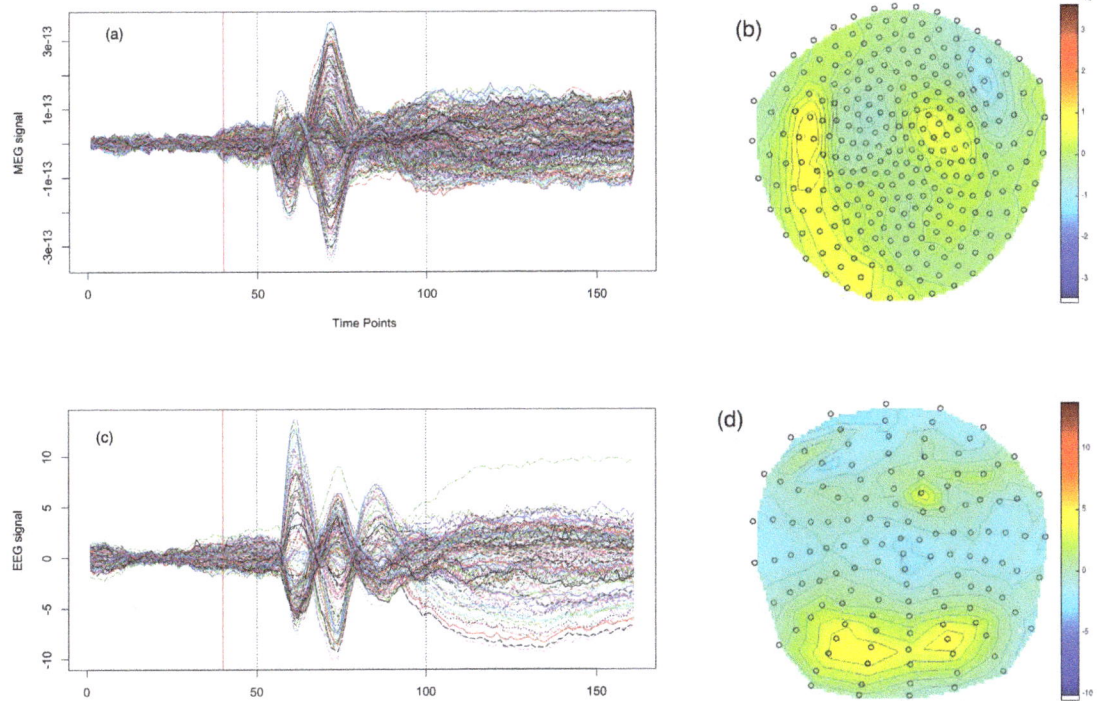

Figure 4. The Magnetoencephalography (MEG) and Electroencephalography (EEG) data considered in the face perception study: panels (**a**,**c**) show the time series observed at each MEG sensor and EEG sensor, respectively; panels (**b**,**d**) depict the spatially interpolated values of the MEG data and the EEG data, respectively, each observed at $t = 80$, roughly 200 ms after presentation of the stimulus. In panels (**b**,**d**) the black circles correspond to the sensor locations after projecting these locations onto a 2-dimensional grid (for presentation).

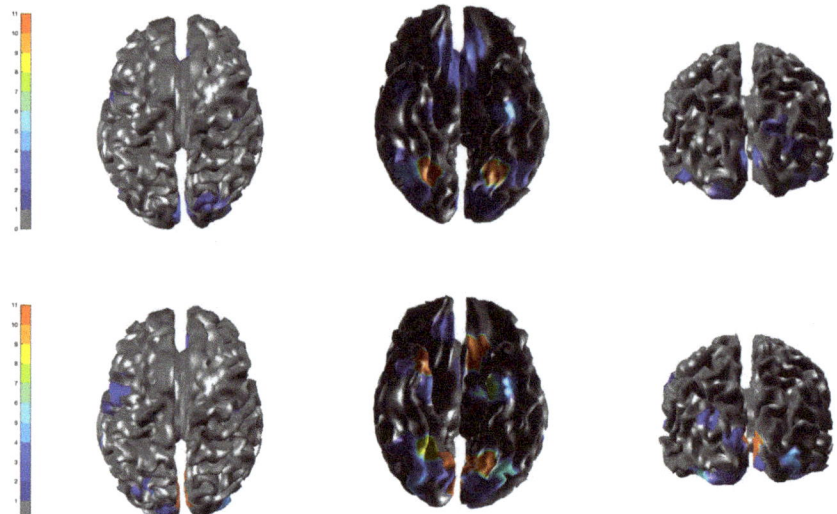

Figure 5. Brain activation for scrambled faces—the power of the estimated source activity $\sum_{t=1}^{T} \hat{S}_j(t)^2$ at each location j of the cortical surface. **Row 1** displays results from our ICM algorithm applied to the combined MEG and EEG data; **Row 2** displays results from ACS-ICM applied to the combined MEG and EEG data.

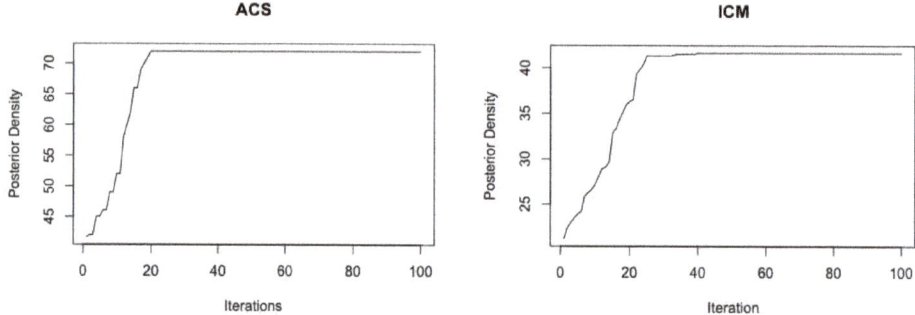

Figure 6. Objective function values obtained from the data with the ACS-ICM (**left**) and ICM (**right**) algorithms.

In addition, we present the temporal summary from our bootstrap replicates representing the interval estimation for the estimated temporal profile of brain activity at the peak location of the T-map. The interval estimate represents a 95% confidence interval depicted in Figure 8. One of the key components of our work is varying the inverse temperature parameter for sensitivity analysis. We fixed the inverse temperature at $\beta = (0.1, 0.44)$ and run the ACS-ICM algorithm to convergence. We run our algorithm together with $K = 10$, $n_v = 560$ and a cluster size of J = 250. For $\beta = 0.1$, the corresponding results obtained are depicted in the first row of Figure 9. The results indicate activation on the bilateral ventral occipital cortex. Additionally, at $\beta = 0.44$, the power map results from ACS-ICM, depicted in the second row of Figure 9, differ when compared with results from ACS-ICM at $\beta = 0.1$. In particular, the highest power signals occured in the right ventral temporal region where there was low activation for using $\beta = 0.1$.

For our real data application we applied both algorithms with J = 500 clusters so as to evaluate how the performance varies as this tuning parameter changes. The results are displayed in Appendix B. The corresponding results obtained from ACS-ICM are displayed in the second row of Figure A5. Examining Figure A5, ACS-ICM seems to indicate similar results to those obtained from using a tuning parameter of J = 250, in particular with respect to activation on the bilateral ventral occipital cortex. For our sensitivity analysis, we present results obtained from using inverse temperature ($\beta = 0.1$ and $\beta = 0.44$) displayed in Figure A6. We observe that from ACS-ICM, the spatial spread of the high power occurs on the bilateral ventral occipital cortex. In addition, source estimates obtained from ACS-ICM indicate bilateral activation in the occipital cortex, and activation in the right temporal and right frontal regions of the brain. These estimated source locations reveal activation in areas known to be involved in the processing of visual stimuli. More interestingly, ACS-ICM also detected high power in a region on the corpus callosum; given that the inverse problem is ill-posed with an infinite number of possible configurations this may be the reason.

In our real data analysis, the required computation time for ICM was 3 min on a single core (Intel Skylake 2.4 GHz, AVX512) with R software, whereas the computation time for the ACS-ICM was roughly 7 h. The choice of cluster size will have an impact on the computational time required by the algorithm. With regards to ACS-ICM, the required computing time for a cluster size of 250 was approximately 7 h, whereas for a cluster size of 500, ACS-ICM required 12 h of computing time. While there is a substantially increase in computation, the paper has demonstrated uniform improvements in the quality of the solutions, in terms of both source estimation and model selection. Furthermore, the bootstrap can be implemented in parallel on a computing cluster to obtain standard errors with no increase to the required computation time.

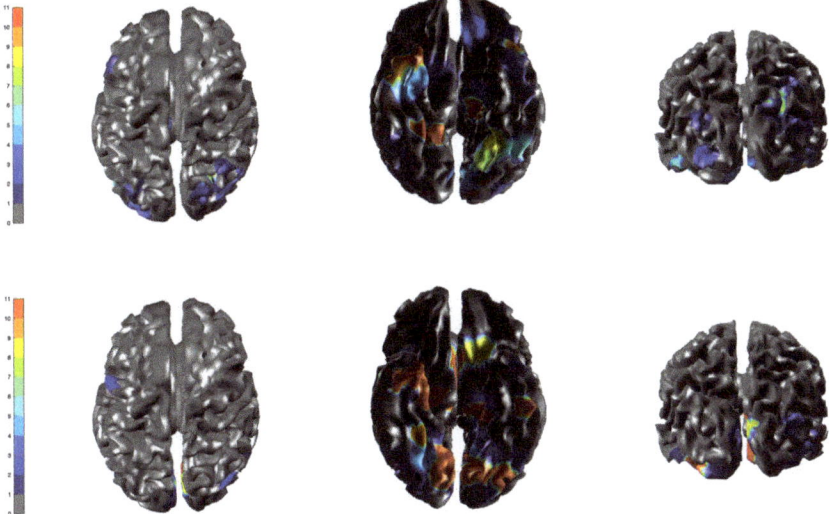

Figure 7. The spatial profile of brain activity from ACS-ICM based on our bootstrap replicates. **Row 1** displays standard deviations of the total power of the estimated source activity; **Row 2** displays the T-map.

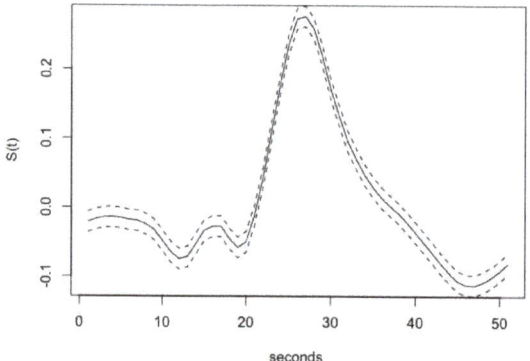

Figure 8. The 95% confidence interval for the estimated temporal profile of brain activity at the peak location of the T-map from the bootstrap replicates.

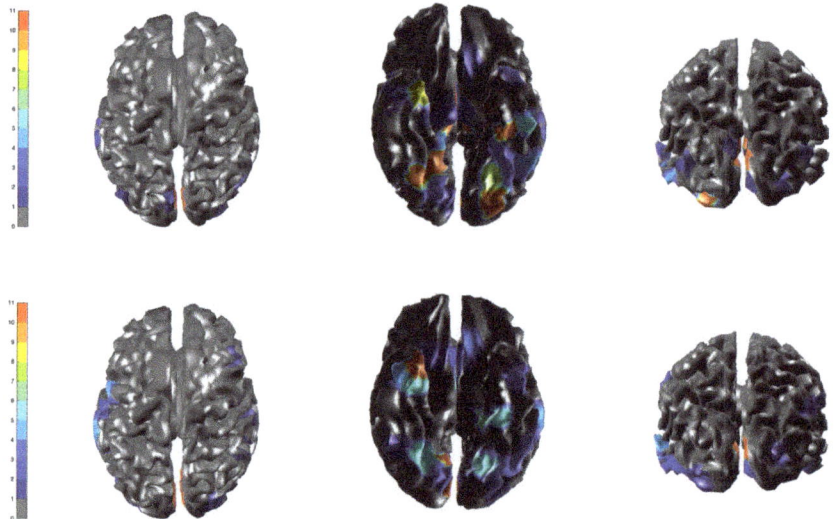

Figure 9. Brain activation for scrambled faces—the power of the estimated source activity $\sum_{t=1}^{T} \hat{S}_j(t)^2$ at each location j of the cortical surface. **Row 1** displays results from our ACS-ICM algorithm applied to the combined MEG and EEG data with $\beta = 0.1$; **Row 2** displays results from ACS-ICM applied to the combined MEG and EEG data with $\beta = 0.44$.

Residual Diagnostics for the Scrambled Faces MEG and EEG Data

We assessed the goodness of fit of the model by checking the residual time series plot, normal quantile–quantile plot and residuals versus fitted values after running the ACS-ICM and ICM algorithms. This was done by computing the residuals for both EEG and MEG after applying both algorithms. The residuals were computed as $\hat{e}_M(t) = M(t) - X_M \hat{S}(t)$ and $\hat{e}_E(t) = E(t) - X_E \hat{S}(t)$ at each time point $t = 1, \ldots, T$. The assumption made for the residuals was that they should be draws from a mean-zero Gaussian distribution if the assumed model generated the observed data. The residual time series plot for EEG and MEG from the ACS-ICM algorithm is displayed in Figure 10, panels (a) and (b). The plots from Figure 11, panels (a) and (b), also depicts residuals time series plots obtained from

ICM for EEG and MEG, respectively. Examining the plots, the residual time series plots obtained from both algorithms exhibit similar patterns for MEG and EEG. However, there are significant improvements seen in estimates from ACS-ICM. Specifically for the EEG data, there are sensors with relatively large peaks remaining from the ICM but significant improvements from ACS-ICM as we observe no fewer residuals patterns relative to ICM. In the case of MEG data, we observe that the residuals obtained from ACS-ICM reveal few sensors with peaks remaining as compared to ICM, where there are more sensors with large peaks and residuals.

In Figures 10 and 11, panels (c) and (d), we show plots of the residuals versus fitted values from ACS-ICM and ICM. For the EEG data, the ACS-ICM residuals reveal fewer extreme values with smaller residual patterns but more outliers are seen in the residuals obtained from ICM comparably. The residuals obtained from ICM are characterized by higher values to the left of zero and lower values to the right of zero. In the case of MEG data, the residuals obtained from ACS-ICM also show fewer extreme values with a smaller residual pattern but a similar resemblance for residuals obtained from the ICM algorithm with few extreme values outside the zero band. We observe more extreme values in the residual plot obtained from ICM than that obtained from ACS-ICM. This signifies improvements of the ACS-ICM algorithm over ICM. Inspecting Figure 10, panels (e) and (f), reveals normal quantile–quantile plots for the EEG and MEG residuals obtained from the ACS-ICM algorithm. There is no deviation from normality observed from the EEG and MEG data. Hence, the Gaussian assumption holds from using the ACS-ICM algorithm. In the case of the ICM, in Figure 11, panels (e) and (f) depict the normal quantile–quantile plots for the EEG and MEG data. In this case we observe a clear divergence from the normal distribution for the EEG and MEG residuals. In particular, we see a strong deviation from normality in the left and right tail of the distribution for the EEG data. There is also a deviation from normality in the right tail of the distribution for the MEG data.

In summary, the residual analysis revealed the use of the ACS-ICM algorithm resulted in estimates with a better fit of the spatial mixture model for the EEG and MEG data relative to ICM. Thus our proposed approach leads to improvements in point estimation and model selection uniformly in all settings in simulation studies and in our application with larger objective function values and improved model fit based on residual analysis.

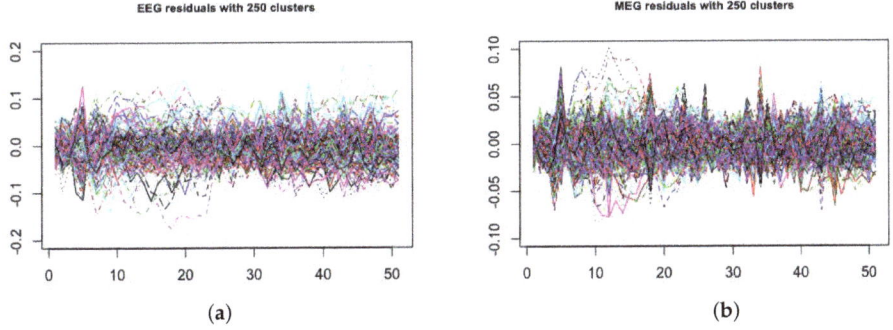

Figure 10. *Cont.*

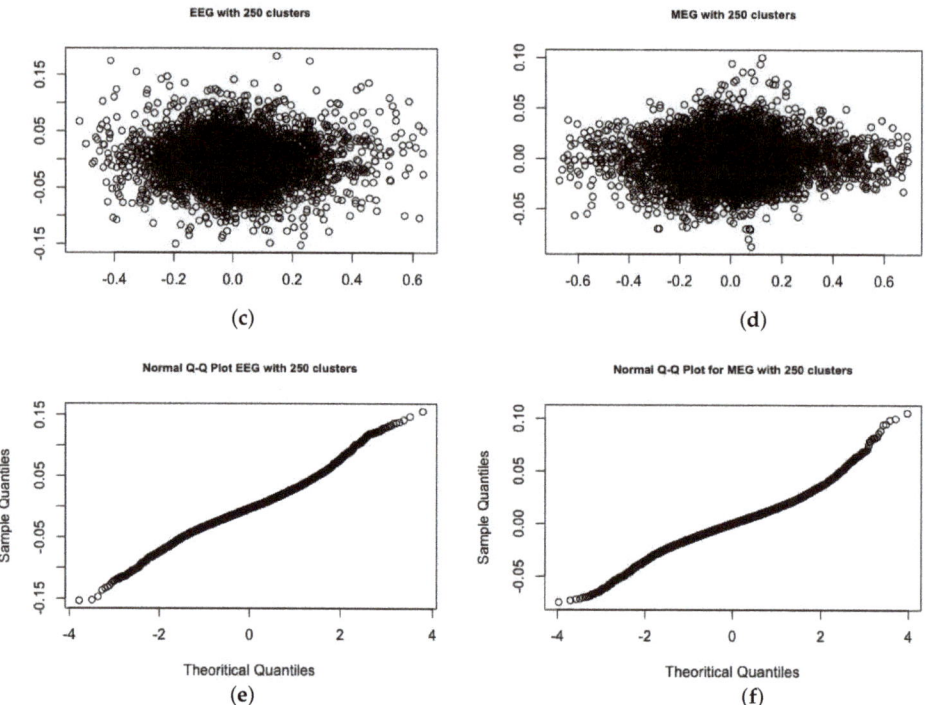

Figure 10. Brain activation for scrambled faces using the ACS-ICM algorithm—residual diagnostics: time series of residuals, (**a**) EEG, (**b**) MEG; residuals versus fitted values, (**c**) EEG, (**d**) MEG; residual normal quantile–quantile plots, (**e**) EEG, (**f**) MEG.

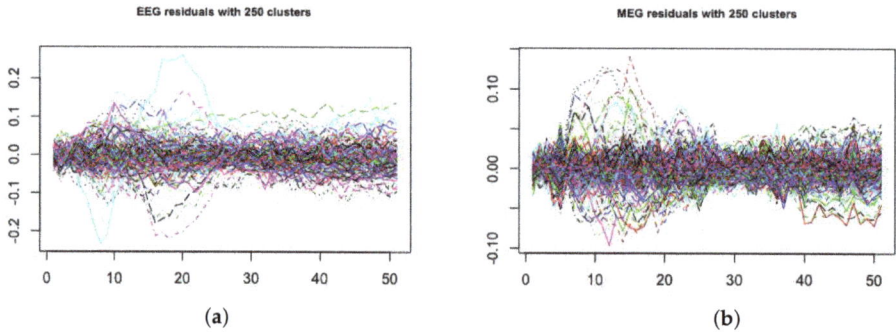

Figure 11. *Cont.*

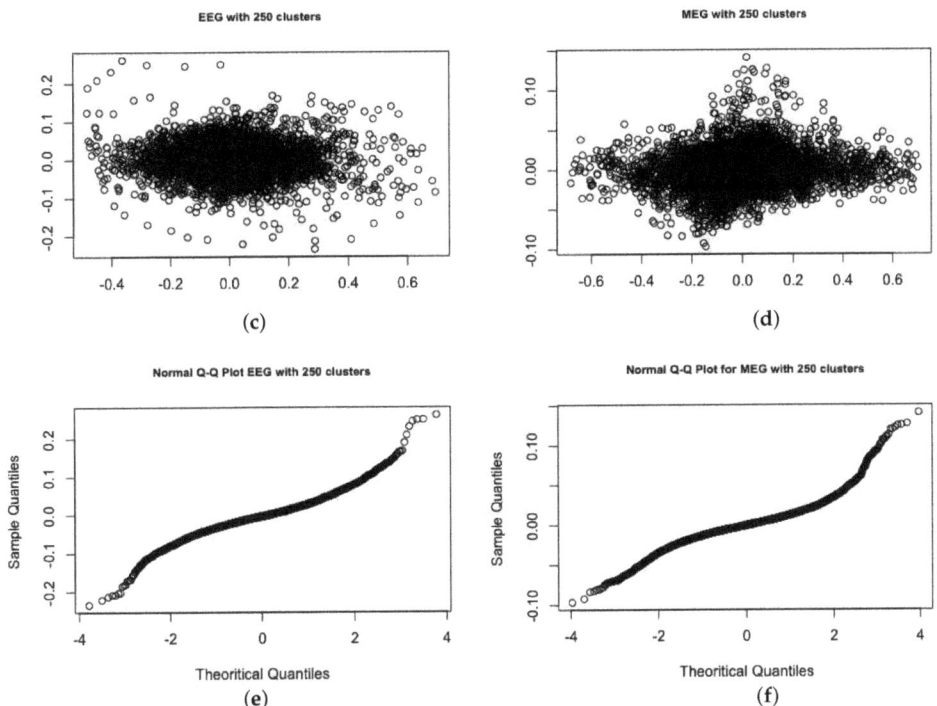

Figure 11. Brain activation for scrambled faces using icm algorithm—residual diagnostics: time series of residuals, (**a**) EEG, (**b**) MEG; residuals versus fitted values, (**c**) EEG, (**d**) MEG; residual normal quantile–quantile plots, (**e**) EEG, (**f**) MEG.

5. Discussion and Conclusions

In this section, we provide numerical results obtained in the data analysis, limitations of the proposed approach and the prospects for future research. We have developed an ACS-ICM algorithm for spatiotemporal modeling of combined MEG/EEG data for solving the neuroelectromagnetic inverse problem. Adopting a Bayesian finite mixture model with a Potts model as a spatial prior, the focus of our work has been to improve source localization estimates, model selection and model fit. The primary contribution is the design and implementation of the ACS-ICM algorithm as an approach for source localization that result in better performance over ICM, which is very positive uniformly in every setting on simulation studies and real data application. Another key development is the technique implemented in choosing the tuning parameters for the ACS-ICM by using an outer level optimization that numerically optimizes the choice of the tuning parameters for this algorithm. This strategy ensures that the optimal tuning parameters based on the data and problem complexity are selected.

5.1. Numerical Results

In our simulation studies, we observed four significant improvements associated with ACS-ICM over ICM: (1) ACS-ICM neural source estimates provided improved correlation between estimated and truth sources uniformly across all settings considered; (2) the objective function values obtained from the posterior density values for ACS-ICM were larger than those obtained from ICM uniformly across all settings considered; (3) ACS-ICM showed significant improvement with respect to the total mean square error for all cluster sizes considered compared to ICM; (4) ACS-ICM exhibited improved performance in terms of both bias and mean square error for the non-regular problem of estimating number of mixture components. Moreover, the application of ACS-ICM to real data led to higher quality estimates with larger maximized posterior density values. These improvements have demonstrated the advantage of the ACS-ICM algorithm when compared with ICM in both the face perception analysis as well as the simulation studies. In addition to implementing the ACS-ICM algorithm for point estimation, we demonstrated how a nonparametric bootstrap can be used to obtain standard errors, confidence intervals and T-maps for the proposed methodology. This was done to account for uncertainty in our point estimates of the neural source activity.

5.2. Limitations of the Proposed Approach

An important limitation of the simulation studies is the use of white noise added to the signals. This is because MEG/EEG data would have structured noise that arise from, e.g., motion, and such noise would be spatially correlated. The spatially correlated noise will make the simulation scenarios more challenging, which we expect to result in a decline in performance. We did not pursue this scenario in our simulation and we will consider it in our future studies.

5.3. Prospects for Future Research

In our current work, we are implementing ACS-ICM for the spatial mixture model developed in [13]. We hope in the future to extend the model by considering a robust error structure in the MEG/EEG model. The model currently assumes that the errors are independent in time. This will be extended by allowing for an autoregressive structure. A second extension would be to relax the assumption that the errors have a Gaussian distribution by incorporating a multivariate t distribution for the error terms. Integrating these extensions, we will develop a new joint model for the MEG/EEG data and implement the ACS-ICM and ICM algorithms for the neuroelectromagnetic inverse problem.

Furthermore, when we obtained the source estimates from the ACS-ICM algorithm, we mapped a function of them (the total power) on the cortex and in that map we used no thresholding. That is to say, the locations were not thresholded so we can see all the locations with estimated power. For our future studies we hope to map the total power on the cortex with a threshold so that we can see the locations with highest power. In a better way to choose the threshold, our next objective is to extend this work by implementing thresholding of cortical maps using random field theory [41]. Random field theory is mainly applied in dealing with thresholding problems encountered in functional imaging. This is used to solve the problem of finding the height threshold for a smooth statistical map, which gives the required family-wise error rate. In going forward with our current work, the idea is to take the point estimate obtained from ACS-ICM and standard errors (obtained from bootstrap) to provide estimates of p-values for t-statistics pertaining to the number of activated voxels comprising a particular region.

It should be noted that the ACS-ICM algorithm and spatial model developed can also be applied to studies involving multiple subjects. Expanding from a single subject model to a model developed for multiple subjects would be of great interest for the MEG/EEG inverse problem. This will be based on developing a fully Bayesian analysis based on a divide and conquer Markov Chain Monte Carlo (MCMC) method [42]. This approach for Bayesian computation with multiple subjects is to partition the data into partitions, perform local inference for each piece separately, and combine the results to obtain a global posterior approximation.

Author Contributions: Conceptualization, E.A.O. and F.S.N.; methodology, E.A.O., F.S.N. and Y.S.; formal analysis, E.A.O.; writing—original draft preparation, E.A.O.; writing—review and editing, E.A.O., S.E.A., Y.S. and F.S.N.; supervision, F.S.N. and S.E.A.; funding acquisition, F.S.N. and S.E.A. All authors have read and agreed to the published version of the manuscript.

Funding: This research was funded by the Natural Sciences and Engineering Research Council of Canada (NSERC) and supported by infrastructure provided by Compute Canada (www.computecanada.ca (accessed on 14 November 2020) and data sourced from [13].

Institutional Review Board Statement: Not applicable.

Informed Consent Statement: Not applicable.

Data Availability Statement: Publicly available datasets were analyzed in this study. This data can be found here https://www.fil.ion.ucl.ac.uk/spm/data/mmfaces (accessed on 26 January 2021).

Acknowledgments: This research is supported by the Visual and Automated Disease Analytics (VADA) graduate training program.

Conflicts of Interest: The authors declare no conflict of interest.

Appendix A

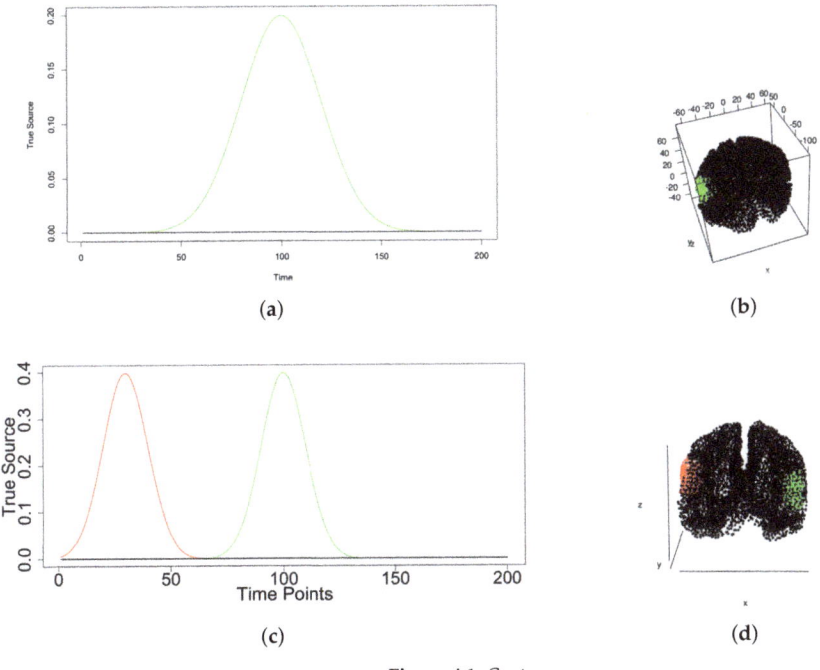

Figure A1. *Cont.*

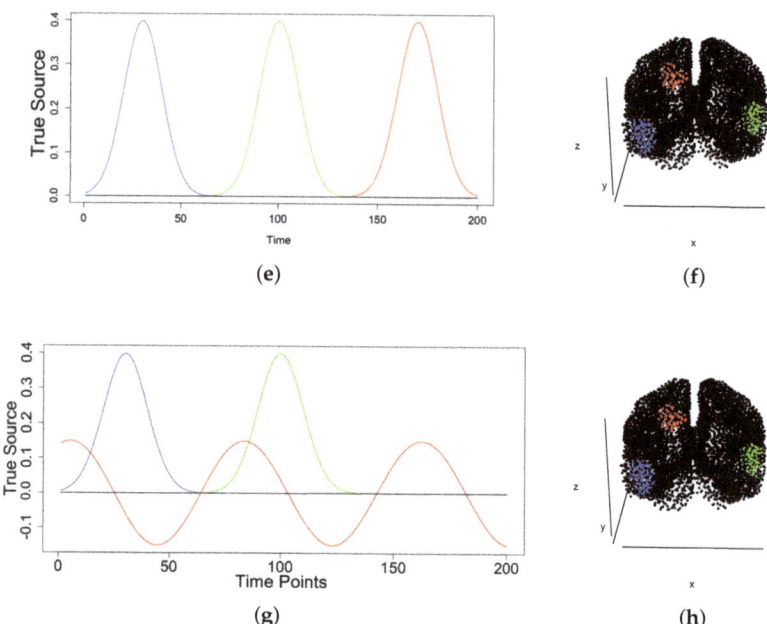

Figure A1. The true signal $S_j(t)$ used in each of the distinct active and inactive regions in the simulation studies of Section 3 for $K = 2$; panel (**a**), $K = 3$; panel (**c**) and $K = 4$; panel (**e**) & (**g**) are depicted in the left column. The right column presents the true partition of the cortex into active and inactive states for the corresponding states for $K = 2$; panel (**b**), $K = 3$; panel (**d**) and $K = 4$; panel (**f**) & (**h**).

Table A1. Simulation study I—Average (Ave.) correlation between the neural source estimates and the true values for the ICM and ACS-ICM algorithms. The simulation study is based on $R = 500$ simulation replicates where each replicate involves the simulation of MEG and EEG data based on a known configuration of the neural activity depicted in Appendix A, Figures A1 and A2.

| | | $K = 2$ | $K = 3$ | $K = 4$ | $K = 9$ |
Algorithm	Clusters	Ave. Corr.	Ave. Corr.	Ave. Corr.	Ave. Corr.
ICM	250	0.60	0.63	0.62	0.54
ACS-ICM	250	0.64	0.67	0.63	0.59
ICM	500	0.53	0.55	0.49	0.44
ACS-ICM	500	0.56	0.61	0.53	0.46
ICM	1000	0.41	0.43	0.40	0.37
ACS-ICM	1000	0.46	0.47	0.45	0.43

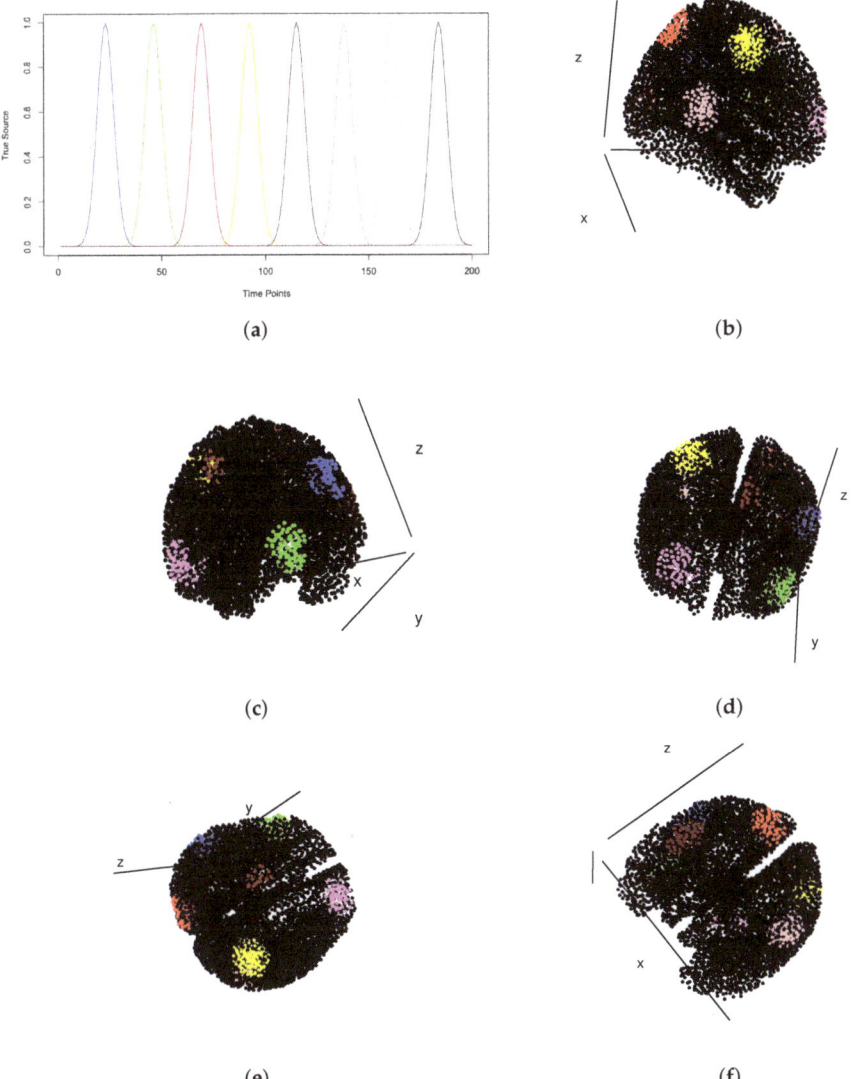

Figure A2. The true signal $S_j(t)$ in panel (**a**) and true partition of the cortex into active and inactive states for the case of $K = 9$ states (panel **b**–**f**) used in simulation studies of Section 3.

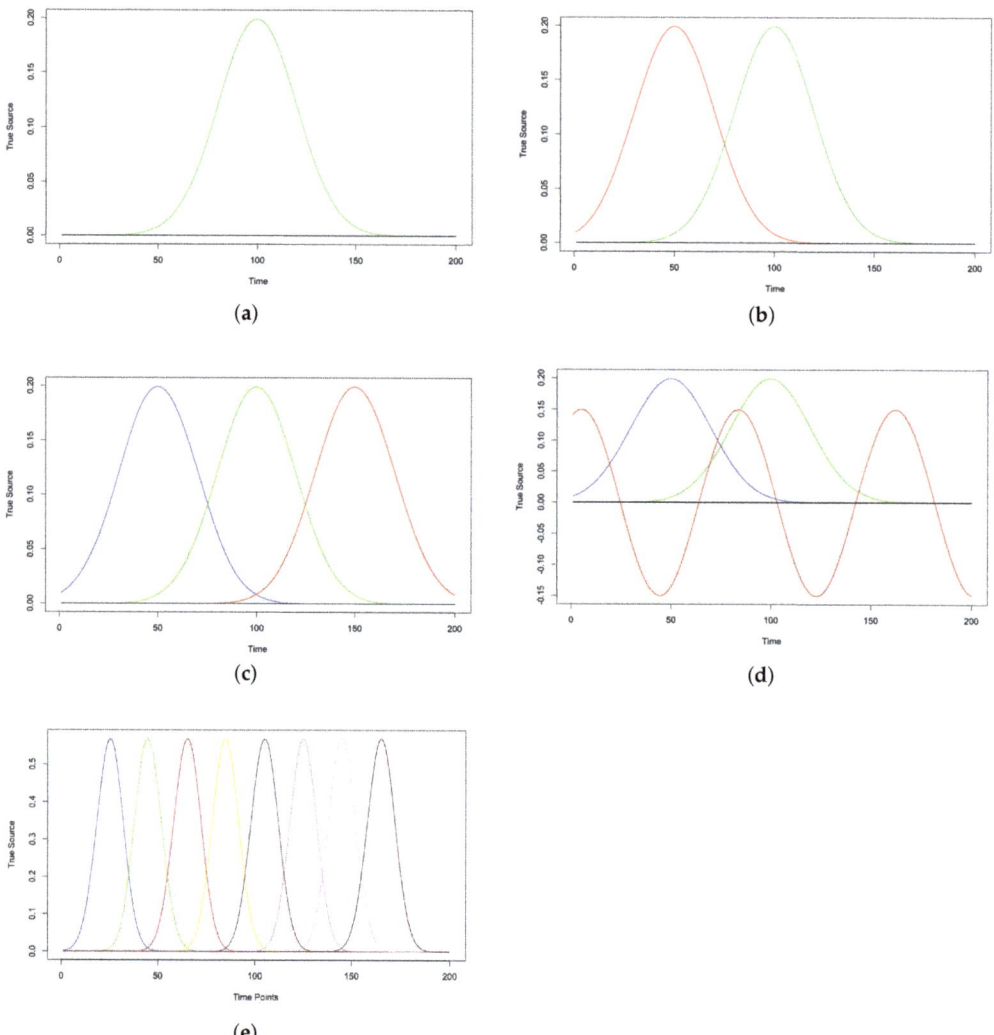

Figure A3. The true signal $S_j(t)$ used in in each of the distinct active and inactive regions for $K = 2$; panel (**a**), $K = 3$; panel (**b**), $K = 4$; panel (**c**) & (**d**), $K = 9$; panel (**e**) in the simulation study of Section 3.2, in the second part of the study where the mixture components were less well separated.

Table A2. Simulation study I—Total Mean-Squared Error (TMSE) of the neural source estimators decomposed into variance and squared bias for the ICM and ACS-ICM algorithms. This total was obtained separately for locations in active regions and then for the inactive region.

Algorithm	Clusters	Active Region			Inactive Region		
		TMSE	(Bias)^2	Variance	TMSE	(Bias)^2	Variance
				$K=2$			
ICM	250	92	36	56	141	65	76
ACS-ICM	250	83	32	51	127	61	66
ICM	500	196	91	105	211	103	108
ACS-ICM	500	183	87	96	191	93	98
ICM	1000	271	147	124	285	127	158
ACS-ICM	1000	263	144	119	274	125	149
				$K=3$			
ICM	250	490	237	253	523	255	268
ACS-ICM	250	465	219	246	497	244	253
ICM	500	1203	582	621	705	345	360
ACS-ICM	500	904	434	470	593	282	311
ICM	1000	1657	817	840	674	321	353
ACS-ICM	1000	1051	379	672	601	289	312
				$K=4$			
ICM	250	776	378	396	674	289	385
ACS-ICM	250	681	336	345	614	248	366
ICM	500	1404	651	753	804	387	417
ACS-ICM	500	1152	555	597	781	375	406
ICM	1000	2493	1100	1393	796	359	437
ACS-ICM	1000	1763	797	966	774	346	428
				$K=9$			
ICM	250	2100	918	1182	1541	727	814
ACS-ICM	250	1446	709	737	1303	632	671
ICM	500	2515	1246	1269	1260	618	642
ACS-ICM	500	2142	1104	1038	1046	492	554
ICM	1000	3561	1720	1839	1549	740	809
ACS-ICM	1000	2714	1281	1433	1415	688	727

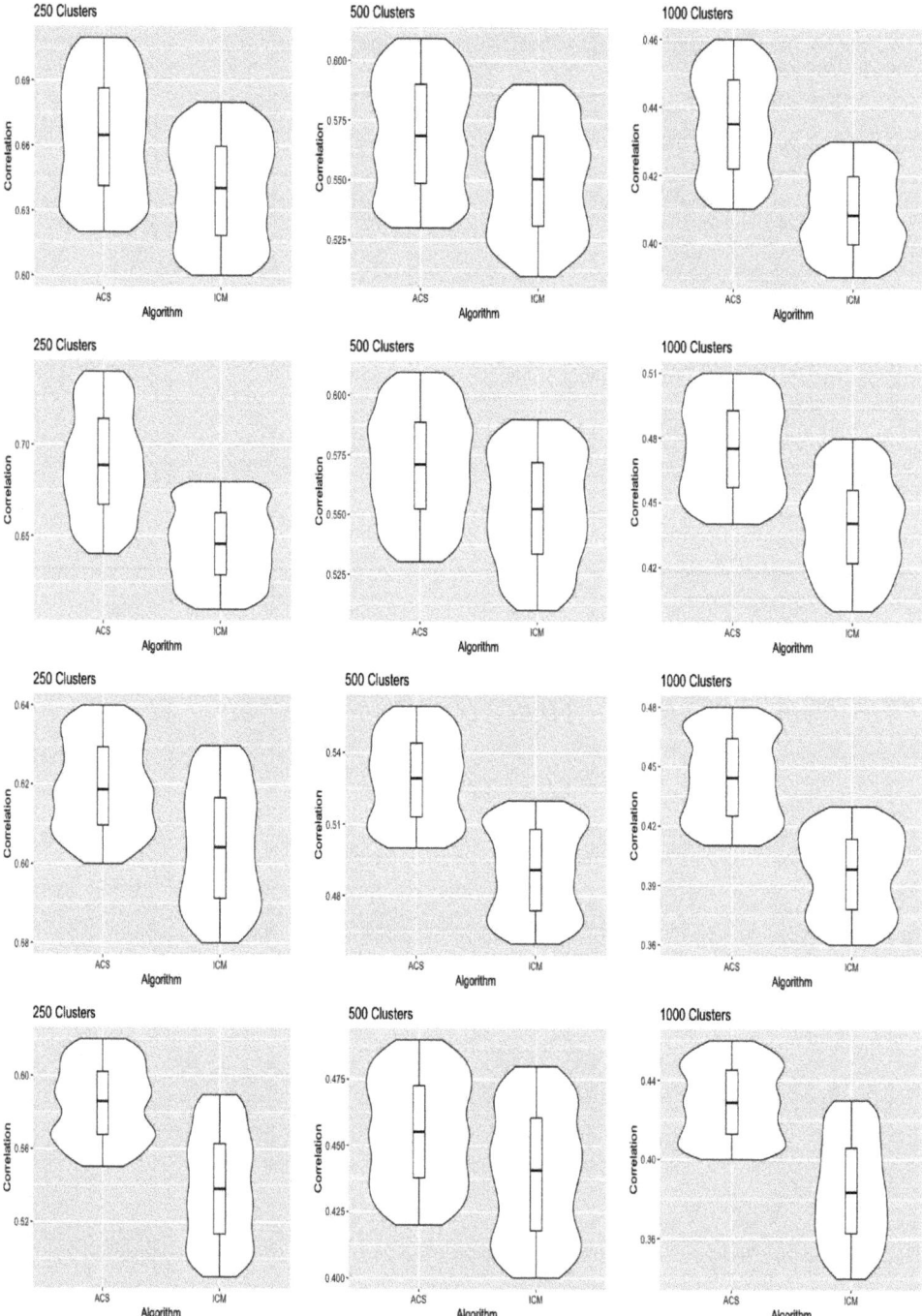

Figure A4. Violin plots comparing the correlation values obtained in the simulation studies for the ICM and ACS-ICM algorithms. The first row corresponds to the case when $K = 2$, the second row corresponds to when $K = 3$, the third row is when $K = 4$ and the last row is when $K = 9$.

Appendix B

We present results from the ACS-ICM and ICM algorithms with tuning parameters for the cluster size of 500. The cortical maps displaying the spatial patterns of the total power for estimated sources are represented below.

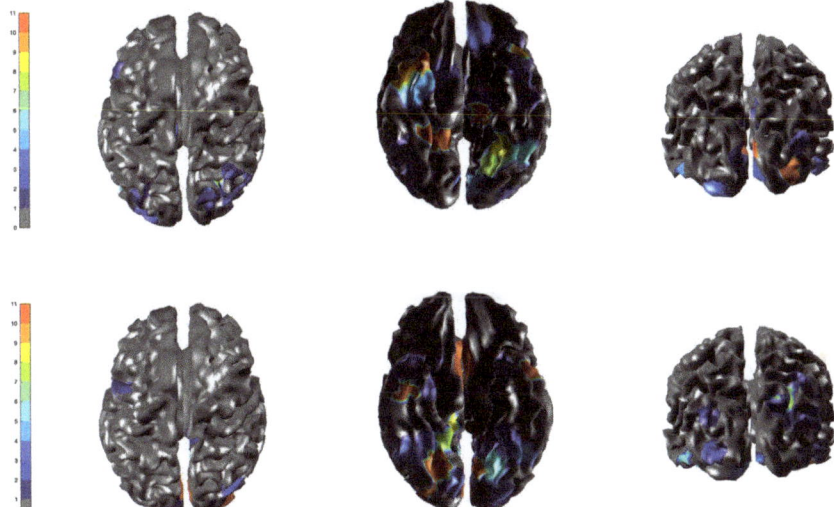

Figure A5. Brain activation for scrambled faces—the power of the estimated source activity $\sum_{t=1}^{T} \hat{S}_j(t)^2$ at each location j of the cortical surface. **Row 1** displays results from our ICM algorithm applied to the combined MEG and EEG data; **Row 2** displays results from ACS applied to the combined MEG and EEG data.

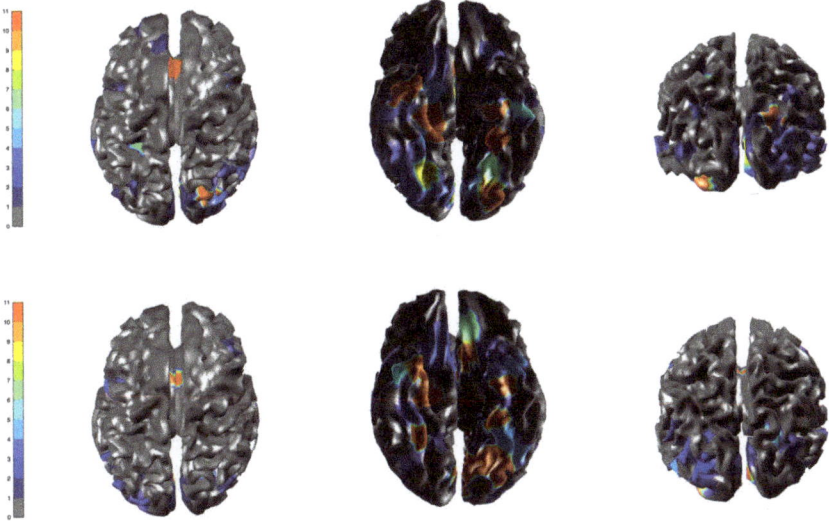

Figure A6. Brain activation for scrambled faces—the power of the estimated source activity $\sum_{t=1}^{T} \hat{S}_j(t)^2$ at each location j of the cortical surface. **Row 1** displays results from our ACS applied to the combined MEG and EEG data with $\beta = 0.1$; **Row 2** displays results from ACS applied to the combined MEG and EEG data with $\beta = 0.44$.

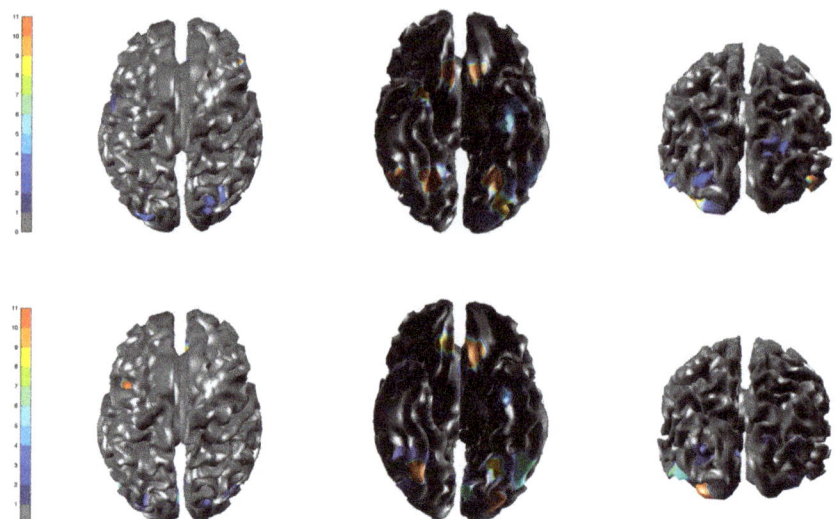

Figure A7. The spatial profile of brain activity from ACS-ICM based on our bootstrap replicates. **Row 1** displays standard deviations of the total power of the estimated source activity; **Row 2** displays the T-map.

References

1. Chowdhury, R.A.; Zerouali, Y.; Hedrich, T.; Heers, M.; Kobayashi, E.; Lina, J.M.; Grova, C. MEG–EEG information fusion and electromagnetic source imaging: From theory to clinical application in epilepsy. *Brain Topogr.* **2015**, *28*, 785–812. [CrossRef]
2. Zhang, X.; Lei, X.; Wu, T.; Jiang, T. A review of EEG and MEG for brainnetome research. *Cogn. Neurodyn.* **2014**, *8*, 87–98. [CrossRef]
3. Baillet, S.; Mosher, J.C.; Leahy, R.M. Electromagnetic brain mapping. *IEEE Signal Process. Mag.* **2001**, *18*, 14–30. [CrossRef]
4. Sorrentino, A.; Piana, M. Inverse Modeling for MEG/EEG data. In *Mathematical and Theoretical Neuroscience*; Springer: Berlin/Heidelberg, Germany, 2017; pp. 239–253.
5. Giraldo-Suarez, E.; Martínez-Vargas, J.D.; Castellanos-Dominguez, G. Reconstruction of neural activity from EEG data using dynamic spatiotemporal constraints. *Int. J. Neural Syst.* **2016**, *26*, 1650026. [CrossRef]
6. Long, C.J.; Purdon, P.L.; Temereanca, S.; Desai, N.U.; Hämäläinen, M.S.; Brown, E.N. State-space solutions to the dynamic magnetoencephalography inverse problem using high performance computing. *Ann. Appl. Stat.* **2011**, *5*, 1207. [CrossRef]
7. Rytsar, R.; Pun, T. EEG source reconstruction using global optimization approaches: Genetic algorithms versus simulated annealing. *Int. J. Tomogr. Simul.* **2010**, *14*, 83–94.
8. Wipf, D.; Nagarajan, S. A unified Bayesian framework for MEG/EEG source imaging. *NeuroImage* **2009**, *44*, 947–966. [CrossRef] [PubMed]
9. Henson, R.N.; Mouchlianitis, E.; Friston, K.J. MEG and EEG data fusion: Simultaneous localisation of face-evoked responses. *Neuroimage* **2009**, *47*, 581–589. [CrossRef] [PubMed]
10. Friston, K.; Harrison, L.; Daunizeau, J.; Kiebel, S.; Phillips, C.; Trujillo-Barreto, N.; Henson, R.; Flandin, G.; Mattout, J. Multiple sparse priors for the M/EEG inverse problem. *NeuroImage* **2008**, *39*, 1104–1120. [CrossRef] [PubMed]
11. Daunizeau, J.; Friston, K.J. A mesostate-space model for EEG and MEG. *NeuroImage* **2007**, *38*, 67–81. [CrossRef]
12. Olier, I.; Trujillo-Barreto, N.J.; El-Deredy, W. A switching multi-scale dynamical network model of EEG/MEG. *Neuroimage* **2013**, *83*, 262–287. [CrossRef] [PubMed]
13. Song, Y.; Nathoo, F.; Babul, A. A Potts-mixture spatiotemporal joint model for combined magnetoencephalography and electroencephalography data. *Can. J. Stat.* **2019**, *47*, 688–711. [CrossRef]
14. Dorigo, M.; Gambardella, L.M. Ant colony system: A cooperative learning approach to the traveling salesman problem. *IEEE Trans. Evol. Comput.* **1997**, *1*, 53–66. [CrossRef]
15. İnkaya, T.; Kayalıgil, S.; Özdemirel, N.E. Ant colony optimization based clustering methodology. *Appl. Soft Comput.* **2015**, *28*, 301–311. [CrossRef]
16. Parpinelli, R.S.; Lopes, H.S.; Freitas, A.A. Data mining with an ant colony optimization algorithm. *IEEE Trans. Evol. Comput.* **2002**, *6*, 321–332. [CrossRef]

17. Sharma, S.; Buddhiraju, K.M. Spatial–spectral ant colony optimization for hyperspectral image classification. *Int. J. Remote Sens.* **2018**, *39*, 2702–2717. [CrossRef]
18. Opoku, E.A.; Ahmed, S.E.; Nelson, T.; Nathoo, F.S. Parameter and Mixture Component Estimation in Spatial Hidden Markov Models: A Comparative Analysis of Computational Methods. In *International Conference on Management Science and Engineering Management*; Springer: Berlin/Heidelberg, Germany, 2020; pp. 340–355.
19. Ouadfel, S.; Batouche, M. Ant colony system with local search for Markov random field image segmentation. In Proceedings of the 2003 International Conference on Image Processing (Cat. No. 03CH37429), Barcelona, Spain, 14–17 September 2003; Volume 1, pp. 1–133.
20. Shweta, K.; Singh, A. An effect and analysis of parameter on ant colony optimization for solving travelling salesman problem. *Int. J. Comput. Sci. Mob. Comput* **2013**, *2*, 222–229.
21. Guan, B.; Zhao, Y.; Li, Y. An Ant Colony Optimization Based on Information Entropy for Constraint Satisfaction Problems. *Entropy* **2019**, *21*, 766. [CrossRef]
22. Lecaignard, F.; Bertrand, O.; Caclin, A.; Mattout, J. Empirical Bayes evaluation of fused EEG-MEG source reconstruction: Application to auditory mismatch evoked responses. *NeuroImage* **2021**, *226*, 117468. [CrossRef]
23. Kuznetsova, A.; Nurislamova, Y.; Ossadtchi, A. Modified covariance beamformer for solving MEG inverse problem in the environment with correlated sources. *NeuroImage* **2021**, *228*, 117677. [CrossRef]
24. Samuelsson, J.G.; Peled, N.; Mamashli, F.; Ahveninen, J.; Hämäläinen, M.S. Spatial fidelity of MEG/EEG source estimates: A general evaluation approach. *Neuroimage* **2021**, *224*, 117430. [CrossRef]
25. Cai, C.; Sekihara, K.; Nagarajan, S.S. Hierarchical multiscale Bayesian algorithm for robust MEG/EEG source reconstruction. *NeuroImage* **2018**, *183*, 698–715. [CrossRef]
26. Hindriks, R. A methodological framework for inverse-modeling of propagating cortical activity using MEG/EEG. *NeuroImage* **2020**, *223*, 117345. [CrossRef]
27. Janati, H.; Bazeille, T.; Thirion, B.; Cuturi, M.; Gramfort, A. Multi-subject MEG/EEG source imaging with sparse multi-task regression. *NeuroImage* **2020**, *220*, 116847. [CrossRef]
28. Grova, C.; Aiguabella, M.; Zelmann, R.; Lina, J.M.; Hall, J.A.; Kobayashi, E. Intracranial EEG potentials estimated from MEG sources: A new approach to correlate MEG and iEEG data in epilepsy. *Hum. Brain Mapp.* **2016**, *37*, 1661–1683. [CrossRef] [PubMed]
29. Pellegrino, G.; Hedrich, T.; Chowdhury, R.; Hall, J.A.; Lina, J.M.; Dubeau, F.; Kobayashi, E.; Grova, C. Source localization of the seizure onset zone from ictal EEG/MEG data. *Hum. Brain Mapp.* **2016**, *37*, 2528–2546. [CrossRef]
30. Pellegrino, G.; Hedrich, T.; Chowdhury, R.A.; Hall, J.A.; Dubeau, F.; Lina, J.M.; Kobayashi, E.; Grova, C. Clinical yield of magnetoencephalography distributed source imaging in epilepsy: A comparison with equivalent current dipole method. *Hum. Brain Mapp.* **2018**, *39*, 218–231. [CrossRef] [PubMed]
31. Pellegrino, G.; Hedrich, T.; Porras-Bettancourt, M.; Lina, J.M.; Aydin, Ü.; Hall, J.; Grova, C.; Kobayashi, E. Accuracy and spatial properties of distributed magnetic source imaging techniques in the investigation of focal epilepsy patients. *Hum. Brain Mapp.* **2020**, *41*, 3019–3033. [CrossRef] [PubMed]
32. Chowdhury, R.A.; Pellegrino, G.; Aydin, Ü.; Lina, J.M.; Dubeau, F.; Kobayashi, E.; Grova, C. Reproducibility of EEG-MEG fusion source analysis of interictal spikes: Relevance in presurgical evaluation of epilepsy. *Hum. Brain Mapp.* **2018**, *39*, 880–901. [CrossRef]
33. Sarvas, J. Basic mathematical and electromagnetic concepts of the biomagnetic inverse problem. *Phys. Med. Biol.* **1987**, *32*, 11. [CrossRef] [PubMed]
34. Reimann, M.; Doerner, K.; Hartl, R.F. D-ants: Savings based ants divide and conquer the vehicle routing problem. *Comput. Oper. Res.* **2004**, *31*, 563–591. [CrossRef]
35. Shmygelska, A.; Hoos, H.H. An ant colony optimisation algorithm for the 2D and 3D hydrophobic polar protein folding problem. *BMC Bioinform.* **2005**, *6*, 30. [CrossRef]
36. Singer, S.; Nelder, J. Nelder-mead algorithm. *Scholarpedia* **2009**, *4*, 2928. [CrossRef]
37. Gutjahr, W.J. A graph-based ant system and its convergence. *Future Gener. Comput. Syst.* **2000**, *16*, 873–888. [CrossRef]
38. Gutjahr, W.J. ACO algorithms with guaranteed convergence to the optimal solution. *Inf. Process. Lett.* **2002**, *82*, 145–153. [CrossRef]
39. Henson, R.; Mattout, J.; Phillips, C.; Friston, K.J. Selecting forward models for MEG source-reconstruction using model-evidence. *Neuroimage* **2009**, *46*, 168–176. [CrossRef]
40. Henson, R.N.; Mattout, J.; Singh, K.D.; Barnes, G.R.; Hillebrand, A.; Friston, K. Population-level inferences for distributed MEG source localization under multiple constraints: Application to face-evoked fields. *Neuroimage* **2007**, *38*, 422–438. [CrossRef] [PubMed]
41. Brett, M.; Penny, W.; Kiebel, S. Introduction to random field theory. *Hum. Brain Funct.* **2003**, *2*. [CrossRef]
42. Vehtari, A.; Gelman, A.; Sivula, T.; Jylänki, P.; Tran, D.; Sahai, S.; Blomstedt, P.; Cunningham, J.P.; Schiminovich, D.; Robert, C.P. Expectation Propagation as a Way of Life: A Framework for Bayesian Inference on Partitioned Data. *J. Mach. Learn. Res.* **2020**, *21*, 1–53.

Article

Evaluation of Survival Outcomes of Endovascular Versus Open Aortic Repair for Abdominal Aortic Aneurysms with a Big Data Approach

Hao Mei, Yaqing Xu, Jiping Wang and Shuangge Ma *

Department of Biostatistics, Yale University, New Haven, CT 06520, USA; hao.mei@yale.edu (H.M.); yaqing.xu@yale.edu (Y.X.); jiping.wang@yale.edu (J.W.)
* Correspondence: shuangge.ma@yale.edu

Received: 10 September 2020; Accepted: 27 November 2020; Published: 30 November 2020

Abstract: Abdominal aortic aneurysm (AAA) is a localized enlargement of the abdominal aorta. Once ruptured AAA (rAAA) happens, repairing procedures need to be applied immediately, for which there are two main options: open aortic repair (OAR) and endovascular aortic repair (EVAR). It is of great clinical significance to objectively compare the survival outcomes of OAR versus EVAR using randomized clinical trials; however, this has serious feasibility issues. In this study, with the Medicare data, we conduct an emulation analysis and explicitly "assemble" a clinical trial with rigorously defined inclusion/exclusion criteria. A total of 7826 patients are "recruited", with 3866 and 3960 in the OAR and EVAR arms, respectively. Mimicking but significantly advancing from the regression-based literature, we adopt a deep learning-based analysis strategy, which consists of a propensity score step, a weighted survival analysis step, and a bootstrap step. The key finding is that for both short- and long-term mortality, EVAR has survival advantages. This study delivers a new big data strategy for addressing critical clinical problems and provides valuable insights into treating rAAA using OAR and EVAR.

Keywords: abdominal aortic aneurysm; emulation; deep learning; Medicare data

1. Introduction

Abdominal aortic aneurysm (AAA) is a balloon-like dilatation of the aorta that supplies blood to the body and happens below the chest. Each year, it is estimated that 200,000 people in the U.S. are diagnosed with AAA, and ruptured AAA (rAAA) poses significant clinical and public health challenges [1]. rAAA is associated with an overall mortality rate of over 80%, which causes more than 5000 deaths in the country each year [2,3]. Once rAAA occurs, repairing procedures need to be conducted immediately. In the current clinical practice, there are two main approaches: emergent open aortic repair (OAR) and endovascular aortic repair (EVAR). OAR has a relatively longer history and is still considered as the standard procedure for AAA repair, during which large incisions are unavoidable [4]. EVAR was first successfully conducted and reported in 1994, and only small incisions in the groins are needed [5]. However, this circumvented procedure makes EVAR require more intense monitoring and probable reintervention [6]. Moreover, preoperative imaging and specific anatomic requirements make EVAR less well suitable for emergent rAAA. As suggested in multiple studies [7–11], the preferred minimum invasion but awaited long-term postoperative complications may account for the favorable 30-day mortality but similar or even inferior late survival of EVAR compared to OAR. With the criticalness of rAAA and prevalence of EVAR and OAR, it is of significant interest to objectively evaluate and directly compare their survival outcomes.

In general, to compare the effects of two treatments, the gold-standard approach is to conduct a randomized controlled clinical trial. However, most of the existing clinical trials have focused on patients who have elective/intact AAA (eAAA/iAAA) and excluded those who have rAAA and require emergent care (e.g., OVER [7], DREAM [8]). This is sensible as patients with rAAA cannot bear the prolonged process of eligibility examination, treatment assignment, and finally, surgical procedure, which are non-negligible steps in a clinical trial for bias control but unacceptable for saving lives in a real-world setting.

With the aforementioned concerns, researchers have focused on observational data and analysis to assess the survival outcomes of the two procedures for rAAA patients. Our literature review suggests that quite a few of them have relied on large medical claims databases in particular, including Medicare [6,10,12,13]. In these studies [6,10], regression and other association analysis techniques have been the main tools. It is well recognized that such analyses, even after accounting for confounders, can only lead to conclusions on association, as opposed to the desired cause-and-effect relationship. To overcome such limitations, causal inference techniques [14–16] can be adopted. Here we note that, with extensive examinations and comparisons, no approach has been observed to dominate others—it is expected that such an approach may not exist, and different approaches have different pros and cons. In this article, we adopt the emulation approach, which is relatively new but has already been examined in many publications [17–19]. With this approach, a clinical trial is explicitly designed and assembled using observational data, and statistical analysis approaches designed for clinical trials can be then adopted, bearing the potential of drawing causal conclusions. Comparatively, the biggest advantage of this approach may be its lucid interpretations.

Built on the emulation strategy, we take a big data analysis approach. Here "big data" is manifested in at least two perspectives. The first is that our effort is built on the Medicare data. The Medicare database is massive, covers the dominating majority of the U.S. senior population, and contains comprehensive information. Compared to for example hospital- and community-based data, Medicare data is advantageous with its unbiased sample selection and relatively uniform and detailed data collection. It has served as the basis of a large number of clinical and public health studies, including those that adopt causal inference analysis techniques [19,20]. More details on the analyzed Medicare data are provided below in Section 2.1. The second big data perspective is that in the analysis of the emulated trial, deep learning techniques are adopted. In "standard" emulation analysis (as well as most if not all analysis of real clinical trials) [18,19], regression (e.g., logistic and Cox) techniques have been adopted. For diverse fields including engineering, business, social science, and others [21,22], the superiority of deep learning techniques in prediction has been well established through a myriad of published studies. Relatively recently, deep learning techniques have been applied to biomedical studies on cancer [23], fracture [24], chronic diseases [25], and cardiovascular diseases [26]. The studied outcomes/phenotypes include continuous [25], categorical [24], and, more recently, survival [27]. It is noted that the existing deep learning analyses of biomedical data are mostly in the association analysis domain.

The overarching goal of this study is to directly compare EVAR versus OAR for rAAA patients and draw conclusions as close to causal as possible, so as to further inform clinical practice. This study may advance from the existing literature in the following aspects:

- It strives to compare the treatment effects of EVAR and OAR under the clinical trial framework, as opposed to the commonly adopted observational data analysis framework. The conclusion so generated may have important and direct clinical implications.
- It advances from the existing emulation analyses by investigating a new disease condition and treatments, which may further expand the paradigm of emulation analysis.
- Deep learning techniques, as opposed to "simple" regressions, are adopted. This study may assist in introducing deep learning to the emulation paradigm, as well as further fostering deep learning research. Specifically, this is the first application of deep learning to the emulation analysis and study of rAAA. Building on the existing deep learning components, we assemble an analysis

pipeline that mimics the "propensity score + inverse probability treatment (IPT) weighting Cox regression" approach [18,19].

Looking at a higher level, an "ordinary" clinical trial generates an information set (target), whose most notable characteristic is the balance in information between two treatment arms. In addition, it is usually assumed that such an information set can be sufficiently described using a (semi)parametric model. Information contained in observational data fundamentally differs from the target. As such, a central goal of the emulation approach is to properly carve a subset of information, as large as possible, that mimics the target. With the deep learning analysis approach, the (semi)parametric probabilistic structure can be significantly relaxed. Overall, this study falls into the intersection of information theory and machine learning.

2. Methods

This section details the procedures of conducting an emulation study to compare the treatment effects of EVAR and OAR using a big data approach. First, we introduce the Medicare data used in this study in Section 2.1. In Sections 2.2 and 2.3, we develop protocols of the target randomized clinical trial and the corresponding emulated trial, respectively. Last, we describe the analysis approaches in Section 2.4.

2.1. Data Source

As briefly mentioned above, we analyze the Medicare data in this study. Medicare is a federal health insurance program for adults aged 65 years and above, certain younger people with disabilities, and people with end-stage renal disease (permanent kidney failure requiring dialysis or a transplant). As the single largest payer of health care in the U.S., it covers 98% of adults who are over 65 years old, accounts for 99% of death in the elderly population, and generates a huge amount of medical claims data [28]. The Centers for Medicare & Medicaid Services (CMS) offers a wide range of datasets that follow Medicare beneficiaries across multiple care settings. More specifically, it collects over two billion data points per year through reimbursement to hospital care (Medicare Part A), physician and outpatient services (Medicare Part B), drug prescription (Medicare Part D), and other health care claims. It also collects billions of other data points through enrollment information, beneficiary eligibility checks, quality metrics, and calls to 1-800-MEDICARE [29].

For our study, we first retrieve all inpatient claims between 1 January 2011 to 30 September 2015 from the Medicare provider utilization and payment data: hospital care (Part A), which contains detailed information on health services provided in 54 million inpatient episodes for 23 million Medicare beneficiaries. Information contained in each claim includes beneficiary demographics (e.g., age, sex, race), Medicare enrollment status, services provided (up to 25 diagnosis codes and up to 25 procedure codes), and beneficiary death information. More details on such information and how it is utilized in our analysis are provided below.

It is noted that for research purposes, Medicare data can be viewed as publicly available. We only conduct a secondary analysis of the existing deidentified data. As such, no IRB or other approvals are needed.

2.2. The Target Randomized Clinical Trial

Under the emulation analysis paradigm [30], one of the first and most important steps is the design of a target randomized clinical trial. For treating rAAA, there is a lack of real clinical trials. As such, similar to in some literature [19], we need to design a hypothetical target trial. The following design has been motivated by relevant observational studies [6,10–13] and is clinically well grounded.

The target randomized clinical trial aims to compare the short- and long-term all-cause mortality of rAAA patients treated with EVAR and OAR. More specifically, we enroll participants who are diagnosed with rAAA within the enrollment period and exclude those who meet any of the following criteria: (1) the participant is under 65 years old at enrollment; (2) conversion between EVAR and OAR is necessary after randomization; (3) the participant has concurrent conditions of thoracic aneurysms, thoracoabdominal aneurysms, or aortic dissection; and (4) a repair of the thoracic aorta or visceral or renal bypass is considered necessary for the participant. If a participant develops multiple cases of rAAA during the enrollment period, only the first is considered as the primary case and included in analysis. Such criteria have been motivated by observational studies [6,11] and data availability, and have the same level of rigor as a real clinical trial.

The trial enrolls participants from 1 January 2011 to 30 September 2015. After enrollment, each eligible participant is randomized to receive either EVAR or OAR and followed until death, loss to follow-up, or end of the study (30 June 2019). Such decisions have been made with the considerations that both treatments have been extensively adopted in the study period, the enrollment is long enough to ensure a sufficient sample size, and the follow-up is long enough to ensure a sufficient effective sample size.

To assess both short- and long-term mortality after EVAR and OAR, we define two primary outcomes: time from treatment to short-term perioperative mortality and time from treatment to long-term all-cause mortality. The short-term perioperative mortality is defined as death during the index hospitalization or within 30 days of discharge, for which all participants alive at 30 days after discharge are censored. For the long-term all-cause mortality, a subject is censored at loss to follow-up or end of the study (30 June 2019), whichever comes first. The two survival outcomes have different implications and are both critically important [11].

2.3. The Emulated Trial

To emulate the target randomized clinical trial described above, we develop an emulated trial using the Medicare claims data. The strategy closely follows that developed in the emulation literature [19]. First, we identify Medicare beneficiaries who were diagnosed with rAAA and underwent EVAR or OAR between 1 January 2011 and 30 September 2015. We exclude individuals that met any of the following criteria: (1) the individual was under 65 years old at diagnosis; (2) both EVAR and OAR were present in the same index hospitalization, which indicated conversion; (3) concurrent diagnosis codes of thoracic aneurysms, thoracoabdominal aneurysms, or aortic dissection; (4) concurrent procedure codes of repair of the thoracic aorta or visceral or renal bypass; and (5) less than 12 months of Medicare enrollment before the index hospitalization. If a beneficiary had multiple eligible claims, only the first was considered as the primary case and included in analysis. Additional information on patient selection is provided in the flowchart in Figure 1. The relevant International Classification of Diseases, Ninth Revision, Clinical Modification (ICD-9-CM) codes are provided in Appendix A.

We then classify each eligible subject into one of the two treatment groups: EVAR and OAR, based on the procedure he/she actually received. Follow-up information is then extracted for each subject (to death, loss to follow-up, or end of the study which is 30 June 2019). A loss to follow-up is defined as discontinuation of Medicare enrollment. To identify the primary outcomes, we track each study subject from treatment to his/her documented death. We note that there are 5.35% of the study subjects for whom the date of treatment is missing. For these subjects, we use the date of admission to approximate the date of treatment, since rAAA is an emergent condition that needs immediate treatment, and the average lag time between admission and procedure is 0.53 days in our cohort.

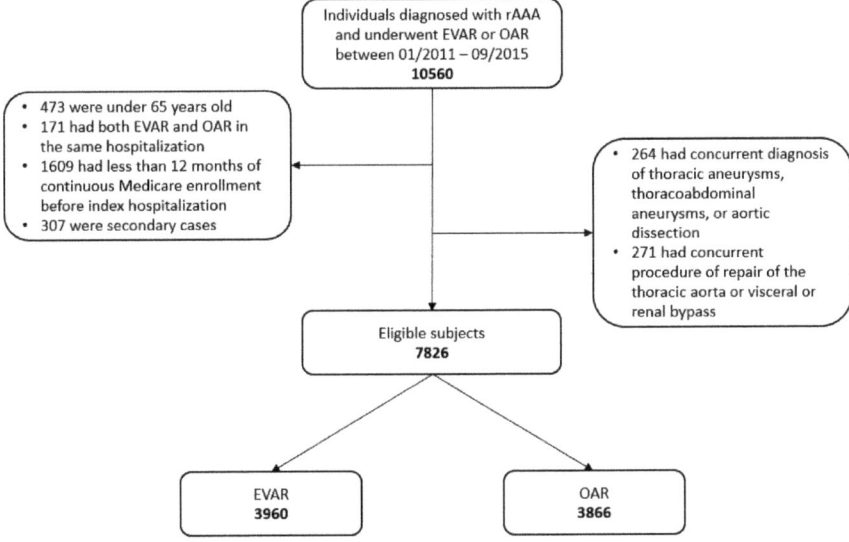

Figure 1. Flowchart of cohort definition.

2.4. Data Analysis

This study has survival outcomes. If this were a real clinical trial, analysis could be conducted using a Cox model. Although balance is expected with proper randomization, to be cautious, in clinical trial analysis, potential confounders are still commonly adjusted. For an emulated trial with a survival outcome, published studies [18,19] suggest the following main analysis steps: (a) conduct a propensity score analysis for treatment using the logistic regression approach, and (b) conduct a Cox regression analysis for survival with IPT weighting. When it can be assumed that all relevant variables are properly included, the first step is a simple parametric regression, and the consistency of parameter estimation can be easily established. The balance in covariate distributions between the two treatment arms for the pseudo sample created by the IPT weighting directly follows. With this balance, the validity of the (weighted) Cox regression follows [31,32].

As briefly mentioned in Section 1, deep learning has demonstrated promising performance with biomedical data. It is of significant interest to apply it to emulation. Equally importantly, the analysis presented in the Supplementary Materials shows that the Cox proportionality assumption is not satisfied. The deep learning approach described below, although has some connections with the Cox model, can be more flexible and less dependent on model assumptions, with its "built-in" flexibility. It consists of the following steps: (a) estimate propensity scores for treatment using a single-layer neural network. This is the counterpart of the logistic regression mentioned above; (b) construct a multi-layer neural network for survival. Advancing from the "standard" deep learning survival, we incorporate weights generated in Step (a), which is the counterpart of the IPT weighted Cox regression mentioned above; and (c) advancing from the existing deep learning literature, we also conduct a bootstrap-type procedure to gain insights into the variation of the neural network weight estimation, which is analogous to the regression coefficient estimation and reveals the treatment effects.

Denote n as the number of independent subjects. For subject i, denote C_i as the censoring time and T_i as the event time. We observe the right-censored survival outcome $Y_i = \min(T_i, C_i)$ and event indicator $d_i = I(T_i \leq C_i)$ with $I(\cdot)$ being the indicator function. Denote $X_i = (X_{i1}, X_{i2}, \ldots, X_{ip})$ as the baseline covariates and Z_i as the binary treatment assignment.

Step 1: We employ a single-layer neural network to estimate the propensity score, which is the probability of treatment assignment conditional on the baseline covariates. In particular, the input includes the covariates described in Table 1, with standardization for the continuous variables and coding for the categorical variables. The labels in the data are the binary treatment assignment variables. For the neural network architecture, we use the Rectified Linear Units (ReLU) as the activation function, sigmoid activation function to produce the probability output, and logarithmic loss function (binary cross-entropy). For optimization, a stochastic gradient descent algorithm with Nesterov momentum is used, and a grid search is conducted to tune the learning rate. For such tasks, we adopt the open-source python module keras (https://keras.io). With the outputted propensity score, we compute the IPT weight as its inverse for a subject in one treatment group and the inverse of one minus propensity score for a subject in the other group.

Table 1. Descriptive characteristics of the study cohort.

	EVAR (N = 3930)	OAR (N = 3866)	*p*-Value *
Demographic			
Age, mean(sd)	78.03 (7.52)	76.59 (6.90)	<0.0001
Male	3023 (76.34)	2786 (72.06)	<0.0001
Race			0.0030
White	3588 (90.77)	3550 (92.02)	
Black	249 (6.30)	178 (4.61)	
Other	116 (2.93)	130 (3.37)	
Medical conditions			
Congestive heart failure	464 (11.72)	299 (7.73)	<0.0001
Cardiac arrhythmia	596 (15.05)	438 (11.33)	<0.0001
Valvular disease	199 (5.03)	172 (4.45)	0.2304
Coronary disease	758 (19.14)	603 (15.60)	<0.0001
Diabetes	329 (8.31)	256 (6.62)	0.0045
Hypertension	1250 (31.25)	1078 (27.88)	0.0004
Chronic obstructive pulmonary diseases	707 (17.85)	584 (15.11)	0.0011
Clinically significant lower extremity vascular diseases	26 (0.66)	27 (0.70)	0.8215
Renal atherosclerosis	20 (0.51)	27 (0.70)	0.2684
Vascular intestine disease	7 (0.18)	2 (0.05)	0.1027
Renal failure	493 (12.45)	358 (9.26)	<0.0001
Other renal diseases	3 (0.08)	1 (0.03)	0.3289
Kidney transplant	4 (0.10)	3 (0.08)	0.7291
Liver disease	33 (0.83)	30 (0.78)	0.7766
Cerebrovascular diseases and paralysis	93 (2.35)	67 (1.73)	0.0544
Other neurological diseases	153 (3.86)	114 (2.95)	0.0258
Hyperlipidemia	817 (20.63)	687 (17.77)	0.0013
Cancer	132 (3.33)	87 (2.25)	0.0037
Rheumatoid arthritis	76 (1.92)	39 (1.01)	0.0008
Prior intact AAA diagnosis	511 (12.90)	440 (11.38)	0.0393
Other			
Year in which repair was performed			<0.0001
2011	808 (20.40)	1013 (26.20)	
2012	869 (21.94)	913 (23.62)	
2013	819 (20.68)	785 (20.31)	
2014	837 (21.14)	701 (18.13)	
2015	627 (15.83)	454 (11.74)	
Outcome (followed until death, loss to follow-up, or 06/30/2019)			
All-cause mortality	2430 (61.36)	2542 (65.75)	<0.0001
Perioperative mortality (in-hospital or 30 days after discharge)	1107 (27.95)	1704 (44.08)	<0.0001

* *p*-values based on *t*-tests for continuous variables and Chi-squared test for categorical variables.

Step 2: Here we conduct the IPT weighted survival analysis. The input includes the same set of covariates and treatment indicator as in Step 1, as well as the IPT weights computed above. For subject i, denote w_i as the IPT weight and $R_i = \{j : T_j > T_i\}$ as the at-risk set (at time T_i). We consider a neural network with two hidden layers and the number of nodes determined by tuning. Denote θ as the weights that characterize the network (note that they are not the IPT weights), and $g_\theta(X_i, Z_i)$ as the output for subject i. Partly motivated by the loss function under the Cox regression as well as recent deep learning studies, such as DeepSurv, we consider the objective function:

$$l(\theta) = -\frac{1}{\sum_i d_i} \sum_{i=1}^{n} d_i w_i \left[g_\theta(X_i, Z_i) - \log \sum_{j \in R_i} \exp\left(g_\theta(X_j, Z_j)\right) \right].$$

For optimization, we adopt a gradient descent approach. ReLU is used as the activation function, and the adaptive moment estimation algorithm (Adam) for gradient descent optimization with a cyclical learning rate method is adopted. We perform a grid search for hyper-parameter tuning. The computational program is developed based on the open-source python module pycox (https://github.com/havakv/pycox).

Step 3: A procedure similar to the 0.632 bootstrap for regression analysis [33] is conducted. In particular, $0.632n$ samples are randomly selected from the original data without replacement. With the bootstrapped samples, the above analysis is conducted, and the neural network weight estimates are extracted. This is repeated multiple (e.g., 1000) times to assess the variability of estimates. For regression, the 0.632 bootstrap is equivalent to the "n-out-of-n with replacement" bootstrap. By sampling without replacement, it can reduce ties and computational cost.

In Appendix B, we sketch the algorithm for conducting the Step 3 bootstrap type analysis. The analysis of the whole data amounts to skipping the bootstrapping step and otherwise applying the same procedures.

The above analysis can deliver the following. The first is a propensity score estimate for each subject. If needed, the weights of the neural network can be extracted to help assess the relative contributions of covariates. The second is the survival neural network. For a subject with a set of known confounder values and treatment assignment, it can generate the (relative) survival risk. Most of the existing deep learning studies have treated neural networks as black boxes. As we conduct a clinical trial analysis, the effect of the treatment is of the most essential interest. As such, we retrieve the estimated weights for the treatment indicator and confounders. With the presence of hidden layers, the weight matrices need to be multiplied across layers to obtain the overall contributions. The third product is that, for the (overall) weight of the treatment indicator, the bootstrap type analysis can generate an evaluation of its variability. The same is also applicable to the confounders.

Remarks

For a large number of binary data analysis problems, the superiority of neural networks over logistic and other regressions has been established [23,24]. Several recent publications, such as DeepSurv [27] and Cox-nnet [34] and others, seem to suggest similar superiority for survival data. As our goal is to take advantage of the recent deep learning developments, we choose not to "re-establish" the merit of deep learning. We also note that there are multiple "base techniques" for building neural networks. The adopted ones have been shown in recent studies as having a strong mathematical/statistical ground and competitive numerical performance. For example, it has been proved that stochastic gradient descent algorithms can find global minima on the training objective of deep neural networks in polynomial time under mild assumptions [35]. Farrell et al. (2020) established novel nonasymptotic high probability bounds for the fully connected feedforward neural networks with ReLU activations [36]. On the other hand, our literature review suggests that, compared to regression analysis, theoretical research on deep learning remains very rare. Consistency properties (for example, for the weights and bootstrapped estimation) remain unclear. Published literature seems to suggest tremendous challenges. It is beyond our scope to conduct such theoretical investigation.

3. Results

3.1. Patient Characteristics and Unadjusted Incidences

Our analysis includes 7826 eligible subjects, with 3960 in the EVAR arm and 3866 in the OAR arm. The summary statistics are shown in Table 1. It is observed that the study subjects were slightly younger in the OAR arm and more likely to be white males in both arms. Participants in the OAR arm were healthier with lower percentages of almost all medical conditions (except for two rare conditions: clinically significant lower extremity vascular diseases and renal atherosclerosis). It is also observed that, as time passed by (from year 2011 to 2015), the rAAA patients were more and more likely to receive EVAR. Here we note that, without the IPT weighting, all demographic variables and most medical condition variables are significantly unbalanced between the two treatment arms, highlighting a significant difference between real clinical trials and observational studies. Table 1 also shows the unadjusted incidence rates by treatment. The EVAR arm has a slightly lower unadjusted incidence rate for long-term all-cause mortality and a significantly lower unadjusted incidence rate for short-term perioperative mortality.

3.2. Analysis of the Emulated Trial

Prior to analysis, we deleted 15 records with missing measurements (7 in the EVAR arm and 8 in the OAR arm). Analysis was conducted using the approach described in Section 2.4. For the propensity score analysis, the baseline covariates include age, gender, race, year in which repair was performed (this variable has been considered in the published observational studies [6,10]; it is also motivated by the changing rates of EVAR and OAR), and 20 medical conditions, as shown in Table 1 (and with related ICD-9-CM codes in Appendix A). For survival analysis, the same baseline covariates and the treatment indicator are included.

For both the propensity score and survival analysis, the obtained fully connected neural network architectures are available from the authors. For the propensity score analysis, the learning rate is tuned as 0.008. The distributions of propensity scores are shown in Figure 2. Minor differences between the two arms are observed.

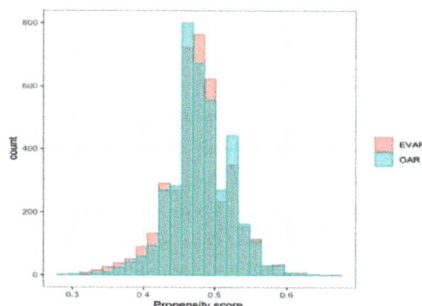

Figure 2. Distribution of propensity score.

For the analysis of short-term survival, the learning rate for Adam optimizer is tuned as 0.016. The analysis results are summarized in Figure 3. The left panel shows the estimated survival curves, after accounting for IPT weights, for the two treatments separately. With the bootstrap procedure, we are also able to obtain the pointwise 90% confidence intervals. It is noted that this analysis mimics the "familiar" regression analysis and differs from most of the existing deep learning studies, however, it lacks rigorous theoretical justifications that are available for regression analysis. EVAR is observed to have a modest survival advantage, with the lower bounds of its confidence intervals almost coinciding with the upper bounds of OAR's confidence intervals. Based on the estimated survival curves, we compute the expected survival under EVAR as 83.5 days, compared to 79.2 days under OAR. In the right panel of Figure 3, the forest plot, which shows the medians as well as the 25% and 75% quantile values of the overall estimated weights ("accumulated" over layers), again suggests the survival advantage of EVAR. The right panel of Figure 3 also contains weight information for confounders that demonstrate considerable and "persistent" effects (across the bootstrapped datasets), including race and seven medical conditions.

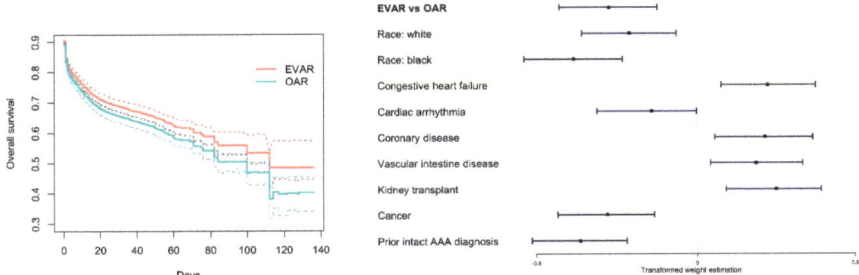

Figure 3. Analysis of short-term mortality. **Left**: estimated survival curves with pointwise 90% confidence intervals. **Right**: forest plot of the estimated weights.

For the analysis of long-term survival, the learning rate for Adam optimizer is tuned as 0.036. The analysis results are summarized in Figure 4, which are parallel to those in Figure 3. The findings are similar to those for short-term survival. Briefly, the left panel suggests some advantages of EVAR, but the pointwise confidence intervals overlap. We compute the expected survival as 1464.2 days under EVAR and 1348.0 days under OAR. The forest plot in the right panel shows that the advantage of EVAR is smaller than that for short-term survival. Confounders that demonstrate considerable and "persistent" effects include race, sex, and six medical conditions.

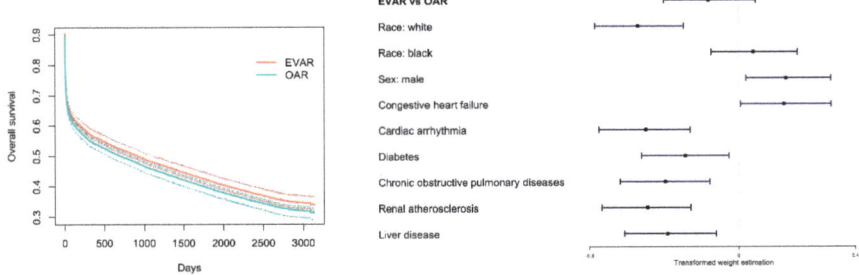

Figure 4. Analysis of long-term mortality. **Left**: estimated survival curves with pointwise 90% confidence intervals. **Right**: forest plot of the estimated weights.

For comprehensiveness, we also conduct a regression-based analysis. The results are presented in the Supplementary Materials. As the Cox model assumption is violated in both survival analyses, the results cannot be sensibly utilized.

4. Discussion

As fully discussed in published literature, the Medicare data has multiple unique advantages. With its broad coverage of the U.S. elderly population, our findings can be applied to this population with high confidence. Although there is no evidence that the relative treatment effects of EVAR and OAR differ by age, sex, and race [11–13], application of the findings to the younger U.S. population and populations in other countries/regions should be conducted with cautions. Besides, we have analyzed the Medicare inpatient claims data from 1 January 2011 to 30 June 2019. Both the enrollment and follow-up times are long enough, especially compared to many peer studies [11–13].

Recently, deep learning has become increasingly popular in biomedical studies, especially including in the analysis of Medicare and other healthcare data. Beyond those mentioned in the first section, other examples also include Ali et al. (2020), which develops a novel information framework using ensemble deep learning and a feature fusion technique to analyze wearable sensor and electronic medical test data for heart disease prediction [37]. Additional examples include Ali et al. (2020) [38], Jain et al. (2020) [39], and Selden (2020) [40]. The deep learning techniques adopted in this article can fully meet our needs, and it is beyond our scope to comprehensively review/compare existing techniques. We do recognize that it is of interest to explore the applications of other deep learning techniques to the emulation setting.

The emulation strategy has been developed and adopted in quite a few studies. Its pros and cons have been well documented. It is especially noted that, first, emulated trials, although resembling real clinical trials in multiple perspectives, still have notable limitations and cannot replace real clinical trials. Second, there is still a lack of objective comparison and definitive conclusion on its relative performance with respect to other causal inference approaches. Although important, this is beyond the scope of this study. The adopted deep learning methods have been based on certain well-developed components and software programs. Nevertheless, their "combination" and application to the emulation setting and rAAA treatment problem are new and novel. Our analysis has demonstrated how to "replace" regression using deep learning under settings more sophisticated than in the literature. As the "propensity score + survival analysis" strategy and individual components of the deep learning analysis have been more or less developed in the literature, we choose not to methodologically further discuss or conduct more numerical investigations.

Our main finding is that EVAR has advantageous short- and long-term survival. Although the improvement in expected survival is modest, considering the severity of rAAA, it may still have important clinical implications. In the literature, the short-term survival advantage of EVAR has been suggested in multiple observational analyses [6,10,12]. However, there has been a lack of definitive conclusion on the long-term benefit. For example, Behrendt et al. (2017) suggested early survival benefit of EVAR over OAR, which reversed at ~2.5 years of follow-up, for iAAA and rAAA patients in Germany [11]. Schermerborn et al. (2015) observed similar survival of the two procedures after 3 years from initial surgery for iAAA patients in the Medicare population [6]. And a 15-year follow-up resulted from the EVAR-1 trial indicated that EVAR had inferior late survival compared to OAR [9]. Multiple factors can contribute to the differences observed in the aforementioned and other studies. First, the studied populations have different characteristics. Second, the analysis strategies also differ, with our strategy closer to a controlled clinical trial. It is also noted that the study periods are different. Although there is still no indication of temporal variation in treatment effects, related confounders may change over time.

Besides treatment, our analysis also suggests that race, gender, and certain medical conditions are associated with survival after EVAR and OAR among rAAA patients. While most observational studies that compare EVAR and OAR match study subjects or adjust for potential confounders, there is a lack of attention on how these variables may impact survival after rAAA. We have found that compared to other races, the white race is associated with lower short- and long-term mortality, and the black race is associated with lower short-term mortality. This race difference has been insufficiently studied in the literature. It can be caused by genetic effects (considering that genetic factors contribute to many cardiovascular diseases), lifestyle, cultural factors, access to care, and other factors that may confound survival. While Egorova et al. (2011) observed significantly worse outcomes after EVAR and OAR for female patients, we have found no gender difference in short-term mortality and male associated with higher long-term mortality [12]. One contributing factor is the difference in analysis technique: Egorova et al. (2011) compared the observed survival with expected survival in a life table [12], while we have conducted a more comprehensive adjusted analysis. Lastly, we have identified certain medical conditions as associated with survival. What may seem counterintuitive is that some medical conditions are found as negatively associated with mortality. For example, it is found that prior intact AAA diagnosis decreases short-term mortality risk after rAAA, and the presence of cardiac arrhythmia increases both short- and long-term survival. One plausible explanation is that patients with related medical conditions are more likely to have regular hospital visits and more access to healthcare services, which may lead to more timely detection of emergent rAAA. For example, it is noted in Edwards et al. (2014) that patients who had a prior diagnosis of intact AAA were less commonly admitted through the emergency department, and were more commonly transferred between hospitals before treatment, which was associated with better survival [10]. Dardic et al. (1998) also found that the presence of hypertension, diabetes, and COPD was correlated with a statistically significant lower mortality rate, whereas the presence of smoking, heart disease, and renal disease was correlated with a statistically insignificant lower mortality rate after the diagnosis of rAAA [41].

This study inevitably has limitations. In particular, there is limited information covered by the Medicare data [28,29]. Therefore, the treatment arms may be imbalanced on unmeasured confounders such as over-the-counter drug uses and patients' socioeconomic information. We note that this limitation is shared by other emulation studies and analysis of observational data. Moreover, due to limited data access, this study examines rAAA patients' inpatient treatments from 1 January 2011 to 30 September 2015. With the special nature of rAAA, inpatient claims should be able to catch the dominating majority of the cases. However, it remains unclear whether utilization of other clinical settings (e.g., emergency room or outpatient) affects the treatment effects of EVAR and OAR. Also, although there is no indication that the treatment effects have temporal variations, it may still be of interest to examine more extensive data. In addition to data limitations, it is well documented in the literature that the emulation approach, while being lucidly interpretable, has limitations [14,18,19]. For example, the approach can only emulate target trials without blind assignment. Given the specific natures of EVAR and OAR, lack of blinding is not necessarily a limitation for this study. However, future studies using the emulation approach should be cautious of these limitations.

With the possibility of more extensive data, it is of future interest to investigate the aforementioned potential confounders that are unmeasured in this study, the effects of other clinical settings on the treatment effects of EVAR and OAR after rAAA, and the potential temporal variations. Moreover, it is also postponed to future work to conduct a direct comparison of the emulation approach and other causal inference approaches using the large-scale Medicare data.

5. Conclusions

This study has suggested certain short- and long-term survival advantage of EVAR over OAR for rAAA patients. It has also further advanced the emulation and deep learning techniques for analyzing data mined from large medical record databases. Both the medical findings and analytic developments can complement the existing literature and be of interest to stakeholders at multiple levels.

Supplementary Materials: The following are available online at http://www.mdpi.com/1099-4300/22/12/1349/s1.

Author Contributions: Conceptualization, H.M., Y.X., J.W., and S.M.; methodology, H.M., Y.X., J.W., and S.M; software, H.M. and Y.X.; validation, H.M., Y.X., J.W., and S.M.; formal analysis, H.M. and Y.X.; investigation, H.M., Y.X., J.W., and S.M.; resources, Y.X.; data curation, H.M.; writing—original draft preparation, H.M., Y.X., J.W., and S.M.; writing—review and editing, H.M., Y.X., J.W., and S.M.; visualization, Y.X.; supervision, S.M.; project administration, S.M. All authors have read and agreed to the published version of the manuscript.

Funding: This research received no external funding.

Acknowledgments: We thank the reviewers for their careful review and insightful comments, which have led to a significant improvement of this article.

Conflicts of Interest: The authors declare no conflict of interest.

Appendix A

Table A1. International Classification of Disease, 9th edition, Clinical Modification (ICD-9-CM) codes for identifying eligible individuals and defining confounders (one year of medical history).

Variable	ICD-9-CM Code *
Inclusion	
Ruptured abdominal aortic aneurysm	441.3
Endovascular aortic repair	39.71
Open aortic repair	38.44, 39.25, 39.52, 38.34, 38.64, 38.40, 38.60
Exclusion	
Thoracic aneurysms	441.1, 441.2
Thoracoabdominal aneurysms	441.6, 441.7
Aortic dissection	441.00-441.03
Repair of the thoracic aorta	38.35, 38.45, 39.73
Visceral or renal bypass	38.46, 39.24, 39.26
Medical history	
Congestive heart failure	398.91, 402.01, 402.11, 402.91, 404.01, 404.03, 404.11, 404.91, 404.13, 404.93, 425.4, 425.5, 425.7, 425.8, 425.9, 428.0, 428.1, 428.20, 428.22, 428.30, 428.32, 428.40, 428.42, 428.9
Cardiac arrhythmia	426.0, 426.10, 426.11, 426.12, 426.13, 426.7, 426.9, 427.0, 427.1, 427.2, 427.3, 427.9, V45.0, V53.3
Valvular disease	093.2, 394, 395, 396, 397, 424, V42.2, V43.3
Coronary disease	412, 413, 414, 429.2
Diabetes	250
Hypertension	401, 402, 403, 404, 405
Chronic Obstructive Pulmonary diseases	416, 417.9, 490, 491, 492, 493, 494, 495.0, 495.1, 495.2, 495.3, 495.4, 495.5, 495.6, 495.8, 495.9, 496, 500, 501, 502, 503, 504, 505, 506.0, 506.2, 506.4, 506.9, 508.1, 508.8, 508.9
Clinically significant lower extremity vascular diseases	440.22, 440.23, 440.24, 440.3, 444.22, V43.4,
Renal atherosclerosis	440.1
Vascular intestine disease	557.1
Renal failure w dialysis	V45.1, V56.0, V56.1, V56.2, V56.3, V56.8, 585.6, 39.95 (w/o 586)
Renal failure without dialysis	403.01, 403.11, 403.91, 404.02, 404.03, 404.12, 404.13, 404.92, 404.93, 585 (w/o 585.6), 588.0
Other renal diseases	582, 583.0, 583.1, 583.2, 583.4
Kidney transplant	V420
Liver disease	070.22, 070.23, 070.32, 070.33, 070.44, 070.54, 070.9, 456.0, 456.1, 571, 572.1, 572.2, 572.3, 572.4, 572.8, 573.0, 573.1, 573.8, 573.9
Cerebrovascular diseases and paralysis	342, 344.1, 344.3, 344.4, 344.5, 344.9, 437.0, 438
Other neurological diseases	330, 331, 332, 333, 334.0, 334.1, 334.2, 334.4, 334.8, 335.0, 335.1, 335.2, 335.8, 335.9, 336.0, 336.2, 343, 344.0, 348.1, 348.3, 344.2, 344.6, 345, 437.3, 437.4, 437.5, 437.6, 437.7
Hyperlipidemia	272
Cancer	140, 141, 142, 143, 144,145, 146, 147, 148, 149, 150, 151, 152, 153, 154, 155, 156, 157, 158,159, 160, 161, 162, 163, 164, 165, 170, 171,172, 174, 175, 176, 179, 180, 181, 182, 183, 184, 185, 186, 187, 188, 189, 190, 191, 192, 193, 194, 195, 196, 197, 198, 199, 200, 201, 202, 203.0, 238.6
Rheumatoid arthritis	446, 701.0, 710.0, 710.1, 710.2, 710.3, 710.4, 710.8, 710.9, 711.2, 719.3, 714,720, 725, 728.5, 728.89
Prior intact AAA diagnosis	441.4, without mention 441.3

* Primary or any secondary diagnosis/procedure code.

Appendix B

Algorithm A1. Algorithm for the bootstrap type analysis (Step 3).

1. Initialize b = 0.
2. Update b = b + 1, randomly sample $0.632n$ subjects without replacement from the original data and conduct the following procedure.

   Step 1. Estimate the propensity score.
   - Construct a simple multilayer perceptron with one hidden layer, ReLU and sigmoid activations, and binary cross-entropy as the loss function.
   - For a fixed learning rate from a vector of hyper-parameter values:
   ○ Compile the model using a stochastic gradient descent optimizer with Nesterov momentum.
   ○ Compute the classification accuracy as the metrics for model selection.
   - Select the best model with the highest accuracy from the above grid search, and calculate the IPT weight by transforming the estimated propensity score.

   Step 2. Assess the treatment effects using the IPT weighted survival analysis.
   - Construct a multilayer perceptron with two hidden layers, ReLU activations, and the proposed loss function $l(\theta)$.
   - For a fixed learning rate, the number of nodes in hidden layers from a grid of hyper-parameter values:
   ○ Compile the model using Adam optimizer.
   ○ Estimate the weights θ, and compute the concordance statistic for model selection.
   - Select the best model with the largest concordance, and obtain the effects of treatment and other confounders by multiplying the estimated weights across layers.

3. Repeat until b = B is large enough (e.g., B = 1000).

References

1. Kent, K.C.; Zwolak, R.M.; Egorova, N.N.; Riles, T.S.; Manganaro, A.; Moskowitz, A.J.; Gelijns, A.C.; Greco, G. Analysis of risk factors for abdominal aortic aneurysm in a cohort of more than 3 million individuals. *J. Vasc. Surg.* **2010**, *52*, 539–548. [CrossRef] [PubMed]
2. Heikkinen, M.; Salenius, J.P.; Auvinen, O. Ruptured abdominal aortic aneurysm in a well- defined geographic area. *J. Vasc. Surg.* **2002**, *36*, 291–296. [CrossRef] [PubMed]
3. Minino, A.M.; Murphy, S.H.; Xu, J.; Kochanek, K.D. Division of Vital Statistics. Deaths: Final Data for 2009. Centers for Disease Control and Prevention. *Natl. Vital Stat. Rep.* **2011**, *59*, 10.
4. Sakalihasan, N.; Limet, R.; Defawe, O.D. Abdominal aortic aneurysm. *Lancet* **2005**, *365*, 1577–1589. [CrossRef]
5. MacSweeney, S.T.R.; Ellis, M.; Greenhalgh, R.M.; Powell, J.T. Smoking and growth rate of small abdominal aortic aneurysms. *Lancet* **1994**, *344*, 651–652. [CrossRef]
6. Schermerhorn, M.L.; Buck, D.B.; O'Malley, A.J.; Curran, T.; McCallum, J.C.; Darling, J.; Landon, B.E. Long-Term Outcomes of Abdominal Aortic Aneurysm in the Medicare Population. *N. Engl. J. Med.* **2015**, *373*, 328–338. [CrossRef]
7. Lederle, F.A.; Freischlag, J.A.; Kyriakides, T.C.; Padberg, F.T.; Matsumura, J.S.; Kohler, T.R.; Lin, P.H.; Jean-Claude, J.M.; Cikrit, D.F.; Peduzzi, P.N.; et al. Outcomes Following Endovascular vs Open Repair of Abdominal Aortic Aneurysm: A Randomized Trial. *JAMA* **2009**, *302*, 1535–1542. [CrossRef]
8. De Bruin, J.L.; Baas, A.F.; Buth, J.; Prinssen, M.; Verhoeven, E.L.G.; Cuypers, P.W.M.; van Sambeek, M.R.H.M.; Balm, R.; Grobbee, D.E.; Blankensteijn, J.D. Blankensteijn Long-Term Outcome of Open or Endovascular Repair of Abdominal Aortic Aneurysm. *N. Engl. J. Med.* **2010**, *362*, 1881–1889. [CrossRef]
9. Patel, R. Endovascular versus open repair of abdominal aortic aneurysm in 15-years' follow-up of the UK endovascular aneurysm repair trial 1 (EVAR trial 1): A randomised controlled trial. *Lancet* **2016**, *388*, 2366–2374. [CrossRef]
10. Edwards, S.T.; Schermerhorn, M.L.; O'Malley, A.J.; Bensley, R.P.; Hurks, R.; Cotterill, P.; Landon, B.E. Landon Comparative effectiveness of endovascular versus open repair of ruptured abdominal aortic aneurysm in the Medicare population. *J. Vasc. Surg.* **2014**, *59*, 575–582. [CrossRef]
11. Behrendt, C.-A.; Sedrakyan, A.; Rieß, H.C.; Heidemann, F.; Kölbel, T.; Debus, E.S. Short-term and long-term results of endovascular and open repair of abdominal aortic aneurysms in Germany. *J. Vasc. Surg.* **2017**, *66*, 1704–1711. [CrossRef] [PubMed]

12. Egorova, N.N.; Vouyouka, A.G.; McKinsey, J.F.; Faries, P.L.; Kent, K.C.; Moskowitz, A.J.; Gelijns, A. Effect of gender on long-term survival after abdominal aortic aneurysm repair based on results from the Medicare national database. *J. Vasc. Surg.* **2011**, *54*, 1–12. [CrossRef] [PubMed]
13. Jackson, R.S.; Chang, D.C.; Freischlag, J.A. Comparison of Long-term Survival After Open vs Endovascular Repair of Intact Abdominal Aortic Aneurysm Among Medicare Beneficiaries. *JAMA* **2012**, *307*, 1621–1628. [CrossRef] [PubMed]
14. Hernán, M.A.; Robins, J.M. *Causal Inference: What If*; Chapman Hill/Crc: Boca Raton, FL, USA, 2020.
15. Van der Laan, M.J.; Rose, S. *Targeted Learning: Causal Inference for Observational and Experimental Data*; Springer Science & Business Media: New York, NY, USA, 2011; ISBN 978-1-4419-9782-1.
16. Chipman, H.A.; George, E.I.; McCulloch, R.E. BART: Bayesian additive regression trees. *Ann. Appl. Stat.* **2010**, *4*, 266–298. [CrossRef]
17. Hernán, M.A.; Alonso, A.; Logan, R.; Grodstein, F.; Stampfer, M.J.; Willett, W.C.; Manson, J.E.; Robins, M. Observational studies analyzed like randomized experiments: An application to postmenopausal hormone therapy and coronary heart disease. *Epidemiology (Camb. Mass.)* **2008**, *19*, 766–779. [CrossRef]
18. Dickerman, B.A.; García-Albéniz, X.; Logan, R.W.; Denaxas, S.; Hernán, M.A. Avoidable flaws in observational analyses: An application to statins and cancer. *Nat. Med.* **2019**, *25*, 1601–1606. [CrossRef]
19. Petito, L.C.; García-Albéniz, X.; Logan, R.W.; Howlader, N.; Mariotto, A.B.; Dahabreh, I.J.; Hernán, M.A. Estimates of Overall Survival in Patients With Cancer Receiving Different Treatment Regimens: Emulating Hypothetical Target Trials in the Surveillance, Epidemiology, and End Results (SEER)–Medicare Linked Database. *JAMA Netw Open* **2020**, *3*, e200452. [CrossRef]
20. Zigler, C.M.; Kim, C.; Choirat, C.; Hansen, J.B.; Wang, Y.; Hund, L.; Samet, J.; King, G.; Dominici, F. Causal Inference Methods for Estimating Long-Term Health Effects of Air Quality Regulations. *Europe PMC* **2016**, *187*, 5–49.
21. Deng, L.; Yu, D. Deep Learning: Methods and Applications. *SIG* **2014**, *7*, 197–387. [CrossRef]
22. Lv, Y.; Duan, Y.; Kang, W.; Li, Z.; Wang, F.-Y. Traffic Flow Prediction With Big Data: A Deep Learning Approach—IEEE Journals & Magazine. *IEEE Trans. Intell. Transp. Syst.* **2015**, *16*, 865–873.
23. Wang, D.; Khosla, A.; Gargeya, R.; Irshad, H.; Beck, A.H. Deep Learning for Identifying Metastatic Breast Cancer. *arXiv* **2016**, arXiv:1606.05718.
24. Badgeley, M.A.; Zech, J.R.; Oakden-Rayner, L.; Glicksberg, B.S.; Liu, M.; Gale, W.; McConnell, M.V.; Percha, B.; Snyder, T.M.; Dudley, J.T. Dudley Deep learning predicts hip fracture using confounding patient and healthcare variables. *NPJ Digit. Med.* **2019**, *2*, 1–10. [CrossRef] [PubMed]
25. Chung, K.; Yoo, H.; Choe, D. Ambient context-based modeling for health risk assessment using deep neural network. *J. Ambient Intell. Humaniz. Comput.* **2020**, *11*, 1387–1395. [CrossRef]
26. Hsiao, H.C.; Chen, S.H.; Tsai, J.J. Deep Learning for Risk Analysis of Specific Cardiovascular Diseases Using Environmental Data and Outpatient Records. In Proceedings of the 2016 IEEE 16th International Conference on Bioinformatics and Bioengineering (BIBE), Taichung, Taiwan, 31 October–2 November 2016.
27. Katzman, J.L.; Shaham, U.; Cloninger, A.; Bates, J.; Jiang, T.; Kluger, Y. DeepSurv: Personalized treatment recommender system using a Cox proportional hazards deep neural network. *BMC Med. Res. Methodol.* **2018**, *18*, 24. [CrossRef]
28. Mues, K.E.; Liede, A.; Liu, J.; Wetmore, J.B.; Zaha, R.; Bradbury, B.D.; Collins, A.J.; Gilbertson, D.T. Use of the Medicare database in epidemiologic and health services research: A valuable source of real-world evidence on the older and disabled populations in the US. *Clin. Epidemiol.* **2017**, *9*, 267. [CrossRef]
29. Brennan, N.; Oelschlaeger, A.; Cox, C.; Tavenner, M. Leveraging The Big-Data Revolution: CMS Is Expanding Capabilities To Spur Health System Transformation. *Health Aff.* **2014**, *33*, 1195–1202. [CrossRef]
30. Hernán, M.A.; Robins, J.M. Using Big Data to Emulate a Target Trial When a Randomized Trial Is Not Available. *Am. J. Epidemiol.* **2016**, *183*, 758–764. [CrossRef]
31. Westreich, D.; Cole, S.R.; Schisterman, E.F.; Platt, R.W. A simulation study of finite-sample properties of marginal structural Cox proportional hazards models. *Stat. Med.* **2012**, *31*, 2098–2109. [CrossRef] [PubMed]
32. Robins, J.M. Association, causation, and marginal structural models. *Synthese* **1999**, *121*, 151–179. [CrossRef]
33. Jiang, W.; Simon, R. A comparison of bootstrap methods and an adjusted bootstrap approach for estimating the prediction error in microarray classification. *Stat. Med.* **2007**, *26*, 5320–5334. [CrossRef] [PubMed]
34. Ching, T.; Zhu, X.; Garmire, L.X. Cox-nnet: An artificial neural network method for prognosis prediction of high-throughput omics data. *PLoS Comput. Biol.* **2018**, *14*, e1006076. [CrossRef] [PubMed]

35. Allen-Zhu, Z.; Li, Y.; Song, Z. A convergence theory for deep learning via over-parameterization. In Proceedings of the 36th International Conference on Machine Learning (PMLR 2019), Long Beach, CA, USA, 9–15 June 2019; Volume 97, pp. 242–252.
36. Farrell, M.H.; Liang, T.; Misra, S. Deep neural networks for estimation and inference. *arxiv* **2018**, arXiv:1809.09953.
37. Ali, F.; El-Sappagh, S.; Islam, S.M.R.; Kwak, D.; Ali, A.; Imran, M.; Kwak, K.-S. A smart healthcare monitoring system for heart disease prediction based on ensemble deep learning and feature fusion. *Inf. Fusion* **2012**, *63*, 208–222. [CrossRef]
38. Ali, F.; El-Sappagh, S.; Islam, S.M.R.; Ali, A.; Attique, M.; Imran, M.; Kwak, K.-S. An intelligent healthcare monitoring framework using wearable sensors and social networking data. *Future Gener. Comput. Syst.* **2020**, *114*, 23–43. [CrossRef]
39. Jain, K.; Neelakantan, M.; Key, P. Limitations in the Analysis of Atherectomy Using Medicare Big Data. *J. Endovasc. Ther.* **2020**. [CrossRef] [PubMed]
40. Selden, T.M. Differences Between Public And Private Hospital Payment Rates Narrowed, 2012–2016: A data analysis comparing payment rate differences between private insurance and Medicare for inpatient hospital stays, emergency department visits, and outpatient hospital care. *Health Aff.* **2020**, *39*, 94–99.
41. Dardik, A.; Burleyson, G.P.; Bowman, H.; Gordon, T.A.; Williams, G.M.; Webb, T.H.; Perler, B.A. Surgical repair of ruptured abdominal aortic aneurysms in the state of Maryland: Factors influencing outcome among 527 recent cases. *J. Vasc. Surg.* **1998**, *28*, 413–421. [CrossRef]

Publisher's Note: MDPI stays neutral with regard to jurisdictional claims in published maps and institutional affiliations.

© 2020 by the authors. Licensee MDPI, Basel, Switzerland. This article is an open access article distributed under the terms and conditions of the Creative Commons Attribution (CC BY) license (http://creativecommons.org/licenses/by/4.0/).

Article

Variable Selection Using Nonlocal Priors in High-Dimensional Generalized Linear Models With Application to fMRI Data Analysis

Xuan Cao [1] and Kyoungjae Lee [2,*]

1. Department of Mathematical Sciences, University of Cincinnati, Cincinnati, OH 45221, USA; caox4@ucmail.uc.edu
2. Department of Statistics, Inha University, Incheon 22212, Korea
* Correspondence: leekjstat@gmail.com

Received: 8 June 2020; Accepted: 21 July 2020; Published: 23 July 2020

Abstract: High-dimensional variable selection is an important research topic in modern statistics. While methods using nonlocal priors have been thoroughly studied for variable selection in linear regression, the crucial high-dimensional model selection properties for nonlocal priors in generalized linear models have not been investigated. In this paper, we consider a hierarchical generalized linear regression model with the product moment nonlocal prior over coefficients and examine its properties. Under standard regularity assumptions, we establish strong model selection consistency in a high-dimensional setting, where the number of covariates is allowed to increase at a sub-exponential rate with the sample size. The Laplace approximation is implemented for computing the posterior probabilities and the shotgun stochastic search procedure is suggested for exploring the posterior space. The proposed method is validated through simulation studies and illustrated by a real data example on functional activity analysis in fMRI study for predicting Parkinson's disease.

Keywords: high-dimensional; nonlocal prior; strong selection consistency

1. Introduction

With the increasing ability to collect and store data in large scales, we are facing the opportunities and challenges to analyze data with a large number of covariates per observation, the so-called high-dimensional problem. When this situation arises, variable selection is one of the most commonly used techniques, especially in radiological and genetic research, due to the nature of high-dimensional data extracted from imaging scans and gene sequencing. In the context of regression, when the number of covariates is greater than the sample size, the parameter estimation problem becomes ill posed, and variable selection is usually the first step for dimension reduction.

A good amount of work has recently been done for variable selection from both frequentist and Bayesian perspectives. On the frequentist side, extensive studies on variable selection have emerged ever since the appearance of least absolute shrinkage and selection operator (Lasso) [1]. Other penalization approaches for sparse model selection including smoothly clipped absolute deviation (SCAD) [2], minimum concave penalty (MCP) [3] and many variations have also been introduced. Most of these methods are first considered in the context of linear regression and then extended to generalized linear models. Because all the methods share the basic desire of shrinkage toward sparse models, it has been understood that most of these frequentist methods can be interpreted from a Bayesian perspective and many analogous Bayesian methods have also been proposed. See for example [4–6] that discuss the connection between penalized likelihood-based methods and Bayesian approaches. These Bayesian methods employed local priors, which still preserve positive values at null parameter values, to achieve desirable shrinkage.

In this paper, we are interested in nonlocal densities [7] that are identically zero whenever a model parameter is equal to its null value. Compared to local priors, nonlocal prior distributions have relatively appealing properties for Bayesian model selection. In particular, nonlocal priors discard spurious covariates faster as the sample size grows, while preserving exponential learning rates to detect nontrivial coefficients [7]. Johnson and Rossell [8] and Shin et al. [9] study the behavior of nonlocal densities for variable selection in a linear regression setting. When the number of covariates is much smaller than the sample size, [10] establish the posterior convergence rate for nonlocal priors in a logistic regression model and suggest a Metropolis–Hastings algorithm for computation.

To the best of our knowledge, a rigorous investigation of high-dimensional posterior consistency properties for nonlocal priors has not been undertaken in the context of generalized linear regression. Although [11] investigated the model selection consistency of nonlocal priors in generalized linear models, they assumed a fixed dimension p. Motivated by this gap, our first goal was to examine the model selection property for nonlocal priors, particularly, the product moment (pMOM) prior [8] in a high-dimensional generalized linear model. It is known that the computation problem can arise for Bayesian approaches due to the non-conjugate structure in generalized linear regression. Hence, our second goal was to develop efficient algorithms for exploring the massive posterior space. These were challenging goals of course, as the posterior distributions are not available in closed form for this type of nonlocal priors.

As the main contributions of this paper, we first establish model selection consistency for generalized linear models with pMOM prior on regression coefficients (Theorems 1–3) when the number of covariates grows at a sub-exponential rate of the sample size. Next, n terms of computation, we first obtain the posteriors via Laplace approximation and then implement an efficient shotgun stochastic search (SSS) algorithm for exploring the sparsity pattern of the regression coefficients. In particular, the SSS-based methods have been shown to significantly reduce the computational time compared with standard Markov chain Monte Carlo (MCMC) algorithms in various settings [9,12,13]. We demonstrate that our model can outperform existing state-of-the-art methods including both penalized likelihood and Bayesian approaches in different settings. Finally, the proposed method is applied to a functional Magnetic Resonance Imaging (fMRI) data set for identifying alternative brain activities and for predicting Parkinson's disease.

The rest of paper is organized as follows. Section 2 provides background material regarding generalized linear models and revisits the pMOM distribution. We detail strong selection consistency results in Section 3, and proofs are provided in the Appendix A. The posterior computation algorithm is described in Section 4, and we show the performance of the proposed method and compare it with other competitors through simulation studies in Section 5. In Section 6, we conduct a real data analysis for predicting Parkinson's disease and show our method yields better prediction performance compared with other contenders. To conclude our paper, a discussion is given in Section 7.

2. Preliminaries

2.1. Model Specification for Logistic Regression

We first describe the framework for Bayesian variable selection in logistic regression followed by our hierarchical model specification. Let $y \in \{0,1\}^n$ be the binary response vector and $X = (x_{ij}) \in \mathbb{R}^{n \times p}$ be the design matrix. Without loss of generality, we assume that the columns of X are standardized to have zero mean and unit variance. Let $x_i \in \mathbb{R}^p$ denote the ith row vector of X that contains the covariates for the ith subject. Let β be the $p \times 1$ vector of regression coefficients. We first consider the following standard logistic regression model:

$$\mathbb{P}(y_i = 1 \mid x_i, \beta) = \frac{\exp(x_i^\top \beta)}{1 + \exp(x_i^\top \beta)}, \quad i = 1, 2, \ldots, n, \tag{1}$$

We will work in a scenario where the dimension of predictors, p grows with the sample size n. Thus, we consider the number of predictors is function of n, i.e., $p = p_n$, but we denote it as p for notational simplicity.

Our goal is variable selection, i.e., the correct identification of all non-zero regression coefficients. In light of that, we denote a model by $k = \{k_1, k_2, \ldots, k_{|k|}\} \subseteq [p] =: \{1, 2, \ldots, p\}$ if and only if all the nonzero elements of β are $\beta_{k_1}, \beta_{k_2}, \ldots, \beta_{k_{|k|}}$ and denote $\beta_k = (\beta_{k_1}, \beta_{k_2}, \ldots, \beta_{k_{|k|}})^\top$, where $|k|$ is the cardinality of k. For any $m \times p$ matrix A, let $A_k \in \mathbb{R}^{m \times |k|}$ denote the submatrix of A containing the columns of A indexed by model k. In particular, for $1 \leq i \leq n$, we denote x_{ik} as the subvector of x_i containing the entries of x_i corresponding to model k.

The class of pMOM densities [8] can be used for model selection through the following hierarchical model

$$\pi(\beta_k \mid \tau, k) = d_k (2\pi)^{-\frac{|k|}{2}} (\tau)^{-r|k| - \frac{|k|}{2}} |U_k|^{\frac{1}{2}} \exp\left(-\frac{\beta_k^\top U_k \beta_k}{2\tau}\right) \prod_{i=1}^{|k|} \beta_{k_i}^{2r}, \quad (2)$$

$$\pi(k) \propto I(|k| \leq m_n). \quad (3)$$

Here U is a $p \times p$ nonsingular matrix, r is a positive integer referred to as the order of the density and d_k is the normalizing constant independent of the positive constant τ. Please note that prior (2) is obtained as the product of the density of multivariate normal distribution and even powers of parameters, $\prod_{i=1}^{|k|} \beta_{k_i}^{2r}$. This results in $\pi(\beta_k \mid \tau, k) = 0$ at $\beta_k = 0$, which is desirable because (2) is a prior for the nonzero elements of β. Some standard regularity assumptions on the hyperparameters will be provided later in Section 3. In (3), $m_n \in [p]$ is a positive integer restricting the size of the largest model, and a uniform prior is placed on the model space restricting our analysis to models having size less than or equal to m_n. Similar structure has also been considered in [5,9,14]. An alternative is to use a complexity prior [15] that takes the form of

$$\pi(k) \propto c_1^{-|k|} p^{-c_2|k|},$$

for some positive constants c_1, c_2. The essence is to force the estimated model to be sparse by penalizing dense models. As noted in [9], the model selection consistency result based on the nonlocal priors derives strength directly from the marginal likelihood and does not require strong penalty over model size. This is indeed reflected in the simulation studies in [14], where the authors compare the model selection performance under uniform prior and complexity prior. The result under uniform prior is much better than that under complexity prior, as the complexity prior always tends to prefer the sparse models.

By the hierarchical model (1) to (3) and Bayes' rule, the resulting posterior probability for model k is denoted by,

$$\pi(k|y) = \frac{\pi(k)}{\pi(y)} m_k(y), \quad (4)$$

where $\pi(y)$ is the marginal density of y, and $m_k(y)$ is the marginal density of y under model k given by

$$m_k(y) = \int \exp\{L_n(\beta_k)\} \pi(\beta_k \mid k) d\beta_k$$

$$= \int \exp\{L_n(\beta_k)\} d_k (2\pi)^{-|k|/2} (\tau)^{-r|k| - |k|/2} |U_k|^{\frac{1}{2}} \exp\left(-\frac{\beta_k^\top U_k \beta_k}{2\tau}\right) \prod_{i=1}^{|k|} \beta_{k_i}^{2r} d\beta_k, \quad (5)$$

where

$$L_n(\beta_k) = \log \left(\prod_{i=1}^{n} \left\{ \frac{\exp(x_{ik}^\top \beta_k)}{1 + \exp(x_{ik}^\top \beta_k)} \right\}^{y_i} \left\{ \frac{1}{1 + \exp(x_{ik}^\top \beta_k)} \right\}^{1-y_i} \right) \qquad (6)$$

is the log likelihood function. In particular, these posterior probabilities can be used to select a model by computing the posterior mode defined by

$$\hat{k} = \arg\max_k \pi(k|y). \qquad (7)$$

Of course, the closed form of these posterior probabilities cannot be obtained due to not only the nature of logistic regression but also the structure of nonlocal prior. Therefore, special efforts need to be devoted to both consistency analysis and computational strategy as we shall see in the following sections.

2.2. Extension to Generalized Linear Model

We can easily extend our previous discussion on logistic regression to a generalized linear model (GLM) [16]. Given predictors x_i and an outcome y_i for $1 \leq i \leq n$, a probability density function (or probability mass function) of a generalized linear model has the following form of the exponential family

$$p(y_i|\theta) = \exp\{a(\theta)y_i + b(\theta) + c(y_i)\},$$

in which $a(\cdot)$ is a continuously differentiable function with respect to θ with nonzero derivative, $b(\cdot)$ is also a continuously differentiable function of θ, $c(\cdot)$ is some constant function of y, and θ is also known as the natural parameter that relates the response to the predictors through the linear function $\theta_i = \theta_i(\beta) = x_i^\top \beta$. The mean function is $\mu = E(y_i|x_i) = -b'(\theta_i)/a'(\theta_i) \triangleq \phi(\theta_i)$, where $\phi(\cdot)$ is the inverse of some chosen link function.

The class of pMOM densities specified in (2) can still be used for model selection in this generalized setting by noting that the log likelihood function in (5) and (6) now takes the general form of

$$L_n(\beta_k) = \sum_{i=1}^{n} \{a(\theta_i(\beta_k))y_i + b(\theta_i(\beta_k)) + c(y_i)\}. \qquad (8)$$

After obtaining the posterior probabilities in (4) with the log likelihood substituted as (8), we can select a model by computing the posterior mode. In Section 4, we will adopt some search algorithm that use these posterior probabilities to target the mode in a more efficient way.

3. Main Results

In this section, we show that the proposed Bayesian model enjoys desirable theoretical properties. Let $t \subseteq [p]$ be the true model, which means that the nonzero locations of the true coefficient vector are $t = (j, j \in t)$. We consider t to be a fixed vector. Let $\beta_0 \in \mathbb{R}^p$ be the true coefficient vector and $\beta_{0,t} \in \mathbb{R}^{|t|}$ be the vector of the true nonzero coefficients. In the following analysis, we will focus on logistic regression, but our argument can be easily extended to any other GLM as well. In particular,

$$H_n(\beta_k) = -\frac{\partial^2 L_n(\beta_k)}{\partial \beta_k \partial \beta_k^\top} = \sum_{i=1}^{n} \sigma_i^2(\beta_k) x_i x_i^\top = X_k^\top \Sigma(\beta_k) X_k$$

as the negative Hessian of $L_n(\beta_k)$, where $\Sigma(\beta_k) \equiv \Sigma_k = diag(\sigma_1^2(\beta_k), \ldots, \sigma_n^2(\beta_k))$, $\sigma_i^2(\beta_k) = \mu_i(\beta_k)(1 - \mu_i(\beta_k))$ and

$$\mu_i(\beta_k) = \frac{\exp(x_{ik}^\top \beta_k)}{1 + \exp(x_{ik}^\top \beta_k)}.$$

In the rest of the paper, we denote $\Sigma = \Sigma(\beta_t)$ and $\sigma_i^2 = \sigma_i^2(\beta_t)$ for simplicity.

Before we establish our main results, the following notations are needed for stating our assumptions. For any $a, b \in \mathbb{R}$, $a \vee b$ and $a \wedge b$ mean the maximum and minimum of a and b, respectively. For any sequences a_n and b_n, we denote $a_n \lesssim b_n$, or equivalently $a_n = O(b_n)$, if there exists a constant $C > 0$ such that $|a_n| \leq C|b_n|$ for all large n. We denote $a_n \ll b_n$, or equivalently $a_n = o(b_n)$, if $a_n/b_n \longrightarrow 0$ as $n \to \infty$. Without loss of generality, if $a_n \geq b_n > 0$ and there exist constants $C_1 > C_2 > 0$ such that $C_2 < b_n/a_n \leq a_n/b_n < C_1$, we denote $a_n \sim b_n$. For a given vector $v = (v_1, \ldots, v_p)^\top \in \mathbb{R}^p$, the vector ℓ_2-norm is denoted as $\|v\|_2 = (\sum_{j=1}^p v_j^2)^{1/2}$. For any real symmetric matrix A, $\lambda_{\max}(A)$ and $\lambda_{\min}(A)$ are the maximum and minimum eigenvalue of A, respectively. To attain desirable asymptotic properties of our posterior, we assume the following conditions:

Condition (A1) $\log n \lesssim \log p = o(n^{1/2})$ and $m_n = O\left((n/\log p)^{\frac{1-d'}{2}} \wedge \log p\right)$ for some $0 \leq d < (1+d)/2 \leq d' \leq 1$.

Condition (A1) ensures our proposed method can accommodate high dimensions where the number of predictors grows at a sub-exponential rate of n. Condition (A1) also specifies the parameter m_n in the uniform prior (3) that restricts our analysis on a set of *reasonably large* models. Similar assumptions restricting the model size have been commonly assumed in the sparse estimation literature [4,5,9,17].

Condition (A2) For some constant $C \in (0, \infty)$ and $0 \leq d < (1+d)/2 \leq d' \leq 1$,

$$\max_{i,j} |x_{ij}| \leq C,$$

$$0 < \lambda \leq \min_{k:|k| \leq m_n + |t|} \lambda_{\min}\left(n^{-1} H_n(\beta_{0,k})\right) \leq \Lambda_{m_n + |t|} \leq C^2 (\log p)^d,$$

and $\Lambda_\zeta = \max_{k:|k| \leq \zeta} \lambda_{\max}(n^{-1} X_k^\top X_k)$ for any integer $\zeta > 0$. Furthermore, $\|\beta_{0,t}\|_2^2 = O((\log p)^d)$.

Condition (A2) gives lower and upper bounds of $n^{-1} H_n(\beta_{0,k})$ and $n^{-1} X_k^\top X_k$, respectively, where k is a large model satisfying $|k| \leq m_n + |t|$. The lower bound condition can be regarded as a restricted eigenvalue condition for ℓ_0-sparse vectors. Restricted eigenvalue conditions are routinely assumed in high-dimensional theory to guarantee some level of curvature of the objective function and are satisfied with high probability for sub-Gaussian design matrices [5]. Similar conditions have also been used in the linear regression literature [18–20]. The last assumption in Condition (A2) says that the magnitude of true signals is bounded above $(\log p)^d$ up to some constant, which allows the magnitude of signals to increase to infinity.

Condition (A3) For some constant $c_0 > 0$,

$$\min_{j \in t} \beta_{0,j}^2 \geq c_0 \left(\frac{|t| \Lambda_{|t|} \log p}{n} \vee \frac{1}{\log p} \right). \tag{9}$$

Condition (A3) gives a lower bound for nonzero signals, which is called the *beta-min* condition. In general, this type of condition is necessary for catching every nonzero signal. Please note that due to Conditions (A1) and (A2), the right-hand side of (9) decreases to zero as $n \to \infty$. Thus, it allows the smallest nonzero coefficients to tend to zero as we observe more data.

Condition (A4) For some small constant $\delta > 0$, the hyperparameters τ and r satisfy

$$\tau^{r+1/2} \sim n^{-1/2} p^{2+\delta}.$$

Condition (A4) suggests appropriate conditions for the hyperparameter τ in (2). A similar assumption has also been considered in [9]. The scale parameter τ in the nonlocal prior density reflects the dispersion of the nonlocal prior density around zero, and implicitly determines the size of the regression coefficients that will be shrunk to zero [8,9]. For the below theoretical results, we assume that $U = I$ for simplicity, but our results are still valid for other choices of U as long as $\lambda_{\max}(U) = O(1)$ and $\lambda_{\min}(U) = O(1)$.

Theorem 1 (No super set). *Under conditions (A1), (A2) and (A4),*

$$\pi(k \supsetneq t \mid y) \xrightarrow{P} 0, \text{ as } n \to \infty.$$

Theorem 1 says that, asymptotically, our posterior will not overfit the model, i.e., not include unnecessarily many variables. Of course, it does not guarantee that the posterior will concentrate on the true model. To capture every significant variable, we require the magnitudes of nonzero entries in $\beta_{0,t}$ not to be too small. Theorem 2 shows that with an appropriate lower bound specified in Condition (A3), the true model t will be the mode of the posterior.

Theorem 2 (Posterior ratio consistency). *Under conditions (A1)–(A4) with $c_0 = \{(1-\epsilon_0)\lambda\}^{-1}\{2(3+\delta) + 5\{(1-\epsilon_0)\lambda\}^{-1}\}$ for some small constant $\epsilon_0 > 0$,*

$$\max_{k \neq t} \frac{\pi(k \mid y)}{\pi(t \mid y)} \xrightarrow{P} 0, \text{ as } n \to \infty.$$

Posterior ratio consistency is a useful property especially when we are interested in the point estimation with the posterior mode, but does not provide how large probability the posterior puts on the true model. In the following theorem, we state that our posterior achieves *strong selection consistency*. By strong selection consistency, we mean that the posterior probability assigned to the true model t converges to 1. Since strong selection consistency implies posterior ratio consistency, it requires a slightly stronger condition on the lower bound for the magnitudes of nonzero entries in $\beta_{0,t}$, i.e., a larger value of c_0, compared to that in Theorem 2.

Theorem 3 (Strong selection consistency). *Under conditions (A1)–(A4) with $c_0 = \{(1-\epsilon_0)\lambda\}^{-1}\{2(9+2\delta) + 5\{(1-\epsilon_0)\lambda\}^{-1}\}$ for some small constant $\epsilon_0 > 0$, the following holds:*

$$\pi(t \mid y) \xrightarrow{P} 1, \text{ as } n \to \infty.$$

4. Computational Strategy

In this section, we describe how to approximate the marginal density of the data and to conduct the model selection procedure. The integral formulation in (4) leads to the posterior probabilities not available in closed form. Hence, we use Laplace approximation to compute $m_k(y)$ and $\pi(k|y)$. A similar approach to compute posterior probabilities has been used in [8–10].

Please note that for any model k, when $U_k = I_k$, the normalization constant d_k in (2) is given by $d_k = \{(2r-1)!!\}^{-|k|}$. Let

$$\begin{aligned}
f(\beta_k) &= \log\left(\exp\{L_n(\beta_k)\}\pi(\beta_k \mid k)\right) \\
&= \sum_{i=1}^n \left\{y_i x_{ik}^\top \beta_k - \log(1 + \exp(x_{ik}^\top \beta_k))\right\} - |k|\log((2r-1)!!) - \frac{|k|}{2}\log(2\pi) - \left(r|k| + \frac{|k|}{2}\right)\log\tau \\
&\quad - \frac{\beta_k^\top \beta_k}{2\tau} + \sum_{i=1}^{|k|} 2r\log(|\beta_{k_i}|).
\end{aligned}$$

For any model k, the Laplace approximation of $m_k(y)$ is given by

$$(2\pi)^{\frac{|k|}{2}} \exp\{f(\hat{\beta}_k)\}|V(\hat{\beta}_k)|^{-\frac{1}{2}}, \tag{10}$$

where $\hat{\beta}_k = \arg\max_{\beta_k} f(\beta_k)$ obtained via the optimization function optim in R using a quasi-Newton method and $V(\hat{\beta}_k)$ is a $|k| \times |k|$ symmetric matrix which can be calculated as:

$$-\sum_{i=1}^{n} \frac{x_{ik} x_{ik}^\top \exp(x_{ik}^\top \beta_k)}{\{1 + \exp(x_{ik}^\top \beta_k)\}^2} - \frac{1}{\tau} I_k - \text{diag}\left(\frac{2r}{\beta_{k_1}^2}, \ldots, \frac{2r}{\beta_{k_{|k|}}^2}\right).$$

The above Laplace approximation can be used to compute the log of the posterior probability ratio between any given model k and true model t, and select a model k with the highest probability.

We then adopt the shotgun stochastic search (SSS) algorithm [9,12] to efficiently navigate through the massive model space and identify the global maxima. Using the Laplace approximations of the marginal probabilities in (10), the SSS method aims at exploring "interesting" regions of the resulting high-dimensional model spaces and quickly identifies regions of high posterior probability over models. Let $\text{nbd}(k) = \{\Gamma^+, \Gamma^-, \Gamma^0\}$ containing all the neighbors of model k, in which $\Gamma^+ = \{k \cup \{j\} : j \notin k\}$, $\Gamma^- = \{k \setminus \{j\} : j \in k\}$ and $\Gamma^0 = \{k \setminus \{j\} \cup \{l\} : j \in k, l \notin k\}$. The SSS procedure is described in Algorithm 1.

Algorithm 1 Shotgun Stochastic Search (SSS)

Choose an initial model $k^{(1)}$
for $i = 1$ to $i = N - 1$ do
    Compute $\pi(k|y)$ for all $k \in \text{nbd}(k^{(i)})$
    Sample k^+, k^- and k^0, from Γ^+, Γ^- and Γ^0 with probabilities proportional to $\pi(k|y)$
    Sample the next model $k^{(i+1)}$ from $\{k^+, k^-, k^0\}$ with probability proportional to
    $\{\pi(k^+|y), \pi(k^-|y), \pi(k^0|y)\}$
end for

---

5. Simulation Studies

In this section, we demonstrate the performance of the proposed method in various settings. Let X be the design matrix whose first $|t|$ columns correspond to the active covariates for which we have nonzero coefficients, while the rest correspond to the inactive ones with zero coefficients. In all the simulation settings, we generate $x_i \stackrel{i.i.d.}{\sim} N_p(0, \Sigma)$ for $i = 1, \ldots, n$ under the following two different cases of Σ:

- Case 1: Isotropic design, where $\Sigma = I_p$, i.e., no correlation imposed between different covariates.
- Case 2: Autoregressive correlated design, where $\Sigma_{ij} = 0.3^{|i-j|}$, for all $1 \leq i \leq j \leq p$. The correlations among different covariates are set to different values.

Following the simulation settings in [9,10], we consider the following two designs, each with the same sample size $n = 100$ and number of predictors being either $p = 100$ or 150:

- Design 1 (Dense model): The number of predictors $p = 100$ and $|t| = 8$.
- Design 2 (High-dimensional): The number of predictors $p = 150$ and $|t| = 4$.

We investigate the following two settings for the true coefficient vector $\beta_{0,t}$ to include different combinations of small and large signals.

- Setting 1: All the entries of $\beta_{0,t}$ are set to 3.
- Setting 2: All the entries of $\beta_{0,t}$ are generated from Unif(1.5, 3).

Finally, for given X and $1 \leq i \leq n$, we sample y_i from the following logistic model as in (1)

$$\mathbb{P}(y_i = 1 \mid x_i, \beta_0) = \frac{\exp(x_i^\top \beta_0)}{1 + \exp(x_i^\top \beta_0)}.$$

We will refer to our proposed method as "nonlocal" and its performance will then be compared with other existing methods including Spike and Slab prior-based model selection [21], empirical Bayesian LASSO (EBLasso) [22], Lasso [23] and SCAD [24]. The tuning parameters in the regularization approaches are chosen by 5-fold cross-validation. Spike and slab prior method is implemented via the BoomSpikeSlab package in R. For the nonlocal prior, the hyperparameters are set at $U = I, r = 1$ and we tune $\tau = 10^{-i}n^{-1/2}p^{2+0.05}$ for four different values of $i = 0, 1, 2, 3$. We choose the optimal τ by the mean squared prediction error through 5-fold cross-validation. Please note that this implies that τ is data-dependent and the resulting procedure is similar to an empirical-Bayesian approach in the high-dimensional Bayesian literature given the prior knowledge about the sparse true model [13]. For the SSS procedure, the initial model was set by randomly taking three coefficients to be active and the remaining to be inactive. The detailed steps for our method are coded in R and publicly available at https://github.com/xuan-cao/Nonlocal-Logistic-Selection. In particular, the stochastic search is implemented via the SSS function in the R package BayesS5.

To evaluate the performance of variable selection, the precision, sensitivity, specificity, Matthews correlation coefficient (MCC) [25] and mean-squared prediction error (MSPE) are reported at Tables 1–4, where each simulation setting is repeated for 20 times. The criteria are defined as

$$\text{Precision} = \frac{TP}{TP+FP}, \quad \text{Sensitivitiy} = \frac{TP}{TP+FN}, \quad \text{Specificity} = \frac{TN}{TN+FP},$$

$$\text{MCC} = \frac{TP \times TN - FP \times FN}{\sqrt{(TP+FP)(TP+FN)(TN+FP)(TN+FN)}}, \quad \text{MSPE} = \frac{1}{n_{\text{test}}}\sum_{i=1}^{n_{\text{test}}}(\hat{y}_i - y_{\text{test},i})^2,$$

where TP, TN, FP and FN are true positive, true negative, false positive and false negative, respectively. Here we denote $\hat{y}_i = x_i^\top \hat{\beta}$, where $\hat{\beta}$ is the estimated coefficient based on each method. For Bayesian methods, the usual GLM estimates based on the selected support are used as $\hat{\beta}$. We generated test samples $y_{\text{test},1}, \ldots, y_{\text{test},n_{\text{test}}}$ with $n_{\text{test}} = 50$ to calculate the MSPE.

Table 1. The summary statistics for Design 1 (Dense model design) are represented for each setting of the true regression coefficients under the first isotropic covariance case. Different setting means different choice of the true coefficient β_0.

	Setting 1				
	Precision	Sensitivity	Specificity	MCC	MSPE
Nonlocal	1	1	1	1	0.02
Spike and Slab	1	0.38	1	0.60	0.21
Lasso	0.67	1	0.96	0.80	0.17
EBLasso	1	0.38	1	0.60	0.22
SCAD	0.57	1	0.93	0.73	0.14
	Setting 2				
	Precision	Sensitivity	Specificity	MCC	MSPE
Nonlocal	0.73	1	0.97	0.84	0.18
Spike and Slab	1	0.13	1	0.34	0.23
Lasso	0.54	0.88	0.93	0.65	0.15
EBLasso	1	0.63	1	0.78	0.22
SCAD	0.47	0.88	0.91	0.60	0.13

Table 2. The summary statistics for Design 1 (Dense model design) are represented for each setting of the true regression coefficients under the second autoregressive covariance case. Different setting means different choice of the true coefficient β_0.

	Setting 1				
	Precision	Sensitivity	Specificity	MCC	MSPE
Nonlocal	0.89	1	0.99	0.94	0.13
Spike and Slab	0.71	0.63	0.98	0.64	0.20
Lasso	0.70	0.88	0.98	0.76	0.16
EBLasso	1	0.50	1	0.69	0.23
SCAD	0.67	0.75	0.97	0.68	0.17
	Setting 2				
	Precision	Sensitivity	Specificity	MCC	MSPE
Nonlocal	0.88	0.88	0.99	0.86	0.14
Spike and Slab	0.83	0.63	0.99	0.70	0.13
Lasso	0.63	0.88	0.96	0.72	0.14
EBLasso	1	0.38	1	0.60	0.22
SCAD	0.47	0.88	0.91	0.60	0.13

Table 3. The summary statistics for Design 2 (High-dimensional design) are represented for each setting of the true regression coefficients under the first isotropic covariance case. Different setting means different choice of the true coefficient β_0.

	Setting 1				
	Precision	Sensitivity	Specificity	MCC	MSPE
Nonlocal	1	1	1	1	0.08
Spike and Slab	0.75	0.75	0.99	0.74	0.09
Lasso	0.80	1	0.99	0.89	0.14
EBLasso	1	0.75	1	0.86	0.21
SCAD	0.67	1	0.99	0.81	0.12
	Setting 2				
	Precision	Sensitivity	Specificity	MCC	MSPE
Nonlocal	1	1	1	1	0.10
Spike and Slab	0.75	0.75	0.99	0.74	0.11
Lasso	0.67	1	0.99	0.81	0.14
EBLasso	1	0.75	1	0.86	0.23
SCAD	0.44	1	0.97	0.66	0.12

Table 4. The summary statistics for Design 2 (High-dimensional design) are represented for each setting of the true regression coefficients under the second autoregressive covariance case. Different setting means different choice of the true coefficient β_0.

	Setting 1				
	Precision	Sensitivity	Specificity	MCC	MSPE
Nonlocal	1	0.75	1	0.86	0.11
Spike and Slab	1	0.50	1	0.71	0.10
Lasso	0.57	1	0.98	0.75	0.10
EBLasso	1	0.50	1	0.70	0.18
SCAD	0.44	1	0.97	0.66	0.12
	Setting 2				
	Precision	Sensitivity	Specificity	MCC	MSPE
Nonlocal	1	0.75	1	0.86	0.15
Spike and Slab	0.50	0.50	0.99	0.49	0.14
Lasso	0.44	1	0.97	0.66	0.13
EBLasso	1	0.50	1	0.70	0.21
SCAD	0.40	1	0.96	0.62	0.14

Based on the above simulation results, we notice that under the first isotropic covariance case, the nonlocal-based approach overall works better than other methods especially in the strong signal setting (i.e., Setting 1), where our method is able to consistently achieve perfect estimation accuracy. This is because as signal strength gets stronger, the consistency conditions of our method are easier to satisfy which leads to better performance. When the covariance is autoregressive, our method suffers from lower sensitivity compared with the frequentist approaches in high-dimensional design (Table 4), but still has higher precision, specificity and MCC. The poor precision of the regularization methods has also been discussed in previous literature in the sense that selection of the regularization parameter using cross-validation is optimal with respect to prediction but tends to include too many noise predictors [26]. Again we observe under the autoregressive design, the performance of our method improves as the true signals strengthen. To sum up, the above simulation studies indicate that the proposed method can perform well under a variety of configurations with different data generation mechanisms.

6. Application to fMRI Data Analysis

In this section, we apply the proposed model selection method to an fMRI data set for identifying aberrant functional brain activities to aid the diagnosis of Parkinson's Disease (PD) [27]. Data consists of 70 PD patients and 50 healthy controls (HC). All the demographic characteristics and clinical symptom ratings have been collected before MRI scanning. In particular, we adopt the mini-mental state examination (MMSE) for cognitive evaluation and the Hamilton Depression Scale (HAMD) for measuring the severity of depression.

6.1. Image Feature Extraction

Functional imaging data for all subjects are collected and retrieved from the archive by neuroradiologists. Image preprocessing procedure is carried out via Statistical Parametric Mapping (SPM12) operated on the Matlab platform. For each subject, we first discard the first 5 time points for signal equilibrium and the remaining 135 images underwent slice-timing and head motion corrections. Four subjects with more than 2.5 mm maximum displacement in any of the three dimensions or 2.5° of any angular motion are removed. The functional images are spatially normalized to the Montreal Neurological Institute space with $3 \times 3 \times 3$ mm^3 cubic voxels and smoothed with a 4 mm full width at half maximum (FWHM) Gaussian kernel. We further regress out nuisance covariates and applied temporal filter (0.01 Hz $< f < 0.08$ Hz) to diminish high-frequency noise.

Zang et al. [28] proposed the method of Regional Homogeneity (ReHo) to analyze characteristics of regional brain activity and to reflect the temporal homogeneity of neural activity. Since some preprocessing methods especially spatial smoothing fMRI time series may significantly change the ReHo magnitudes [29], preprocessed fMRI data without the spatial smoothing step are used for calculating ReHo. In particular, we focus on the mReHo maps obtained by dividing the mean ReHo of the whole brain within each voxel in the ReHo map. We further segment the mReHo maps and extract all the 112 ROI signals based on the Harvard-Oxford atlas (HOA) using the Resting-State fMRI Data Analysis Toolkit.

Slow fluctuations in activity are fundamental features of the resting brain for determining correlated activity between brain regions and resting state networks. The relative magnitude of these fluctuations can discriminate between brain regions and subjects. Amplitude of Low Frequency Fluctuations (ALFF) [30] are related measures that quantify the amplitude of these low frequency oscillations. Leveraging the preprocessed data, we retain the standardized mALFF maps after dividing the ALFF of each voxel by the global mean ALFF. Using the HOA, we again obtain 112 mALFF values via extracting the ROI signals based on the mALFF maps. Voxel-Mirrored Homotopic Connectivity (VMHC) quantifies functional homotopy by providing a voxel-wise measure of connectivity between hemispheres [31]. By segmenting the VMHC maps according to HOA, we also extract 112 VHMC values.

6.2. Results

Our candidate features consist of 336 radiomic variables along with all the clinical characteristics. We now consider a standard logistic regression model with the binary disease indicator as the outcome and all the radiomic variables together with five clinical factors as predictors. Various models including the proposed and other competing methods will then be implemented for classifying subjects based on these extracted features. The dataset is randomly divided into a training set (80%) and a testing set (20%) while maintaining the PD:HC ratio in both sets. For Bayesian methods, we first obtain the identified variables, and then evaluate the testing set performance using standard GLM estimates based on the selected features. The penalty parameters in all frequentist methods are tuned via 5-fold cross validation in the training set. The hyperparameters for the proposed method are set as in simulation studies.

The HAMD score and nine radiomic features including five mALFFs, two ReHos, two VHMCs are selected by the SSS procedure under pMOM prior. In Figure 1, we plot the histograms of selected radiomic features with different colors representing different groups. The predictive performance of various methods in the test set is summarized in Table 5. We can tell from Table 5 that the nonlocal prior-based approach has overall better prediction performance compared with other methods. Our nonlocal approach has higher precision and specificity compared with all the other methods, but yields a lower sensitivity than the frequentist approaches. Based on the most comprehensive measure MCC, our method outperforms all the other methods.

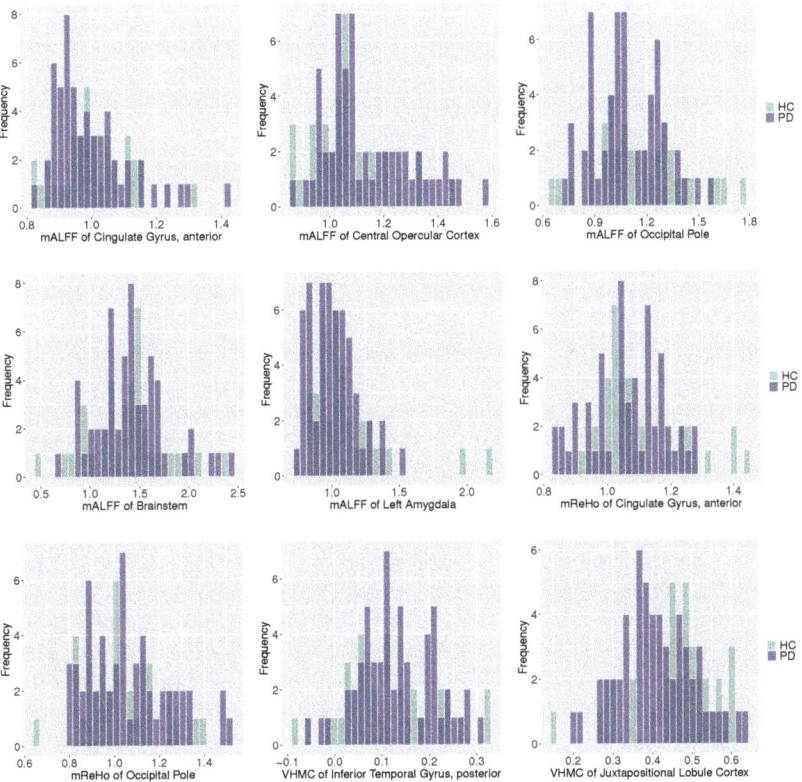

Figure 1. Histograms of selected radiomic features for PD and HC subjects with darker color representing overlapping values. Purple: PD group; Green: HC group.

Table 5. The summary statistics for prediction performance on the testing set for all methods.

	Precision	Sensitivity	Specificity	MCC	MSPE
Nonlocal	0.77	0.83	0.73	0.56	0.21
Spike and Slab	0.53	0.75	0.27	0.40	0.29
Lasso	0.67	1	0.45	0.55	0.18
EBLasso	0.57	1	0.18	0.32	0.28
SCAD	0.58	1	0.37	0.41	0.19

7. Conclusions

In this paper, we propose a Bayesian hierarchical model with a pMOM prior specification over regression coefficients to perform variable selection in high-dimensional generalized linear models. The model selection consistency of our method is established under mild conditions and the shotgun stochastic search algorithm can be used for the implementation of our proposed approach. Our simulation and real data studies indicate that the proposed method has better performance for variable selection compared to a variety of state-of-the-art competing methods. In the fMRI data analysis, our method is able to identify abnormal functional brain activities for PD that occur in the regions of interest including cingulate gyrus, central opercular cortex, occipital pole, brainstem, left amygdala, occipital pole, inferior temporal gyrus, and juxtapositional lobule cortex. These findings suggest disease-related alterations of functional activities that provide physicians sufficient information to get involved with early diagnosis and treatment. Our findings are also coherent with the alternative functional features in cortical regions, brainstem, and limbic regions discovered in previous studies [32–35].

Our fMRI study certainly has limitations. First, we would like to note that fMRI data are typically treated as spatio-temporal objects and a generalized linear model with spatially varying coefficients can be implemented for brain decoding [36]. However, in our application, for each subject, a total of 135 fMRI scans were obtained, each with the dimension of $64 \times 64 \times 31$. If we take each voxel as a covariate to perform the whole-brain functional analysis, it would be computationally challenging and impractical given the extremely high dimension. Hence, we adopt the radiomics approach to extract three different types of features that can summarize the functional activity of the brain, and take these radiomic features as covariates in our generalized linear model. For future studies, we will focus on several regions of interest rather than the entire brain and take the spatio-temporal dependency among voxels into consideration.

Second, although ReHo, ALFF, and VHMC are different types of radiomic features that quantify the functional activity of the brain, it is definitely possible that in some regions, three measures are highly correlated with each other. Our current theoretical and computational strategy can accommodate a reasonable amount of correlations among covariates, but might not work in the presence of high correlation structure. For future studies, we will first carefully examine the potential correlations among features and might only retain one feature for each region if significant correlations are detected.

One possible extension of our methodology is to address the potential misspecification of the hyperparameter τ. The scale parameter τ is of particular importance in the sense that it can reflect the dispersion of the nonlocal density around zero, and implicitly determine the size of the regression coefficients that will be shrunk to zero [8]. Cao et al. [14] investigated the model selection consistency for the hyper-pMOM priors in linear regression setting, where an additional inverse-gamma prior is placed over τ. Wu et al. [11] proved the model selection consistency using hyper-pMOM prior in generalized linear models, but assumed a fixed dimension p. For future study, we will consider this fully Bayesian approach to carefully examine the theoretical and empirical properties for such hyper-pMOM prior in the context of high-dimensional generalized linear regression. We can also extend our method to develop a Bayesian approach for growth models in the context of non-linear regression [37], where the log-transformation is typically used to recover the additive structure.

However, then the model does not fall into the category of GLMs, which is beyond the current setting in this paper. Therefore, we leave it as a future work.

Author Contributions: Conceptualization, X.C.; Methodology, X.C. and K.L.; Software, X.C.; Supervision, K.L.; Validation, K.L.; Writing—original draft, X.C.; Writing—review & editing, K.L. Both authors have read and agreed to the published version of the manuscript.

Funding: This research was funded by Simons Foundation: No.635213; University of Cincinnati: Taft Summer Research Fellowship; National Research Foundation of Korea: No.2019R1F1A1059483; INHA UNIVERSITY Research Grant.

Acknowledgments: We would like to thank two referees for their valuable comments which have led to improvements of an earlier version of the paper. This research was supported by Simons Foundation's collaboration grant (No.635213), Taft Summer Research Fellowship at University of Cincinnati and the National Research Foundation of Korea (NRF) grant funded by the Korea government (MSIT) (No.2019R1F1A1059483).

Conflicts of Interest: The authors declare no conflict of interest.

Abbreviations

The following abbreviations are used in this manuscript:

Lasso	least absolute shrinkage and selection operator
SCAD	smoothly clipped absolute deviation
MCP	minimum concave penalty
pMOM	product moment
SSS	shotgun stochastic search
MCMC	Markov chain Monte Carlo
fMRI	functional Magnetic Resonance Imaging
GLM	generalized linear model
EBLasso	empirical Bayesian LASSO
MCC	Matthews correlation coefficient
MSPE	mean-squared prediction error
PD	Parkinson's Disease
HC	healthy controls
MMSE	mini-mental state examination
HAMD	Hamilton Depression Scale
SPM12	Statistical Parametric Mapping
FWHM	full width at half maximum
HOA	Harvard-Oxford atlas
ALFF	Amplitude of Low Frequency Fluctuations
VMHC	Voxel-Mirrored Homotopic Connectivity

Appendix A

Throughout the Supplementary Material, we assume that for any

$$u \in \{u \in \mathbb{R}^n : u \text{ is in the space spanned by the columns of } \Sigma^{1/2} X_k\}$$

and any model $k \in \{k \subseteq [r] : |k| \leq m_n + |t|\}$, there exists $\delta^* > 0$ such that

$$\mathbb{E}\left[\exp\left\{u^\top \Sigma^{-1/2}(y - \mu)\right\}\right] \leq \exp\left\{\frac{(1+\delta^*) u^\top u}{2}\right\}, \tag{A1}$$

for any $n \geq N(\delta^*)$. However, as stated in [5], there always exists $\delta^* > 0$ satisfying inequality (A1), so it is not really a restriction. Since we will focus on sufficiently large n, δ^* can be considered an arbitrarily small constant, so we can always assume that $\delta > \delta^*$.

Proof of Theorem 1. Let $M_1 = \{k : k \not\supseteq t, |k| \leq m_n\}$ and

$$PR(k,t) = \frac{\pi(k \mid y)}{\pi(t \mid y)},$$

where $t \subseteq [r]$ is the true model. We will show that

$$\sum_{k:k \in M_1} PR(k,t) \xrightarrow{P} 0 \quad \text{as } n \to \infty. \tag{A2}$$

By Taylor's expansion of $L_n(\beta_k)$ around $\widehat{\beta}_k$, which is the MLE of β_k under the model k, we have

$$L_n(\beta_k) = L_n(\widehat{\beta}_k) - \frac{1}{2}(\beta_k - \widehat{\beta}_k)^\top H_n(\widetilde{\beta}_k)(\beta_k - \widehat{\beta}_k)$$

for some $\widetilde{\beta}_k$ such that $\|\widetilde{\beta}_k - \widehat{\beta}_k\|_2 \leq \|\beta_k - \widehat{\beta}_k\|_2$. Furthermore, by Lemmas A.1 and A.3 in [5] and Condition (A2), with probability tending to 1,

$$L_n(\beta_k) - L_n(\widehat{\beta}_k) \leq -\frac{1-\epsilon}{2}(\beta_k - \widehat{\beta}_k)^\top H_n(\beta_{0,k})(\beta_k - \widehat{\beta}_k)$$

for any $k \in M_1$ and β_k such that $\|\beta_k - \beta_{0,k}\|_2 < c\sqrt{|k|\Lambda_{|k|}\log p/n} =: cw_n$, where $\epsilon = \epsilon_n := c'\sqrt{m_n^2 \Lambda_{m_n} \log p/n} = o(1)$, for some constants $c, c' > 0$. Please note that for β_k such that $\|\beta_k - \widehat{\beta}_k\|_2 = cw_n/2$,

$$L_n(\beta_k) - L_n(\widehat{\beta}_k) \leq -\frac{1-\epsilon}{2}\|\beta_k - \widehat{\beta}_k\|_2^2 \lambda_{\min}\{H_n(\beta_{0,k})\}$$

$$\leq -\frac{1-\epsilon}{2}\frac{c^2 w_n^2}{4} n\lambda = -\frac{1-\epsilon}{8}c^2 \lambda |k|\Lambda_{|k|}\log p \longrightarrow -\infty \quad \text{as } n \to \infty,$$

where the second inequality holds due to Condition (A2). It also holds for any β_k such that $\|\beta_k - \widehat{\beta}_k\|_2 > cw_n/2$ by concavity of $L_n(\cdot)$ and the fact that $\widehat{\beta}_k$ maximizes $L_n(\beta_k)$.

Define the set $B := \{\beta_k : \|\beta_k - \widehat{\beta}_k\|_2 \leq cw_n/2\}$, then we have $B \subset \{\beta_k : \|\beta_k - \beta_{0,k}\|_2 \leq cw_n\}$ for some large $c > 0$ and any $k \in M_1$, with probability tending to 1.

$$m_k(y) = \int \exp\{L_n(\beta_k)\} \pi(\beta_k \mid k) d\beta_k$$

$$= \int \exp\{L_n(\beta_k)\} d_k (2\pi)^{-|k|/2} (\tau)^{-r|k|-|k|/2} |U_k|^{\frac{1}{2}} \exp\left(-\frac{\beta_k^\top U_k \beta_k}{2\tau}\right) \prod_{i=1}^{|k|} \beta_{k_i}^{2r} d\beta_k$$

$$\leq d_k (2\pi)^{-|k|/2} (\tau)^{-r|k|-|k|/2} |U_k|^{\frac{1}{2}} \exp\{L_n(\widehat{\beta}_k)\} \tag{A3}$$

$$\times \left[\int_B \exp\left\{-\frac{1-\epsilon}{2}(\beta_k - \widehat{\beta}_k)^\top H_n(\beta_{0,k})(\beta_k - \widehat{\beta}_k) - \frac{\beta_k^\top U_k \beta_k}{2\tau}\right\} \prod_{i=1}^{|k|} \beta_{k_i}^{2r} d\beta_k\right.$$

$$\left. + \exp\left(-\frac{1-\epsilon}{8}c^2 \lambda |k|\Lambda_{|k|}\log p\right) \int_{B^c} \exp\left(-\frac{\beta_k^\top U_k \beta_k}{2\tau}\right) \prod_{i=1}^{|k|} \beta_{k_i}^{2r} d\beta_k\right]$$

Please note that for $A_k = (1-\epsilon)H_n(\beta_{0,k})$ and $\beta_k^* = (A_k + U_k/\tau)^{-1}A_k\widehat{\beta}_k$, we have

$$\int_B \exp\left\{-\frac{1-\epsilon}{2}(\beta_k - \widehat{\beta}_k)^\top H_n(\beta_{0,k})(\beta_k - \widehat{\beta}_k) - \frac{\beta_k^\top U_k \beta_k}{2\tau}\right\} \prod_{i=1}^{|k|} \beta_{k_i}^{2r} d\beta_k$$

$$\leq \int \exp\left\{-\frac{1}{2}(\beta_k - \beta_k^*)^\top (A_k + U_k/\tau)(\beta_k - \beta_k^*)\right\} \prod_{i=1}^{|k|} \beta_{k_i}^{2r} d\beta_k$$

$$\times \exp\left\{-\frac{1}{2}\widehat{\beta}_k^\top (A_k - A_k(A_k + U_k/\tau)^{-1}A_k)\widehat{\beta}_k\right\}$$

$$= (2\pi)^{|k|/2} \det(A_k + U_k/\tau)^{-1/2} \exp\left\{-\frac{1}{2}\widehat{\beta}_k^\top (A_k - A_k(A_k + U_k/\tau)^{-1}A_k)\widehat{\beta}_k\right\} E_k\left(\prod_{i=1}^{|k|} \beta_{k_i}^{2r}\right),$$

where $E_k(.)$ denotes the expectation with respect to a multivariate normal distribution with mean β_k^* and covariance matrix $(A_k + U_k/\tau)^{-1}$. It follows from Lemma 6 in the supplementary material for [8] that

$$E_k\left(\prod_{i=1}^{|k|} \beta_{k_i}^{2r}\right) \leq \left(\frac{n\Lambda_{|k|} + \tau^{-1}}{n\lambda + \tau^{-1}}\right)^{|k|/2} \left\{\frac{4V}{|k|} + \frac{4[(2r-1)!!]^{\frac{1}{r}}}{n(\lambda + \tau^{-2})}\right\}^{r|k|}$$

$$\leq \left(\frac{n\Lambda_{|k|} + \tau^{-1}}{n\lambda + \tau^{-1}}\right)^{|k|/2} 2^{r|k|-1} \left\{\left(\frac{4V}{|k|}\right)^{r|k|} + \left(\frac{4[(2r-1)!!]^{\frac{1}{r}}}{n(\lambda + \tau^{-1})}\right)^{r|k|}\right\},$$

where $V = \|\beta_k^*\|_2^2$, and

$$\int_{B^c} \exp\left\{-\frac{\beta_k^\top U_k \beta_k}{2\tau}\right\} \prod_{i=1}^{|k|} \beta_{k_i}^{2r} d\beta_k \leq (C\tau)^{|k|/2},$$

for some constant $C > 0$. Further note that

$$\det\left\{H_n(\beta_{0,k})(1-\epsilon) + \tau^{-1}I\right\}^{1/2} \leq (n\Lambda_{|k|} + \tau^{-1})^{|k|/2}$$

$$\leq \exp\{C|k|\log n\}$$

$$\ll \exp\left\{\frac{1-\epsilon}{8}c^2\lambda|k|\Lambda_{|k|}\log p\right\}$$

for some constant $C > 0$ and some large constant $c > 0$, by Conditions (A1), (A2) and (A4). Therefore, it follows from (A3) that

$$m_k(y) \leq C^{-|k|/2}(\tau)^{-r|k|-|k|/2} \exp\{L_n(\widehat{\beta}_k)\} \det\left\{H_n(\beta_{0,k})(1-\epsilon) + \tau^{-1}U_k\right\}^{-1/2}$$

$$\times \left[\Lambda_{|k|}^{|k|/2}\left\{\left(\frac{V}{|k|}\right)^{r|k|} + n^{-r|k|}\right\} + o(1)\right] \quad (A4)$$

$$\leq C^{-|k|/2}(\tau)^{-r|k|-|k|/2} \exp\{L_n(\widehat{\beta}_k)\} \det\left\{H_n(\beta_{0,k})(1-\epsilon) + \tau^{-1}U_k\right\}^{-1/2}$$

$$\times \Lambda_{|k|}^{|k|/2}\left\{\left(\frac{V}{|k|}\right)^{r|k|} + n^{-r|k|}\right\},$$

for some constant $C > 0$. Next, note that it follows from Lemma A.3 in the supplementary material for [5] that

$$\begin{aligned} V = \|\beta_k^*\|_2^2 \leq \|\widehat{\beta}_k\|_2^2 &\leq \left(\|\widehat{\beta}_k - \beta_{0,k}\|_2 + \|\beta_{0,k}\|_2\right)^2 \\ &\leq \left(\sqrt{\frac{|k|\Lambda_{|k|}\log p}{n}} + \sqrt{\log p}\right)^2 \\ &\leq 2\left(\frac{|k|\Lambda_{|k|}\log p}{n} + \log p\right). \end{aligned}$$

Therefore,

$$\begin{aligned} \left(\frac{V}{|k|}\right)^{r|k|} &\leq \left(\frac{2\log p}{|k|}\right)^{r|k|} \exp\left(r|k|^2 \frac{\Lambda_{|k|}}{n}\right) \\ &\lesssim \left(\frac{2\log p}{|k|}\right)^{r|k|} \end{aligned}$$

by Conditions (A1) and (A2). Combining with (A4), we obtain the following upper bound for $m_k(y)$,

$$\begin{aligned} m_k(y) &\leq (C_1\tau)^{-r|k|-|k|/2} \exp\left\{L_n(\widehat{\beta}_k)\right\} \det\left\{H_n(\beta_{0,k})(1-\epsilon) + \tau^{-1}U_k\right\}^{-1/2} \\ &\quad \times \Lambda_{|k|}^{|k|/2} \left(\frac{\log p}{|k|}\right)^{r|k|}, \end{aligned} \tag{A5}$$

for any $k \in M_1$ and some constant $C_1 > 0$. Similarly, by Lemma 4 in the supplementary material for [8] and the similar arguments leading up to (A5), with probability tending to 1, we have

$$\begin{aligned} m_t(y) &\gtrsim (C_1\tau)^{-r|t|-|t|/2} \exp\left\{L_n(\widehat{\beta}_t)\right\} \det\left\{H_n(\beta_{0,t})(1+\epsilon) + \tau^{-1}U_t\right\}^{-1/2} \\ &\quad \times \exp\left\{-\frac{1}{2}\widehat{\beta}_t^\top \left(A_t - A_t(A_t + \tau^{-2}I_t)^{-1}A_t\right)\widehat{\beta}_t\right\}(\log p)^{-r|t|} \\ &\gtrsim (C_1\tau)^{-r|t|-|t|/2} \exp\left\{L_n(\widehat{\beta}_t)\right\} \det\left\{H_n(\beta_{0,t})(1+\epsilon) + \tau^{-1}U_t\right\}^{-1/2} (\log p)^{-r|t|} \end{aligned}$$

by Lemma A1, where $A_t = (1+\epsilon)H_n(\beta_{0,t})$. Therefore, with probability tending to 1,

$$\begin{aligned} \frac{m_k(y)}{m_t(y)} &\lesssim \left\{C_1 n^{1/2}\tau^{r+1/2}\right\}^{-(|k|-|t|)} \frac{\det\left\{n^{-1}H_n(\beta_{0,t})(1+\epsilon) + (n\tau)^{-1}U_t\right\}^{1/2}}{\det\left\{n^{-1}H_n(\beta_{0,k})(1-\epsilon) + (n\tau)^{-1}U_k\right\}^{1/2}} \\ &\quad \times \exp\left\{L_n(\widehat{\beta}_k) - L_n(\widehat{\beta}_t)\right\} \Lambda_{|k|}^{\frac{|k|}{2}} \left(\frac{\log p}{|k|}\right)^{r|k|} (\log p)^{r|t|} \\ &\lesssim \left\{C_1 p^{2+\delta}\right\}^{-(|k|-|t|)} \left(\frac{2}{\lambda}\right)^{|k|-|t|} \exp\left\{L_n(\widehat{\beta}_k) - L_n(\widehat{\beta}_t)\right\}(\log p)^{r(2|k|+|t|)} \\ &\lesssim (C_1 p)^{-(2+\delta)(|k|-|t|)} \left(\frac{2}{\lambda}\right)^{|k|-|t|} p^{(1+\delta^*)(1+2w)(|k|-|t|)}(\log p)^{r(2|k|+|t|)} \end{aligned} \tag{A6}$$

for any $k \in M_1$, where the second inequality holds by Lemma 2 in [38], Conditions (A2) and (A4), and the third inequality follows from Lemma 3 in [38], which implies

$$L_n(\widehat{\beta}_k) - L_n(\widehat{\beta}_t) \leq b_n(|k| - |t|) \tag{A7}$$

for any $k \in M_1$ with probability tending to 1, where $b_n = (1+\delta^*)(1+2w)\log p$ with some small constant $w > 0$ satisfying $1 + \delta > (1+\delta^*)(1+2w)$.

Hence, with probability tending to 1, it follows from (A6) that

$$\sum_{k \in M_1} PR(k,t) = \sum_{k \supsetneq t} \frac{\pi(k) m_k(y)}{\pi(t) m_t(y)} \leq \sum_{k \supsetneq t} \frac{m_k(y)}{m_t(y)}$$

$$\leq \sum_{|k|-|t|=1}^{m_n-|t|} \binom{p-|t|}{|k|-|t|} p^{-(1+c)(|k|-|t|)}. \tag{A8}$$

for some constant $c > 0$. Using $\binom{p-|t|}{|k|-|t|} \leq p^{|k|-|t|}$ and (A8), we get

$$\sum_{k \in M_1} PR(k,t) = o(1).$$

Thus, we have proved the desired result (A2). □

Proof of Theorem 2. Let $M_2 = \{k : k \not\supseteq t, |k| \leq m_n\}$. For any $k \in M_2$, let $k^* = k \cup t$, so that $k^* \in M_1$. Let β_{k^*} be the $|k^*|$-dimensional vector including β_k for k and zeros for $t \setminus k$. Then by Taylor's expansion and Lemmas A.1 and A.3 in [5], with probability tending to 1,

$$\begin{aligned} L_n(\beta_{k^*}) &= L_n(\widehat{\beta}_{k^*}) - \frac{1}{2}(\beta_{k^*} - \widehat{\beta}_{k^*})^\top H_n(\tilde{\beta}_{k^*})(\beta_{k^*} - \widehat{\beta}_{k^*}) \\ &\leq L_n(\widehat{\beta}_{k^*}) - \frac{1-\epsilon}{2}(\beta_{k^*} - \widehat{\beta}_{k^*})^\top H_n(\beta_{0,k^*})(\beta_{k^*} - \widehat{\beta}_{k^*}) \\ &\leq L_n(\widehat{\beta}_{k^*}) - \frac{n(1-\epsilon)\lambda}{2}\|\beta_{k^*} - \widehat{\beta}_{k^*}\|_2^2 \end{aligned}$$

for any β_{k^*} such that $\|\beta_{k^*} - \beta_{0,k^*}\|_2 \leq c\sqrt{|k^*|\Lambda_{|k^*|}\log p/n} = cw_n$ for some large constant $c > 0$. Please note that

Let $B_k = n(1-\epsilon)\lambda I_k$ and $\beta_k^* = (B_k + U_k/\tau)^{-1} B_k \widehat{\beta}_k$,

$$\int \exp\left\{-\frac{n(1-\epsilon)\lambda}{2}\|\beta_{k^*} - \widehat{\beta}_{k^*}\|_2^2\right\} \exp\left(-\frac{\beta_k^\top U_k \beta_k}{2\tau}\right) \prod_{i=1}^{|k|} \beta_{k_i}^{2r} d\beta_k$$

$$= \int \exp\left\{-\frac{n(1-\epsilon)\lambda}{2}\|\beta_k - \widehat{\beta}_k\|_2^2 - \frac{1}{2\tau}\beta_k^\top U_k \beta_k\right\} \prod_{i=1}^{|k|} \beta_{k_i}^{2r} d\beta_k \exp\left\{-\frac{n(1-\epsilon)\lambda}{2}\|\widehat{\beta}_{t\setminus k}\|_2^2\right\}$$

$$= (2\pi)^{\frac{|k|}{2}} |B_k + U_k/\tau|^{-1/2} \exp\left\{-\frac{1}{2}\widehat{\beta}_k^\top (B_k - B_k(B_k + U_k/\tau)^{-1}B_k)\widehat{\beta}_k\right\} E_k\left(\prod_{i=1}^{|k|} \beta_{k_i}^{2r}\right)$$

$$\times \exp\left\{-\frac{n(1-\epsilon)\lambda}{2}\|\widehat{\beta}_{t\setminus k}\|_2^2\right\}.$$

where $E_k(.)$ denotes the expectation with respect to a multivariate normal distribution with mean β_k^* and covariance matrix $(B_k + U_k/\tau)^{-1}$. It follows from Lemma 6 in the supplementary material for [8] that

$$\begin{aligned} E_k\left(\prod_{i=1}^{|k|} \beta_{k_i}^{2r}\right) &\leq \left(\frac{n\lambda + \tau^{-1}}{n\lambda + \tau^{-1}}\right)^{|k|/2} \left\{\frac{4V}{|k|} + \frac{4[(2r-1)!!]^{\frac{1}{r}}}{n(\lambda + \tau^{-1})}\right\}^{r|k|} \\ &\leq \left(\frac{n\lambda + \tau^{-1}}{n\lambda + \tau^{-1}}\right)^{|k|/2} 2^{r|k|-1} \left\{\left(\frac{4V}{|k|}\right)^{r|k|} + \left(\frac{4[(2r-1)!!]^{\frac{1}{r}}}{n(\lambda + \tau^{-1})}\right)^{r|k|}\right\}, \end{aligned}$$

where $V = \|\beta_k^*\|_2^2$. Define the set $B_* := \{\beta_k : \|\beta_{k^*} - \widehat{\beta}_{k^*}\|_2 \leq cw_n/2\}$, for some large constant $c > 0$, then by similar arguments used for super sets, with probability tending to 1,

$$\begin{aligned}
\pi(k \mid y) &= d_k (2\pi)^{-|k|/2} (\tau)^{-r|k|-|k|/2} |U_k|^{\frac{1}{2}} \int_{B_* \cup B_*^c} \exp\{L_n(\beta_{G_{k^*}})\} \exp\left(-\frac{\beta_k^\top U_k \beta_k}{2\tau}\right) \prod_{i=1}^{|k|} \beta_{k_i}^{2r} d\beta_k \\
&\lesssim (C_1 \tau)^{-r|k|-|k|/2} \exp\{L_n(\widehat{\beta}_{k^*})\} \det(B_k + U_k/\tau)^{-1/2} \\
&\quad \times \left[\exp\left\{-\frac{n(1-\epsilon)\lambda}{2}\|\widehat{\beta}_{t\setminus k}\|_2^2\right\}\left(\frac{\log p}{|k|}\right)^{r|k|} + \exp\{-cC|k^*|\Lambda_{|k^*|}\log p\}\right]
\end{aligned}$$

for any $k \in M_2$ and for some constant $C > 0$.

Since the lower bound for $\pi(t \mid y)$ can be derived as before, it leads to

$$\begin{aligned}
PR(k,t) &\lesssim \{C_1 n^{1/2} \tau^{r+1/2}\}^{-(|k|-|t|)} \frac{\det\{(1+\epsilon)n^{-1} H_n(\beta_{0,t}) + (n\tau)^{-1} U_t\}^{1/2}}{\det\{(1-\epsilon)\lambda I_k + (n\tau)^{-1} U_k\}^{1/2}} \\
&\quad \times \exp\{L_n(\widehat{\beta}_{k^*}) - L_n(\widehat{\beta}_t)\} \exp\left\{-\frac{n(1-\epsilon)\lambda}{2}\|\widehat{\beta}_{t\setminus k}\|_2^2\right\}\left(\frac{\log p}{|k|}\right)^{r|k|} (\log p)^{r|t|} \quad (A9) \\
&+ \{C_1 n^{1/2} \tau^{r+1/2}\}^{-(|k|-|t|)} \det\{(1+\epsilon)n^{-1} H_n(\beta_{0,t}) + (n\tau)^{-1} U_t\}^{1/2} \\
&\quad \times \exp\{L_n(\widehat{\beta}_{k^*}) - L_n(\widehat{\beta}_t)\} \exp\{-cC|k^*|\Lambda_{|k^*|}\log p\}(\log p)^{r|t|} \quad (A10)
\end{aligned}$$

for any $k \in M_2$ with probability tending to 1.

We first focus on (A9). Please note that

$$\begin{aligned}
&\frac{\det\{(1+\epsilon)n^{-1} H_n(\beta_{0,t}) + (n\tau)^{-1} U_t\}^{1/2}}{\det\{(1-\epsilon)\lambda I_k + (n\tau)^{-1} U_k\}^{1/2}} \\
&\leq \frac{\{(1+\epsilon)\Lambda_{|t|} + (n\tau)^{-1}\}^{|t|/2}}{\{(1-\epsilon)\lambda + (n\tau)^{-1}\}^{|k|/2}} \\
&= \left\{\frac{(1+\epsilon)\Lambda_{|t|} + (n\tau)^{-1}}{(1-\epsilon)\lambda + (n\tau)^{-1}}\right\}^{|t|/2} \left\{\frac{1}{(1-\epsilon)\lambda + (n\tau)^{-1}}\right\}^{(|k|-|t|)/2} \\
&\lesssim \exp\{C|t|\log \Lambda_{|t|}\}\left\{\frac{1}{(1-\epsilon)\lambda + (n\tau)^{-1}}\right\}^{(|k|-|t|)/2}
\end{aligned}$$

for some constant $C > 0$. Furthermore, by the same arguments used in (A7), we have

$$\begin{aligned}
L_n(\widehat{\beta}_{k^*}) - L_n(\widehat{\beta}_t) &\lesssim C_*(|k^*| - |t|)\log p \\
&= C_*|t \setminus k|\log p + C_*(|k| - |t|)\log p
\end{aligned}$$

for some constant $C_* > 0$ and for any $k \in M_2$ with probability tending to 1. Here we choose $C_* = (1+\delta^*)(1+2w)$ if $|k| > |t|$ or $C_* = 3+\delta$ if $|k| \leq |t|$ so that

$$\begin{aligned}
&\{C_1 n^{1/2} \tau^{r+1/2}\}^{-(|k|-|t|)} \left\{\frac{1}{(1-\epsilon)\lambda + (n\tau)^{-1}}\right\}^{(|k|-|t|)/2} p^{C_*(|k|-|t|)} \\
&\quad \times \exp\left\{r|k|\log\left(\frac{\log p}{|k|}\right) + r|t|\log(\log p)\right\} \\
&\lesssim p^{(C_*-2-\delta)(|k|-|t|)} = o(1),
\end{aligned}$$

where the inequality holds by Condition (A4). To be more specific, we divide M_2 into two disjoint sets $M_2' = \{k : k \in M_2, |t| < |k| \leq m_n\}$ and $M_2^* = \{k : k \in M_2, |k| \leq |t|\}$, and will show that

$\sum_{k \in M_2'} PR(k,t) + \sum_{k \in M_2^*} PR(k,t) \longrightarrow 0$ as $n \to \infty$ with probability tending to 1. Thus, we can choose different C_* for M_2' and M_2^* as long as $C_* \geq (1+\delta^*)(1+2w)$. On the other hand, with probability tending to 1, by Condition (A3),

$$\exp\left\{-\frac{n(1-\epsilon)\lambda}{2}\|\widehat{\beta}_{t\setminus k}\|_2^2\right\} \leq \exp\left[-\frac{n(1-\epsilon)\lambda}{2}\left\{\|\beta_{0,t\setminus k}\|_2^2 - \|\widehat{\beta}_{t\setminus k} - \beta_{0,t\setminus k}\|_2^2\right\}\right]$$

$$\leq \exp\left[-\frac{n(1-\epsilon)\lambda}{2}\left\{|t\setminus k|^2 \min_{j \in t}\beta_{0,j}^2 - c'w_n'^2\right\}\right]$$

$$\leq \exp\left\{-\frac{(1-\epsilon)\lambda}{2}(c_0|t\setminus k|^2|t| - c'|t\setminus k|)\Lambda_{|t|}\log p\right\}$$

$$\leq \exp\left\{-\frac{(1-\epsilon)\lambda}{2}(c_0 - c')|t\setminus k|^2|t|\Lambda_{|t|}\log p\right\}$$

for any $k \in M_2$ and some large constants $c_0 > c' > 0$, where $w_n'^2 = |t\setminus k|\Lambda_{|t\setminus k|}\log p/n$. Here, $c' = 5\lambda^{-2}(1-\epsilon)^{-2}$ by the proof of Lemma A.3 in [5].

Hence, (A9) for any $k \in M_2$ is bounded above by

$$\exp\left\{|t|\log\Lambda_{|t|} + C_*|t\setminus k|\log p - \frac{(1-\epsilon)\lambda}{2}(c_0 - c')|t\setminus k|^2|t|\Lambda_{|t|}\log p\right\}$$

$$\lesssim \exp\left\{-\left(\frac{(1-\epsilon)\lambda}{2}(c_0 - c') - C_* - o(1)\right)|t\setminus k|^2|t|\Lambda_{|t|}\log p\right\}$$

$$\leq \exp\left\{-\left(\frac{(1-\epsilon)\lambda}{2}(c_0 - c') - C_* - o(1)\right)|t\setminus k|^2|t|\Lambda_{|t|}\log p\right\}$$

$$\leq \exp\left\{-\left(\frac{(1-\epsilon)\lambda}{2}(c_0 - c') - C_* - o(1)\right)|t|\Lambda_{|t|}\log p\right\}$$

with probability tending to 1, where the last term is of order $o(1)$ because we assume $c_0 = \frac{1}{(1-\epsilon_0)\lambda}\{2(3+\delta) + \frac{5}{(1-\epsilon_0)\lambda}\} > \frac{2}{(1-\epsilon)\lambda}(C_* + o(1)) + c'$ for some small $\epsilon_0 > 0$.

It is easy to see that the maximum (A10) over $k \in M_2$ is also of order $o(1)$ with probability tending to 1 by the similar arguments. Since we have (A2) in the proof of Theorem 1, it completes the proof. □

Proof of Theorem 3. Let $M_2 = \{k : k \not\supseteq t, |k| \leq m_n\}$. Since we have Theorem 1, it suffices to show that

$$\sum_{k:k \in M_2} PR(k,t) \xrightarrow{P} 0 \quad \text{as } n \to \infty. \tag{A11}$$

By the proof of Theorem 2, the summation of (A9) over $k \in M_2$ is bounded above by

$$\sum_{k \in M_2} p^{(C_* - 2 - \delta)(|k| - |t|)} \exp\left\{-\left(\frac{(1-\epsilon)\lambda}{2}(c_0 - c') - C_* - o(1)\right)|t\setminus k|^2|t|\Lambda_{|t|}\log p\right\}$$

$$\leq \sum_{|k|=0}^{r} \sum_{v=0}^{(|t|-1)\wedge|k|} \binom{|t|}{v}\binom{r-|t|}{|k|-v} p^{-(|k|-|t|)} \exp\left\{-\left(\frac{(1-\epsilon)\lambda}{2}(c_0 - c') - C_* - o(1)\right)(|t|-v)^2|t|\Lambda_{|t|}\log p\right\}$$

$$\leq \sum_{|k|=0}^{r} \sum_{v=0}^{(|t|-1)\wedge|k|} \binom{|t|}{v}|t|r|^{|t|-v} \exp\left\{-\left(\frac{(1-\epsilon)\lambda}{2}(c_0 - c') - C_* - o(1)\right)(|t|-v)^2|t|\Lambda_{|t|}\log p\right\}$$

$$\leq \exp\left\{-\left(\frac{(1-\epsilon)\lambda}{2}(c_0 - c') - C_* - o(1)\right)|t|\Lambda_{|t|}\log p + 2(|t|+2)\log p\right\}$$

$$\leq \exp\left\{-\left(\frac{(1-\epsilon)\lambda}{2}(c_0 - c') - C_* - 6 - o(1)\right)|t|\Lambda_{|t|}\log p\right\}$$

with probability tending to 1, where $C_* \leq 3 + \delta$ is defined in the proof of Theorem 2. Please note that the last term is of order $o(1)$ because we assume $c_0 = \frac{1}{(1-\epsilon_0)\lambda}\{2(9+2\delta) + \frac{5}{(1-\epsilon_0)\lambda}\} > \frac{2}{(1-\epsilon)\lambda}(C_* + 6 + o(1)) + c'$ for some small $\epsilon_0 > 0$. It is easy to see that the summation of (A10) over $k \in M_2$ is also of order $o(1)$ with probability tending to 1 by the similar arguments. □

Lemma A1. *Under Condition (A2), we have*

$$\exp\left\{\frac{1}{2}\widehat{\beta}_k^\top (A_k - A_k(A_k + \tau^{-1}U_k)^{-1}A_k)\widehat{\beta}_k\right\} \lesssim 1$$

for any $k \in M_1$ with probability tending to 1.

Proof. Please note that by Condition (A2),

$$(A_k + \tau^{-1}U_k)^{-1} \geq (A_t + (n\tau\lambda)^{-1}A_k)^{-1} = \frac{n\tau\lambda}{n\tau\lambda + 1}A_k^{-1},$$

which implies that

$$\frac{1}{2}\widehat{\beta}_k^\top (A_k - A_k(A_k + \tau^{-1}U_k)^{-1}A_k)\widehat{\beta}_k \leq \frac{1}{2(n\tau\lambda + 1)}\widehat{\beta}_k^\top A_k \widehat{\beta}_k.$$

Thus, we complete the proof if we show that

$$\frac{1}{n\tau\lambda}\widehat{\beta}_k^\top H_n(\beta_{0,k})\widehat{\beta}_k \leq C$$

for some constant $C > 0$ and any $k \in M_1$ with probability tending to 1. By Lemma A.3 in [5] and Condition (A2),

$$\frac{1}{n\tau}\widehat{\beta}_k^\top H_n(\beta_{0,k})\widehat{\beta}_k \leq \frac{1}{\tau}\lambda_{\max}\{n^{-1}H_n(\beta_{0,k})\}\|\widehat{\beta}_k\|_2^2$$

$$\leq \frac{1}{\tau}(\log p)^d(\|\beta_{0,k}\|_2^2 + o(1)) = O(1)$$

for any $k \in M_1$ with probability tending to 1. □

References

1. Tibshirani, R. Regression shrinkage and selection via the lasso. *J. R. Stat. Soc. Ser. (Methodol.)* **1996**, *58*, 267–288. [CrossRef]
2. Fan, J.; Li, R. Variable selection via nonconcave penalized likelihood and its oracle properties. *J. Am. Stat. Assoc.* **2001**, *96*, 1348–1360. [CrossRef]
3. Zhang, C. Nearly unbiased variable selection under minimax concave penalty. *Ann. Stat.* **2010**, *38*, 894–942. [CrossRef]
4. Liang, F.; Song, Q.; Yu, K. Bayesian subset modeling for high-dimensional generalized linear models. *J. Am. Stat. Assoc.* **2013**, *108*, 589–606. [CrossRef]
5. Narisetty, N.; Shen, J.; He, X. Skinny gibbs: A consistent and scalable gibbs sampler for model selection. *J. Am. Stat. Assoc.* **2018**, 1–13. [CrossRef]
6. Ročková, V.; Georg, E. The spike-and-slab lasso. *J. Am. Stat. Assoc.* **2018**, *113*, 431–444. [CrossRef]
7. Johnson, V.; Rossell, D. On the use of non-local prior densities in bayesian hypothesis tests hypothesis. *J. R. Statist. Soc. B* **2010**, *72*, 143–170. [CrossRef]
8. Johnson, V.; Rossell, D. Bayesian model selection in high-dimensional settings. *J. Am. Stat. Assoc.* **2012**, *107*, 649–660. [CrossRef]
9. Shin, M.; Bhattacharya, A.; Johnson, V. Scalable bayesian variable selection using nonlocal prior densities in ultrahigh-dimensional settings. *Stat. Sin.* **2018**, *28*, 1053–1078.
10. Shi, G.; Lim, C.; Maiti, T. Bayesian model selection for generalized linear models using non-local priors. *Comput. Stat. Data Anal.* **2019**, *133*, 285–296. [CrossRef]
11. Wu, Ho.; Ferreira, M.R.; Elkhouly, M.; Ji, T. Hyper nonlocal priors for variable selection in generalized linear models. *Sankhya A* **2020**, *82*, 147–185. [CrossRef]
12. Hans, C.; Dobra, A.; West, M. Shotgun stochastic search for "large p" regression. *J. Am. Stat. Assoc.* **2007**, *102*, 507–516. [CrossRef]

13. Yang, X.; Narisetty, N. Consistent group selection with bayesian high dimensional modeling. *Bayesian Anal.* **2018**. [CrossRef]
14. Cao, X.; Khare, K.; Ghosh, M. High-dimensional posterior consistency for hierarchical non-local priors in regression. *Bayesian Anal.* **2020**, *15*, 241—262. [CrossRef]
15. Castillo, I.; Schmidt-Hieber, J.; Van der Vaart, A. Bayesian linear regression with sparse priors. *Ann. Stat.* **2015**, *43*, 1986–2018. [CrossRef]
16. McCullagh, P.; Nelder, J.A. *Generalized Linear Models*, 2nd ed.; Chapman & Hall: London, UK, 1989.
17. Lee, K.; Lee, J.; Lin, L. Minimax posterior convergence rates and model selection consistency in high-dimensional dag models based on sparse cholesky factors. *Ann. Stat.* **2019**, *47*, 3413–3437. [CrossRef]
18. Ishwaran, H.; Rao, J. Spike and slab variable selection: Frequentist and bayesian strategies. *Ann. Stat.* **2005**, *33*, 730–773. [CrossRef]
19. Song, Q.; Liang, F. Nearly optimal bayesian shrinkage for high dimensional regression. *arXiv* **2017**, arXiv:1712.08964.
20. Yang, Y.; Wainwright, M.; Jordan, M. On the computational complexity of high-dimensional bayesian variable selection. *Ann. Stat.* **2016**, *44*, 2497–2532. [CrossRef]
21. Tüchler, R. Bayesian variable selection for logistic models using auxiliary mixture sampling. *J. Comput. Graph. Stat.* **2008**, *17*, 76–94. [CrossRef]
22. Cai, X.; Huang, A.; Xu, S. Fast empirical bayesian lasso for multiple quantitative trait locus mapping. *BMC Bioinform.* **2011**, *12*, 211. [CrossRef] [PubMed]
23. Friedman, J.; Hastie, T.; Tibshirani, R. Regularization paths for generalized linear models via coordinate descent. *J. Stat. Softw.* **2010**, *33*, 1–22. [CrossRef] [PubMed]
24. Breheny, P.; Huang, J. Coordinate descent algorithms for nonconvex penalized regression, with applications to biological feature selection. *Ann. Appl. Stat.* **2011**, *5*, 232–253. [CrossRef] [PubMed]
25. Matthews, B.W. Comparison of the predicted and observed secondary structure of T4 phage lysozyme. *Biochim. Biophys. Acta (BBA) Protein Struct.* **1975**, *405*, 442–451. [CrossRef]
26. Meinshausen, N.; Bühlmann, P. High-dimensional graphs and variable selection with the lasso. *Ann. Stat.* **2006**, *34*, 1436–1462. [CrossRef]
27. Wei, L.; Hu, X.; Zhu, Y.; Yuan, Y.; Liu, W.; Chen, H. Aberrant intra-and internetwork functional connectivity in depressed Parkinson's disease. *Sci. Rep.* **2017**, *7*, 1–12. [CrossRef]
28. Zang, Y.; Jiang, T.; Lu, Y.; He, Y.; Tian, L. Regional homogeneity approach to fmri data analysis. *NeuroImage* **2004**, *22*, 394–400. [CrossRef]
29. Zuo, Xi.; Xu, T.; Jiang, L.; Yang, Z.; Cao, X.; He, Y.; Zang, Y.; Castellanos, F.; Milham, M. Toward reliable characterization of functional homogeneity in the human brain: Preprocessing, scan duration, imaging resolution and computational space. *NeuroImage* **2013**, *65*, 374–386. [CrossRef]
30. Zang, Y.; He, Y.; Zhu, C.; Cao, Q.; Sui, M.; Liang, M., Tian, L.; Jiang, T.; Wang, Y. Altered baseline brain activity in children with adhd revealed by resting-state functional mri. *Brain Dev.* **2007**, *29*, 83–91.
31. Zuo, Xi.; Kelly, C.; Di Martino, A.; Mennes, M.; Margulies, D.; Bangaru, S.; Grzadzinski, R.; Evans, A.; Zang, Y.; Castellanos, F.; et al. Growing together and growing apart: Regional and sex differences in the lifespan developmental trajectories of functional homotopy. *J. Neurosci.* **2010**, *30*, 15034–15043. [CrossRef]
32. Liu, Y.; Li, M.; Chen, H.; Wei, X.; Hu, G.; Yu, S.; Ruan, X.; Zhou, J.; Pan, X.; Ze Li; et al. Alterations of regional homogeneity in parkinson's disease patients with freezing of gait: A resting-state fmri study. *Front. Aging Neurosci.* **2019**, *11*, 276. [CrossRef] [PubMed]
33. Mi, T.; Mei, S.; Liang, P.; Gao, L.; Li, K.; Wu, T.; Chan, P. Altered resting-state brain activity in parkinson's disease patients with freezing of gait. *Sci. Rep.* **2017**, *7*, 16711. [CrossRef] [PubMed]
34. Prell, T. Structural and functional brain patterns of non-motor syndromes in parkinson's disease. *Front. Neurol.* **2018**, *9*, 138. [CrossRef] [PubMed]
35. Wang, J.; Zhang, J.; Zang, Y.; Wu, T. Consistent decreased activity in the putamen in Parkinson's disease: A meta-analysis and an independent validation of resting-state fMRI. *GigaScience* **2018**, *7*, 6. [CrossRef]
36. Zhang, F.; Jiang, W.; Wong, P.; Wang, J. A Bayesian probit model with spatially varying coefficients for brain decoding using fMRI data. *Stat. Med.* **2016**, *35*, 4380–4397. [CrossRef]

37. Quintero, F.O.L.; Contreras-Reyes, J.E.; Wiff, R.; Arellano-Valle, R.B. Flexible Bayesian analysis of the von Bertalanffy growth function with the use of a log-skew-t distribution. *Fish. Bull.* **2017**, *115*, 13–26. [CrossRef]
38. Lee, K.; Cao, X. Bayesian group selection in logistic regression with application to mri data analysis. In *Biometrics, to Appear*; Wiley: Hoboken, NJ, USA, 2020. [CrossRef]

© 2020 by the authors. Licensee MDPI, Basel, Switzerland. This article is an open access article distributed under the terms and conditions of the Creative Commons Attribution (CC BY) license (http://creativecommons.org/licenses/by/4.0/).

Article

Segmentation of High Dimensional Time-Series Data Using Mixture of Sparse Principal Component Regression Model with Information Complexity

Yaojin Sun and Hamparsum Bozdogan *

Department of Business Analytics and Statistics, University of Tennessee, Knoxville, TN 37996, USA; ysun52@vols.utk.edu
* Correspondence: bozdogan@utk.edu

Received: 29 August 2020; Accepted: 13 October 2020; Published: 17 October 2020

Abstract: This paper presents a new and novel hybrid modeling method for the segmentation of high dimensional time-series data using the mixture of the sparse principal components regression (*MIX-SPCR*) model with information complexity (ICOMP) criterion as the fitness function. Our approach encompasses dimension reduction in high dimensional time-series data and, at the same time, determines the number of component clusters (i.e., number of segments across time-series data) and selects the best subset of predictors. A large-scale Monte Carlo simulation is performed to show the capability of the *MIX-SPCR* model to identify the correct structure of the time-series data successfully. *MIX-SPCR* model is also applied to a high dimensional Standard & Poor's 500 (S&P 500) index data to uncover the time-series's hidden structure and identify the structure change points. The approach presented in this paper determines both the relationships among the predictor variables and how various predictor variables contribute to the explanatory power of the response variable through the sparsity settings cluster wise.

Keywords: high dimensional time-series; segmentation; mixture regression; sparse PCA; entropy-based robust EM; information complexity criteria

1. Introduction

This paper presents a new and novel method for the segmentation and dimension reduction in high dimensional time-series data. We develop hybrid modeling between *mixture-model cluster analysis* and *sparse principal components regression* (MIX-SPCR) model as an expert unsupervised classification methodology with *information complexity* (ICOMP) criterion as the fitness function. This new approach performs dimension reduction in high dimensional time-series data and, at the same time, determines the number of component clusters.

The research of time-series segmentation and change point positioning has been a hot topic of research for a long time. Different research groups have provided solutions with various approaches in this area, including, but not limited to, Bayesian methods Barber et al. [1], fuzzy systems Abonyi and Feil [2], and complex system modeling Spagnolo and Valenti [3], Valenti et al. [4], S Lima [5], Ding et al. [6]. We group these approaches into two branches, one based on complex systems modeling and the other on the statistical model through parameter estimation and inference. Among the complex systems-based modeling approaches, it is worth noting a series of papers that use the stochastic volatility model by Spagnolo and Valenti [3]. For example, these authors used a nonlinear Hestone model to analyze 1071 stocks on the New York Stock Exchange (1987–1998). After accounting for the stochastic nature of volatility, the model is well suited to extracting the escape time distribution from financial time-series data. The authors also identified the NES (Noise Enhanced Stability) effect to measure market dynamics' stabilizing effect. The approach we propose in this paper belongs to another branch of using a statistical model on time scales. Along with the empirical analysis, we show a broader view of how different companies/sectors behaved across different periods. In particular, we use a mixture-model based statistical methodology to segment the time-series and determine change points.

The mixture-model cluster analysis of regression models is not new. These models are also known as *"cluster-wise regression"*, *"latent models"*, and *"latent structure models of choice"*. These models have been well-studied among statisticians, machine learning researchers, and econometricians in the last several decades to construct time-series segmentation models and identify change points. They have many useful theoretical and applied properties. Mixture-model cluster analysis of regression models is a natural extension of the standard multivariate Gaussian mixture-model cluster analysis. These models are beneficial to study heterogeneous data sets that involve not just one response variable but can have several responses or target-dependent variables simultaneously with a given set of independent variables. Recently, they have been proven to be a precious class of models in various disciplines in *behavioral and economic research, ecology, financial engineering, process control, and monitoring, market research, transportation systems*. Additionally, we also witness the mixture model's usage in the *analysis of scanner panel, survey, and other choice data to study consumer choice behavior and dynamics* Dillon et al. [7].

In reviewing the literature, we note that Quandt and Ramsey [8] and Kiefer [9] studied data sets by applying a mixture of two regression models using moment generating function techniques to estimate the unknown model parameters. Later, De Veaux [10] developed an EM algorithm to fit a mixture of two regression models. DeSarbo and Cron [11] used similar estimating equations and extended the earlier work done on a mixture of two regression models to a mixture of K-component regression models. For an excellent review article on this problem, we refer the reviewers to Wedel and DeSarbo [12].

In terms of these models' applications in the segmentation of time-series, they can be seen in the early work of Sclove [13], where the author applied the mixture model to the segmentation of US gross national product, a high dimensional time-series data. Specifically, Sclove [13] used the statistical model selection criteria to choose the number of classes.

With the currently existing challenges in mind in the segmentation of time-series data, in this paper, our objective and goal are to develop a new methodology which can:

- Identify and select variables that are sparse in the *MIX-SPCR* model.
- Treat each time segment continuously in the process with some specified probability density function (pdf).
- Determine the number of time-series segments and the number of sparse variables and estimate the structural change points simultaneously.
- Develop a robust and efficient algorithm for estimating model parameters.

We aim to achieve these objectives by developing the information complexity (ICOMP) criteria as our fitness function throughout the paper for the segmentation of high-dimensional time-series data.

Our approach involves a two-stage procedure. We first make a variable selection by using SPCA with the benefit of sparsity. We then fit the sparse principal component regression (SPCR) model by transforming the original high dimensional data into several main principal components and estimating relationships between the sparse component loadings and the response variable. In this way, the mixture model not only handles the curse of dimensionality but also maintains the model's excessive explanatory power. In this manner, we choose the best subset of predictors and determine the number of time-series segments in the *MIX-SPCR* model simultaneously using ICOMP.

The rest of the paper is organized as follows. In Section 2, we present the model and methods. In particular, we first briefly explain sparse principal component analysis (SPCA) due to Zou et al. [14] in Section 2.1. In Section 2.2, we modify SPCA and develop mixtures of the sparse principal component regression (*MIX-SPCR*) model for the segmentation of time-series data. In Section 3, we present a regularized entropy-based Expectation and Maximization (EM) clustering algorithm. As is well known, the EM algorithm performs through maximizing the likelihood of the mixture models. However, to make the conventional EM algorithm robust (not sensitive to initial values) and converge to global optimum, we use the robust version of the EM algorithm for the *MIX-SPCR* model based on the work of Yang et al. [15]. These authors addressed the robustness issue by adding an entropy term of mixture proportions to the conventional EM algorithm's objective function. While our EM algorithm is in the same spirit of the Yang et al. [15] approach, there are significant differences between our approach and theirs. Yang's robust EM algorithm merely deals with the usual clustering problem without involving any response (or dependent) variable or time factor in the data. We extend it to the case of the *MIX-SPCR* model in the context of time-series data. In Section 4, we discuss various information criteria, specifically the information complexity based criteria (ICOMP). We derive the ICOMP for the *MIX-SPCR* model based on Bozdogan's previous research ([16–20]). In Section 5, we present our Monte Carlo simulation study. Section 5.2 involves an experiment on the detection of structural points, and Section 5.3 presents a large scale Monte Carlo simulation verifying the advantage of the *MIX-SPCR* with statistical information criteria. We provide a real data analysis in Section 6 using the daily adjusted closing S&P 500 index and stock prices from the Yahoo Finance database that spans the period from January 1999 to December 2019. Finally, our conclusion and discussion are presented in Section 7.

2. Model and Methods

In this section, we briefly present the *sparse principal component analysis* (*SPCA*), *sparse principal component regression* (*SPCR*) as a background. Then, by hybridizing these two methods within the mixture model, we propose the *mixture-model cluster analysis of sparse principal component regression* (abbreviated as *MIX-SPCR* model hereafter), for segmentation of high dimensional time-series datasets. Compared with a simple linear combination of all explanatory variables (i.e., the dense PCA model), the new approach interprets better because it maintains a sparsity specification.

Referring to Figure 1, we first show the overall structure of the model in this paper. The overall processing flow is that we clean and standardize the data after obtaining the time-series data. Subsequently, we specify the number of time-series segments and how many Sparse Principal Components (SPCs) each segment contains. Using the Robust EM algorithm (Section 3), we estimate the model parameters, especially the boundaries (also known as *change points*) of each time segment. The information criterion values are calculated using the method of Section 4. By testing different numbers of time segments/SPCs, we obtain multiple criterion values. According to the calculated information criterion values, we choose the most appropriate model with the estimated parameters.

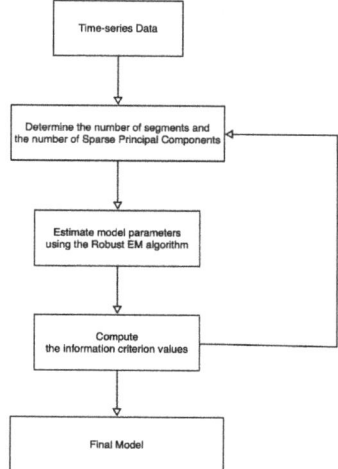

Figure 1. The flowchart of the MIX-SPCR method.

2.1. Sparse Principal Component Analysis (SPCA)

Given the input data matrix, **X** with n number of observations and p variables, we decompose **X** using the singular value decomposition (SVD). We write the decomposition procedure as $\mathbf{X} = \mathbf{UDV}^T$, where **D** is a diagonal matrix of singular values and orthogonal columns **U** and **V** as the left and right singular vectors. When we perform SVD of a data matrix **X** that has been centered, by subtracting each column's mean, the process is the well-known *principal component analysis (PCA)*. As discussed by Zou et al. [14], PCA has several advantages as compared with other dimensionality reduction techniques. For example, the PCA can sequentially identify the source of variability by considering the linear combination of all the variables. Because of the orthonormal constraint during the computation, all the calculated *principal components (PCs)* have clear geometrical interpretation corresponding to the original data space as a dimension reduction technique. Because PCA can deal with *"the curse of dimensionality"* of high-dimensional data sets, it has been widely used in real-world scenarios, including biomedical and financial applications.

Even though PCA has excellent properties that are desirable in real-world applications and statistical analysis, the interpretation of PCs is often difficult since it includes all the variables as linear combinations of all the original variables in each of the PCs. In practice, the principal components always have a large number of non-zero coefficient values for corresponding variables. To resolve this drawback, researchers proposed various improvements focusing on PCA's sparsity while maintaining the minimal loss of information. Shen and Huang [21] designed an algorithm to iteratively extract top PCs using the so-called *penalized least sum of square (PLSS)* criterion. Zou et al. [14] utilized the lasso penalty (via Elastic Net) to maintain a sparse loading of the principal components, which is named *sparse principal component analysis (SPCA)*.

In this paper, we use the sparse principal component analysis (SPCA) proposed by Zou et al. [14]. Given the data matrix **X**, we minimize the objective function to obtain the SPCA results:

$$(\widehat{\mathbf{A}}, \widehat{\mathbf{B}}) = \arg\min_{\mathbf{A},\mathbf{B}} \sum_{i=1}^{n} \left\| \mathbf{x}_i^T - \mathbf{A}\mathbf{B}^T \mathbf{x}_i^T \right\|^2 + \sum_{j=1}^{k} \lambda_{1,j} \left\| \mathbf{B}_{(j)} \right\|_1 + \lambda_2 \sum_{j=1}^{k} \left\| \mathbf{B}_{(j)} \right\|_2^2, \tag{1}$$

subject to

$$\mathbf{A}^T \mathbf{A} = I_k. \tag{2}$$

where I_k is the identity matrix. We maintain the hyperparameters $\lambda_{1,j}$ and λ_2 to be non-negative. The **A** and **B** matrices of size $(p \times k)$ are given by

$$\mathbf{B} = \begin{bmatrix} B_{1,1} & \cdots & B_{1,k} \\ \vdots & \ddots & \vdots \\ B_{p,1} & \cdots & B_{p,k} \end{bmatrix} = \begin{bmatrix} \mathbf{B}_{(1)} | \ldots | \mathbf{B}_{(k)} \end{bmatrix} = \begin{bmatrix} \mathbf{B}_1 \\ \vdots \\ \mathbf{B}_p \end{bmatrix}, \tag{3}$$

and

$$\mathbf{A} = \begin{bmatrix} A_{1,1} & \cdots & A_{1,k} \\ \vdots & \ddots & \vdots \\ A_{p,1} & \cdots & A_{p,k} \end{bmatrix} = \begin{bmatrix} \mathbf{A}_{(1)} | \ldots | \mathbf{A}_{(k)} \end{bmatrix} = \begin{bmatrix} \mathbf{A}_1 \\ \vdots \\ \mathbf{A}_p \end{bmatrix}. \tag{4}$$

If we choose the first k principal components from the data matrix **X**, then the estimate $\hat{\mathbf{B}}_{(j)}$ contains the sparse loading vectors, which are no longer orthogonal.

A bigger $\lambda_{1,j}$ means a greater penalty for having non-zero entries in $\hat{\mathbf{B}}_{(j)}$. By using different $\lambda_{1,j}$, we control the number of zeros in the jth loading vector. If $\lambda_{1,j} = 0$ for $j = 1, 2, \ldots, k$, this problem reduces to usual PCA.

Zou et al. [14] proposed a generalized SPCA algorithm to solve the optimization problem in Equation (1). The algorithm applies the Elastic Net (EN) to estimate $\mathbf{B}_{(j)}$ iteratively and update matrix **A**. However, this algorithm is not the only available approach for extracting principal components with sparse loadings. The SPCA could also be computed through dictionary learning by Mairal et al. [22]. By introducing the probability model of principal component analysis, SPCA is equivalent to the *sparse probabilistic principal component analysis (SPPCA)* if the prior is Laplacian distribution for each weight matrix element (Guan and Dy [23], Williams [24]). For further discussion on SPPCA, we refer readers to those related publications for more details.

Next, we introduce the *MIX-SPCR* model for the segmentation of time-series data.

2.2. Mixtures of SPCR Model for Time-Series Data

Suppose the continuous response variable is denoted as $\mathbf{y} = \{y_i | 1 \leq i \leq n\}$, where n represents the number of observations (time points). Similarly, we have the predictors denoted as $\mathbf{X} = \{\mathbf{x}_i | 1 \leq i \leq n\}$. Each observation $\mathbf{x}_i$ has p dimensions and is represented as $\mathbf{x}_i = [x_{1,i}, x_{2,i}, \cdots, x_{p,i}]^T$. Both the response variable and independent variables are collected sequentially labeled by time points $T = [t_1, t_2, \cdots, t_n]$.

The finite mixture model allows applying cluster analysis on conditionally dependent data into several classes. In the time-series data scenario, researchers cluster the data $((t_1, \mathbf{x}_1, y_1), (t_2, \mathbf{x}_2, y_2), \cdots, (t_n, \mathbf{x}_n, y_n))$ into several homogeneous groups where the number of groups G is unknown in general. Within each group, we apply the SPCA to extract top k principal components that each of them has a sparse loading of p variable coefficients. The extracted top k PCs are denoted as matrix $\mathbf{P}_{p \times k}$. We also use $\mathbf{P}_g$ to represent the principal component matrix obtained from the group indexed by $g = 1, 2, \ldots, G$.

The SPCR model assumes that each pair $(\mathbf{x}_i, y_i)$ is independently drawn from a cluster using both the SPCA and the regression model as follows.

$$y_i = \mathbf{x}_i^T \mathbf{P}_g \beta_g + \epsilon_{i,g}, i = 1, 2, \cdots, n, \tag{5}$$

where $\beta_g = \begin{bmatrix} \beta_{g,1}, \beta_{g,2}, \cdots, \beta_{g,k} \end{bmatrix}^T$.

For each group g, the random error is assumed to be Gaussian distributed. That is, $\epsilon_{i,g} \sim \mathcal{N}(0, \sigma_g^2)$. If the response variable is multivariate, then the random error is usually also assumed to be a multivariate Gaussian distribution. Thus the probability density function (pdf) of the SPCR model is

$$f(y_i|\mathbf{x}_i, \mathbf{P}_g, \beta_g) = \mathcal{N}\left(y_i|\mathbf{x}_i^T \mathbf{P}_g \beta_g, \sigma_g^2\right). \tag{6}$$

We emphasize here that the noise (i.e., the error term) included in the statistical model is drawn from a normal distribution independent for each time-series segment, with different values of σ_g^2 for each period. Since we use the EM algorithm to estimate the parameters of the model, the noise parameter σ_g^2 can be estimated accurately as well. Future studies will consider introducing different noise distributions, such as α-stable Lévy noise [25], and other non-Gaussian noise distributions to further extend the current model.

We also consider time factor t_i in the SPCR model of time-series data to be continuous. The pdf of the time factor is

$$f(t_i|v_g, \sigma_g^{2,\text{time}}) = \mathcal{N}\left(t_i|v_g, \sigma_g^{2,\text{time}}\right), \tag{7}$$

where v_g is the mean, and $\sigma_g^{2,\text{time}}$ is the variance of the time segment g. Apart from the normal distribution, our approach can also be generalized to other distributions for the time factor, such as skewed distributions, Student's t-distribution, ARCH, GARCH time-series models, and so on.

As a result, if we use the *MIX-SPCR* model to perform segmentation of time-series data, the likelihood function of the whole data $((t_1, \mathbf{x}_1, y_1), (t_2, \mathbf{x}_2, y_2), \cdots, (t_n, \mathbf{x}_n, y_n))$ with G number of clusters (or segments) is given by

$$L = \prod_{i=1}^{n} \prod_{g=1}^{G} \left[\pi_g f(y_i|\mathbf{x}_i, \mathbf{P}_g, \beta_g) f(t_i|v_g, \sigma_g^{2,\text{time}})\right]^{z_{g,i}}, \tag{8}$$

where the π_g is the mixing proportion with the constraint that $\pi_g \geq 0$ and $\sum_{g=1}^{G} \pi_g = 1$. We follow the definition of missing values by Yang et al. [15] and let $\mathbf{Z} = \{Z_1, Z_2, \cdots, Z_n\}$. If $Z_i = g$, then $z_{g,i} = 1$, otherwise, $z_{g,i} = 0$. Then the log-likelihood function of the *MIX-SPCR* model models is

$$\begin{aligned} \mathcal{L}_{\text{mix}} &= \log(L) \\ &= \sum_{i=1}^{n} \sum_{g=1}^{G} z_{g,i} \log\left[\pi_g f(y_i|\mathbf{x}_i, \mathbf{P}_g, \beta_g) f(t_i|v_g, \sigma_g^{2,\text{time}})\right] \tag{9} \\ &= \sum_{i=1}^{n} \sum_{g=1}^{G} z_{g,i} \left[\log \pi_g + \log f(y_i|\mathbf{x}_i, \mathbf{P}_g, \beta_g) + \log f(t_i|v_g, \sigma_g^{2,\text{time}})\right] \\ &= \underbrace{\sum_{i=1}^{n} \sum_{g=1}^{G} z_{g,i} \log \pi_g}_{\mathcal{L}_\pi} + \underbrace{\sum_{i=1}^{n} \sum_{g=1}^{G} z_{g,i} \log f(y_i|\mathbf{x}_i, \mathbf{P}_g, \beta_g)}_{\mathcal{L}_{\text{SPCR}}} + \underbrace{\sum_{i=1}^{n} \sum_{g=1}^{G} z_{g,i} \log f(t_i|v_g, \sigma_g^{2,\text{time}})}_{\mathcal{L}_{\text{time}}}. \tag{10} \end{aligned}$$

We denote $\mathbf{z} = [z_{g,i}]$ where $g = 1, 2, \cdots, G$ and $i = 1, 2, \cdots, n$.

Given the number of segments, researchers usually apply the EM algorithm to determine the optimal segmentation by setting the objective function as $\mathcal{J}_{\text{EM}} = \mathcal{L}_{\text{mix}}$ (Gaffney and Smyth [26], Esling and Agon [27], Gaffney [28]).

3. Regularized Entropy-Based EM Clustering Algorithm

The EM algorithm is a method for iteratively optimizing the objective function. As discussed in Section 2.2, by setting the objective function as the log-likelihood function, we can use the EM algorithm to identify optimal segmentation of time series.

However, in practice, the EM algorithm is sensitive to model initialization conditions and cannot estimate the number of clusters appropriately. To deal with the initialization problem, in 2012, Yang et al. [15] proposed using an entropy penalty to stabilize the computation of each step. The improved method is called the *robust EM algorithm*. In this paper, we extend the robust EM algorithm to deal with time-series data for the *MIX-SPCR* model.

In Section 3.1, we discuss the entropy term of the robust EM algorithm. Then, we show the extension of the robust EM algorithm for the *MIX-SPCR* model in Sections 3.2 and 3.3.

3.1. The Entropy of EM Mixture Probability

As introduced in Equation (8), the π_g represents the mixture probability of each cluster or segment. In other words, the value of π_g is the probability that a data point belongs to group g. The clustering complexity is determined by the number of clusters and corresponding probability values, which could be obtained using entropy. Given $\{\pi_g | 1 \leq g \leq G\}$, the entropy of Z_i is

$$H(Z_i | \{\pi_g | 1 \leq g \leq G\}) = -\sum_{g=1}^{G} \pi_g \log(\pi_g), \text{ for } i = 1, 2, \cdots, n. \tag{11}$$

Then the entropy of **Z** is written as,

$$H(\mathbf{Z} | \{\pi_g | 1 \leq g \leq G\}) = \sum_{i=1}^{n} H(Z_i | \{\pi_g | 1 \leq g \leq G\})$$

$$= -\sum_{i=1}^{n} \sum_{g=1}^{G} \pi_g \log(\pi_g)$$

$$= -n \sum_{g=1}^{G} \pi_g \log(\pi_g). \tag{12}$$

The objective function of the robust EM algorithm is

$$\mathcal{J}_{\text{Robust-EM}} = \mathcal{L}_{\text{mix}} - \lambda_{\text{Robust-EM}} H(\mathbf{Z} | \{\pi_g | 1 \leq g \leq G\}), \tag{13}$$

where $\lambda_{\text{Robust-EM}} \geq 0$. The log-likelihood term $\mathcal{L}_{\text{mix}}$ is from Equation (9), which gives the goodness-of-fit. Next, we present the steps of the EM algorithm for maximizing the objective function in Equation (13).

3.2. E-Step (Expectation)

From a Bayesian perspective, we let $\hat{z}_{g,i}$ denote the posterior probability of the true cluster membership that a dataset triplet $(t_i, \mathbf{x}_i, y_i)$ is drawn from group g. Using the Bayes theorem, we have

$$\hat{z}_{g,i} = \mathbb{E}(Z_i = g | y_i, \mathbf{x}_i, \mathbf{P}_g, \beta_g) \tag{14}$$

$$= \frac{\pi_g \mathcal{N}\left(y_i; \mathbf{x}_i \mathbf{P}_g \beta_g, \sigma_g^2\right) \mathcal{N}\left(t_i | v_g, \sigma_g^{2,\text{time}}\right)}{\sum_{h=1}^{G} \pi_h \mathcal{N}\left(y_i; \mathbf{x}_i \mathbf{P}_h \beta_h, \sigma_h^2\right) \mathcal{N}\left(t_i | v_h, \sigma_h^{2,\text{time}}\right)}. \tag{15}$$

3.3. M-Step (Maximization)

Using the robustified derivation of $\hat{\pi}_g$, the estimated mixture proportion, we have

$$\hat{\pi}_g^{\text{new}} = \hat{\pi}_g^{\text{EM}} + \hat{\lambda}_{\text{Robust-EM}} \hat{\pi}_g^{\text{old}} \left(\log(\hat{\pi}_g^{\text{old}}) - \sum_{h=1}^{G} \left(\hat{\pi}_h^{\text{old}} \log(\hat{\pi}_h^{\text{old}}) \right) \right), \tag{16}$$

where

$$\hat{\pi}_g^{\text{EM}} = \frac{\sum_{i=1}^{n} \hat{z}_{g,i}}{n}. \tag{17}$$

We follow the recommendation of Yang et al. [15] for the value of $\hat{\lambda}_{\text{Robust-EM}}^{\text{new}}$ as

$$\hat{\lambda}_{\text{Robust-EM}}^{\text{new}} = \min \left\{ \frac{\sum_{h=1}^{G} \exp\left(-\eta n \left| \hat{\pi}_g^{\text{new}} - \hat{\pi}_g^{\text{old}} \right|\right)}{G}, \frac{1 - \max\left\{ \sum_{i=1}^{n} \hat{z}_{h,i}^{\text{old}} / n | h = 1, 2, \cdots, G \right\}}{-\max\left\{ \hat{\pi}_h^{\text{old}} | h = 1, 2, \cdots, G \right\} \sum_{h=1}^{G} \hat{\pi}_h^{\text{old}} \log \hat{\pi}_h^{\text{old}}} \right\}, \tag{18}$$

where

$$\eta = \min \left\{ 1, 0.5^{\lfloor p/2-1 \rfloor} \right\}, \tag{19}$$

and p is the number of variables in the model.

We iterate E-step and M-step several times until convergence to obtain the parameter estimates. In particular, the β_g values get updated by maximizing the $\mathcal{J}_{\text{Robust-EM}}$ from Equation (13). Since we fix the number of segments and principal components during each E-step and M-step, the updated values of β_g and σ_g can be calculated using $\mathcal{L}_{\text{mix}}$ directly. The estimated values of β_g and σ_g are given as follows.

$$\hat{\beta}_g^{\text{new}} = \left[\sum_{i=1}^{n} \hat{z}_{g,i}^{\text{old}} (\mathbf{x}_i^T \mathbf{P}_g)^T (\mathbf{x}_i^T \mathbf{P}_g) \right]^{-1} \sum_{i=1}^{n} \hat{z}_{g,i}^{\text{old}} (\mathbf{x}_i^T \mathbf{P}_g)^T y_i$$

$$= \left[\sum_{i=1}^{n} \hat{z}_{g,i}^{\text{old}} \mathbf{P}_g^T \mathbf{x}_i \mathbf{x}_i^T \mathbf{P}_g \right]^{-1} \sum_{i=1}^{n} \hat{z}_{g,i}^{\text{old}} \mathbf{P}_g^T \mathbf{x}_i y_i, \tag{20}$$

$$\hat{\sigma}_g^{2,\text{new}} = \sum_{i=1}^{n} \hat{z}_{g,i}^{\text{old}} \left\| y_i - \mathbf{x}_i^T \mathbf{P}_g \hat{\beta}_g^{\text{new}} \right\|_2^2 / \sum_{i=1}^{n} \hat{z}_{g,i}^{\text{old}}. \tag{21}$$

For the time factor, the estimated mean $\widehat{v}_g$ and variance $\widehat{\sigma}_g^{2,\text{time}}$ are

$$\widehat{v}_g = \frac{\sum_{i=1}^{n} \widehat{z}_{g,i} t_i}{\sum_{i=1}^{n} \widehat{z}_{g,i}}, \tag{22}$$

$$\widehat{\sigma}_g^{2,\text{time}} = \frac{\sum_{i=1}^{n} \widehat{z}_{g,i} (t_i - \widehat{v}_g)^2}{\sum_{i=1}^{n} \widehat{z}_{g,i}}. \tag{23}$$

As discussed above, our approach is flexible in considering other distributional models for the time-series factor, which we will pursue in separate research work.

4. Information Complexity Criteria

Recently, the statistical literature recognized the necessity of introducing model selection as one of the technical areas. In this area, the entropy and the Kullback–Leibler [29] information (or KL distance) play a crucial role and serve as an analytical basis to obtain the forms of model selection criteria. In this paper, we use information criteria to evaluate a portfolio of competing models and select the best-fitting model with minimum criterion values.

One of the first information criteria for model selection in the literature is due to the seminal work of Akaike [30]. Following the entropy maximization principle (EMP), Akaike developed the Akaike's Information Criterion (AIC) to estimate the expected KL distance or divergence. The form of AIC is

$$\text{AIC} = -2 \log L(\hat{\theta}) + 2k, \tag{24}$$

where $L(\hat{\theta})$ is the maximized likelihood function, and k is the number of estimated free parameters in the model. The model with minimum AIC value is chosen as the best model to fit the data.

Motivated by Akaike's work, Bozdogan [16–20,31] developed a new information complexity (ICOMP) criteria based on Van Emden's [32] entropic complexity index in parametric estimation. Instead of penalizing the number of free parameters directly, ICOMP penalizes the covariance complexity of the model. There are several forms of ICOMP. In this section, we present the two general forms of ICOMP criteria based on the estimated inverse Fisher information matrix (IFIM). The first form is

$$\begin{aligned}\text{ICOMP(IFIM)} &= -2 \log L(\hat{\theta}) + 2C(\hat{\Sigma}_{model}) \\ &= -2 \log L(\hat{\theta}) + 2C_1(\widehat{\mathcal{F}}^{-1}),\end{aligned} \tag{25}$$

where $L(\hat{\theta})$ is the maximized likelihood function, and $C_1(\widehat{\mathcal{F}}^{-1})$ represents the entropic complexity of IFIM. We define $C_1(\widehat{\mathcal{F}}^{-1})$ as

$$C_1(\widehat{\mathcal{F}}^{-1}) = \frac{s}{2} \log\left(\frac{tr \widehat{\mathcal{F}}^{-1}}{s}\right) - \frac{1}{2} \log\left|\widehat{\mathcal{F}}^{-1}\right|, \tag{26}$$

and where $s = \text{rank}(\widehat{\mathcal{F}}^{-1})$. We can also give the form of $C_1(\widehat{\mathcal{F}}^{-1})$ in terms of eigenvalues,

$$C_1(\widehat{\mathcal{F}}^{-1}) = \frac{s}{2} \log\left(\frac{\bar{\lambda}_a}{\bar{\lambda}_g}\right), \tag{27}$$

where $\bar{\lambda}_a$ is the arithmetic mean of the eigenvalues, $\lambda_1, \lambda_2, \ldots, \lambda_s$, and $\bar{\lambda}_g$ is the geometric mean of the eigenvalues.

We note that ICOMP penalizes the lack of parsimony and the profusion of the model's complexity through IFIM. It offers a new perspective beyond counting and penalizing number of estimated parameters in the model. Instead, ICOMP takes into account interaction (i.e., correlation) among the estimated parameters through the model fitting process.

We define the second form of ICOMP as

$$\text{ICOMP(IFIM)}_{C_{1F}} = -2\log L(\hat{\theta}) + 2C_{1F}(\hat{\mathcal{F}}^{-1}), \tag{28}$$

where $C_{1F}(\hat{\mathcal{F}}^{-1})$ is given by

$$C_{1F}(\hat{\mathcal{F}}^{-1}) = \frac{s}{4} \frac{\frac{1}{s}tr\left(\left(\hat{\mathcal{F}}^{-1}\right)^T \left(\hat{\mathcal{F}}^{-1}\right)\right) - \left(\frac{tr(\hat{\mathcal{F}}^{-1})}{s}\right)^2}{\left(\frac{tr(\hat{\mathcal{F}}^{-1})}{s}\right)^2}. \tag{29}$$

In terms of the eigenvalues of IFIM, we write $C_{1F}(\hat{\mathcal{F}}^{-1})$ as

$$C_{1F}(\hat{\mathcal{F}}^{-1}) = \frac{1}{4\bar{\lambda}_a^2} \sum_{j=1}^{s} (\lambda_j - \bar{\lambda}_a)^2. \tag{30}$$

We want to highlight some features of $C_{1F}(\hat{\mathcal{F}}^{-1})$ here. The term $C_{1F}(\hat{\mathcal{F}}^{-1})$ is a second-order equivalent measure of complexity to the original term $C_1(\hat{\mathcal{F}}^{-1})$. Additionally, we note that $C_{1F}(\hat{\mathcal{F}}^{-1})$ is scale-invariant and $C_{1F}(\hat{\mathcal{F}}^{-1}) \geq 0$ with $C_{1F}(\hat{\mathcal{F}}^{-1}) = 0$ only when all $\lambda_j = \bar{\lambda}_a$. Furthermore, $C_{1F}(\hat{\mathcal{F}}^{-1})$ measures the relative variation in the eigenvalues.

These two forms of ICOMP provide us an easy to use computational means in high dimensional modeling. Next, we derive the analytical forms of ICOMP in the *MIX-SPCR* model.

4.1. Derivation of Information Complexity in MIX-SPCR Model for Time-Series Data

We first consider the log-likelihood function of the *MIX-SPCR* model given in Equation (9),

$$\mathcal{L}_{\text{mix}} = \mathcal{L}_\pi + \mathcal{L}_{\text{SPCR}} + \mathcal{L}_{\text{time}}. \tag{31}$$

After some work, the estimated inverse Fisher information matrix (IFIM) of the mixture probabilities is

$$\hat{\mathcal{F}}_\pi^{-1} = \begin{bmatrix} \hat{\pi}_1^{-1} & 0 & 0 & 0 \\ 0 & \hat{\pi}_2^{-1} & 0 & 0 \\ 0 & 0 & \ddots & 0 \\ 0 & 0 & 0 & \hat{\pi}_G^{-1} \end{bmatrix}. \tag{32}$$

Similarly, for each segment g, the estimated IFIM, $\hat{\mathcal{F}}_{g,\text{SPCR}}^{-1}$, is

$$\hat{\mathcal{F}}_{g,\text{SPCR}}^{-1} = \begin{bmatrix} \hat{\sigma}_g^2 \left[\sum_{i=1}^{n} \hat{z}_{g,i} (x_i^T P_g)^T (x_i^T P_g) \right]^{-1} & 0 \\ 0^T & 2\hat{\sigma}_g^4 (\sum \hat{z}_{g,i})^{-1} \end{bmatrix}, \quad g = 1, 2, \ldots, G. \tag{33}$$

Note that the IFIM should include both the SPCR models $\widehat{\mathcal{F}}_{g,\text{SPCR}}^{-1}$ and the time factor $\widehat{\mathcal{F}}_{g,\text{time}}^{-1}$ for each segment.

For each segment g, the time factor is under the univariate Gaussian distribution. As a result, the IFIM of the time factor is

$$\widehat{\mathcal{F}}_{g,\text{time}}^{-1} = \begin{bmatrix} \widehat{\sigma}_g^{2,\text{time}}/n & 0 \\ 0 & \frac{2}{n}\widehat{\sigma}_g^{4,\text{time}} \end{bmatrix}. \tag{34}$$

By combining the two IFIMs for the SPCR model and the time factor, we have the inverse Fisher information

$$\widehat{\mathcal{F}}_g^{-1} = \begin{bmatrix} \widehat{\mathcal{F}}_{g,\text{SPCR}}^{-1} & \mathbf{0} \\ \mathbf{0}^T & \widehat{\mathcal{F}}_{g,\text{time}}^{-1} \end{bmatrix}. \tag{35}$$

Overall, the inverse of the estimated Fisher information matrix (IFIM) for the *MIX-SPCR* model becomes

$$\widehat{\mathcal{F}}^{-1} \cong \begin{bmatrix} \widehat{\mathcal{F}}_\pi^{-1} & 0 & 0 & \cdots & 0 \\ 0 & \widehat{\mathcal{F}}_1^{-1} & 0 & \cdots & 0 \\ 0 & 0 & \widehat{\mathcal{F}}_2^{-1} & \cdots & 0 \\ \vdots & \vdots & \vdots & \ddots & \vdots \\ 0 & 0 & 0 & \cdots & \widehat{\mathcal{F}}_G^{-1} \end{bmatrix}. \tag{36}$$

Using the above definition of ICOMP(IFIM) and the properties of block-diagonal matrices with their trace and determinant, we have

$$\text{ICOMP(IFIM)} = -2\mathcal{L}_{\text{mix}} + 2C_1(\widehat{\mathcal{F}}^{-1}), \tag{37}$$

where

$$C_1(\widehat{\mathcal{F}}^{-1}) = \frac{s}{2}\log\left[\frac{tr(\widehat{\mathcal{F}}_\pi^{-1}) + \sum_{g=1}^G tr(\widehat{\mathcal{F}}_g^{-1})}{s}\right] - \frac{1}{2}\left[\log\left|\widehat{\mathcal{F}}_\pi^{-1}\right| + \sum_{g=1}^G \log\left|\widehat{\mathcal{F}}_g^{-1}\right|\right], \tag{38}$$

and where $s = \text{rank}(\widehat{\mathcal{F}}^{-1}) = r_\pi + \sum_{g=1}^G r_g = \dim(\widehat{\mathcal{F}}^{-1})$.

Similarly, we derive the second equivalent form of ICOMP(IFIM)$_{C_{1F}}$ as

$$\text{ICOMP(IFIM)}_{C_{1F}} = -2\mathcal{L}_{\text{mix}} + 2C_{1F}(\widehat{\mathcal{F}}^{-1}). \tag{39}$$

Using the properties of the block-diagonal matrices, we have

$$tr\left(\left(\widehat{\mathcal{F}}^{-1}\right)^T\left(\widehat{\mathcal{F}}^{-1}\right)\right) = tr\left(\widehat{\mathcal{F}}_\pi^{-1}\right)^2 + \sum_{g=1}^G tr\left(\widehat{\mathcal{F}}_g^{-1}\right)^2. \tag{40}$$

Thus, an open computational form of ICOMP(IFIM)$_{C_{1F}}$ becomes

$$\text{ICOMP(IFIM)}_{C_{1F}} = -2\mathcal{L}_{\text{mix}} + \frac{s}{2} \frac{\frac{1}{s}\left[tr\left(\hat{\mathcal{F}}_\pi^{-1}\right)^2 + \sum_{g=1}^{G} tr\left(\hat{\mathcal{F}}_g^{-1}\right)^2\right] - \left[\frac{tr(\hat{\mathcal{F}}_\pi^{-1}) + \sum_{g=1}^{G} tr(\hat{\mathcal{F}}_g^{-1})}{s}\right]^2}{\left[\frac{tr(\hat{\mathcal{F}}_\pi^{-1}) + \sum_{g=1}^{G} tr(\hat{\mathcal{F}}_g^{-1})}{s}\right]^2}. \qquad (41)$$

We note that in computing both forms of ICOMP above, we do not need to build the full inverse of the estimated Fisher information matrix (IFIM) for the *MIX-SPCR* model given in Equation (36). All one requires is the computation of IFIM for each segment, which is appealing.

We also use AIC and CAIC (Bozdogan [33]) for comparison purposes given by

$$\text{AIC} = -2\mathcal{L}_{\text{mix}} + 2s^*, \text{ and,} \qquad (42)$$

$$\text{CAIC} = -2\mathcal{L}_{\text{mix}} + s^*(\log n + 1), \qquad (43)$$

where $s^* = G(k+3)$ is the number of estimated parameters in the *MIX-SPCR* model and log denotes the natural logarithm of the sample size n.

Next, we show our numerical examples starting with a detailed Monte Carlo simulation study.

5. Monte Carlo Simulation Study

We perform numerical experiments in a unified computing environment: Ubuntu 18.04 operating system, Intel I7-8700, and 32 GB of RAM. We use the programming language Python and the scientific computing package NumPy [34] to build a computational platform. The size of the input data directly affects the running time of the program. At $n = 4000$ time-series observations, the execution time for each EM iteration is about 0.9 s. Parameter estimation can reach convergence within 40 steps of iterations, with a total machine run time of 37 s.

5.1. Simulation Protocol

In this section, we present the performance of the proposed *MIX-SPCR* model using synthetic data generated from a segmented regression model. Our simulation protocol has $p = 12$ variables and four actual latent variables. Two segmented regression models determine the dependent variable y, and each segment is continuous and has its own specified coefficients (β_1 and β_2). Our simulation set up is as follows:

$$\Lambda = \begin{bmatrix} 1.8 & 0 & 0 & 0 \\ 1.8 & 0 & 0 & 0 \\ 1.8 & 0 & 0 & 0 \\ 0 & 1.7 & 0 & 0 \\ 0 & 1.7 & 0 & 0 \\ 0 & 1.7 & 0 & 0 \\ 0 & 0 & 1.6 & 0 \\ 0 & 0 & 1.6 & 0 \\ 0 & 0 & 1.6 & 0 \\ 0 & 0 & 0 & 1.5 \\ 0 & 0 & 0 & 1.5 \\ 0 & 0 & 0 & 1.5 \end{bmatrix}, \quad (44)$$

$$\psi = \text{diag}\,(1.27, 0.61, 0.74, 0.88, 0.65, 0.81, 0.74, 1.3, 1.35, 0.74, 0.92, 1.32), \quad (45)$$

$$\Sigma = \Lambda\Lambda^T + \psi, \quad (46)$$

$$x_t \sim \text{MVN}\,(0, \Sigma), t = 1, 2, \cdots, 4000, \quad (47)$$

$$\beta_1 = (-10, 0.1, 0.1, 0.1, 2.1, 0, 0, 0.1, 0.1, 0, 0, 0), \quad (48)$$

$$\beta_2 = (0, 0, 0, 0, 0, 0.5, 0.3, 0.1, 2.1, 1, 2, 20), \quad (49)$$

$$y_{t,g=1} = x_{1,t}\beta_1 + \varepsilon_{1,t}, t = 1, 2, \cdots, 2800, \quad (50)$$

$$y_{t,g=2} = x_{2,t}\beta_2 + \varepsilon_{2,t}, t = 2801, 2802, \cdots, 4000. \quad (51)$$

We set the total number of time-series observations, $n = 4000$. The first segment has $n_1 = 2800$, and the second segment has $n_2 = 1200$ time-series observations. We randomly draw error term from a Gaussian distribution with zero mean and $\sigma^2 = 9$. Among all the variables, the first six observable variables explain the first segment, and the remaining six explanatory variables primarily determine the second segment. We set the mixing proportions $\pi_1 = 0.7$ and $\pi_2 = 0.3$ for two time-series segments, respectively.

5.2. Detection of Structural Change Point

In the first simulation study, we limit the actual number of segments equal to two, which means that the first segment expands from the starting point to a structural change point, and the second segment expands from the change point to the end. By design, each segment is continuous on the time scale, and different sets of independent variables explain the trending and volatility. We run the *MIX-SPCR* model to see if it can successfully determine the position of the change point using the information criteria. If a change point is correctly selected, we expect that the information criteria is minimized at this change point.

Figures 2 and 3 show our results from the *MIX-SPCR* model. Specifically, it shows the sample path of the information criteria at each time point. We note that all the information criteria values are minimized from $t = 2800$ to $t = 3000$, which covers the time-series's actual change point position. As the *MIX-SPCR* model selects different change points, the penalty term of AIC and CAIC remain the same because both the number of model parameters and the number of observations do not change. In this simulation scenario, the fixed penalty term means that the AIC and CAIC reflect the changes only in the "lack of fit" term of various models without considering model complexity. This indicates that using AIC-type criteria just counting and penalizing the number of parameters may be necessary but not sufficient in model selection.

As a comparison, however, we note that the penalty term of information complexity-based criteria, C_1 and C_{1F}, are adjusted in selecting different change points. They are varying but not fixed.

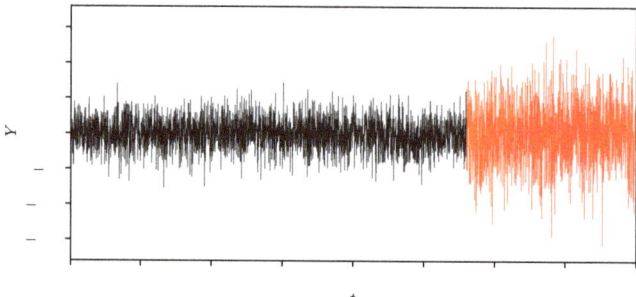

Figure 2. The plot of two-segment simulated time-series data. We show the plot of the simulated time-series data through the whole-time scale. Note that the first segment is from the starting point $t = 1$ to the change point $t = 2800$, and the second time segment expands from the change point $t = 2801$ to the end $t = 4000$.

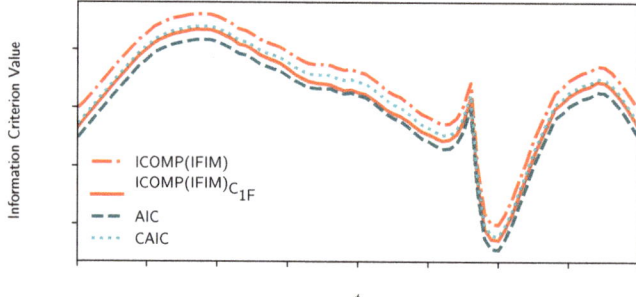

Figure 3. Sample path of information criteria for the simulated time-series data. The horizontal coordinate represents the position of the possible change points, and the vertical coordinate represents the corresponding information criterion (IC) values. The lower the IC value, the more likely the selected position of the change point is the real position. The real change point is $t = 2800$.

5.3. A Large-Scale Monte Carlo Simulation

Next, we perform a large-scale Monte Carlo simulation to illustrate the *MIX-SPCR* model's performance in choosing the correct number of segments and the number of latent variables. A priori, in this simulation, we pretend that we do not know the actual structure of the data and use the information criteria to recover the actual construction of the *MIX-SPCR* model. To achieve this, we follow the above simulation protocol using a different number of time points by varying $n = 1000, 2000, 4000$. As before, there are twelve explanatory variables drawn from four latent variable models generated from a multivariate Gaussian distribution given in Equation (47). The simulated data again consist of two time-series segments with mixing proportions $\pi_1 = 0.7$ and $\pi_2 = 0.3$, respectively. For each data generating process, we replicate the simulation one hundred times and record both information complexity-based criteria (ICOMP(IFIM) & ICOMP(IFIM)$_{C_{1F}}$) and classic AIC-type criteria (AIC & CAIC).

In Table 1, we present how many times the *MIX-SPCR* model selects different models in the one hundred simulations. In this way, we can assess different information criteria by measuring the hit rates.

Looking at Table 1, we see that when the sample size $n = 1000$ (small), AIC selects the correct model ($G = 2, k = 4$) 69 times, CAIC selects 80 times, ICOMP(IFIM) selects 48 times, and ICOMP(IFIM)$_{C_{1F}}$ selects 76 times, respectively, in 100 replications of the Monte Carlo simulation. When the sample size is small, ICOMP(IFIM) tends to choose a sparser regression model sensitive to the sample size. However, as the sample size increases, when $n = 2000$ and $n = 4000$, ICOMP(IFIM) consistently outperforms other information criteria in terms of hit rates. The percentage of the correctly identified model is above 90%, as reported above.

Table 1. Frequency of the choice of the true model with information criteria in 100 replications of the experiment for each sample size (n) of time-series observations. The true model is $G = 2$ and $k = 4$.

		$n = 1000$		$n = 2000$		$n = 4000$	
		$G = 2$	$G = 3$	$G = 2$	$G = 3$	$G = 2$	$G = 3$
AIC	$k = 2$	0	0	0	0	0	0
	$k = 3$	0	6	0	3	0	1
	$k = 4$	69	0	77	0	75	0
	$k = 5$	24	1	20	0	24	0
CAIC	$k = 2$	1	0	0	0	0	0
	$k = 3$	1	3	0	1	0	1
	$k = 4$	80	0	96	0	93	0
	$k = 5$	14	1	3	0	6	0
ICOMP(IFIM)	$k = 2$	31	2	1	0	0	0
	$k = 3$	2	5	0	2	0	1
	$k = 4$	48	0	96	0	96	0
	$k = 5$	11	1	1	0	3	0
ICOMP(IFIM)$_{C_{1F}}$	$k = 2$	2	1	0	0	0	0
	$k = 3$	0	7	0	3	0	1
	$k = 4$	76	0	93	0	93	0
	$k = 5$	13	1	4	0	6	0

Our results show that the *MIX-SPCR* model works well in all settings to estimate the number of time-series segments and the number of latent variables.

Figure 4 illustrates how the *MIX-SPCR* model performs if the number of segments and the number of sparse principal components are unknown beforehand.

The choice of the number of segments (G) has a significant impact on the results. For all the simulation scenarios, the correct choice of the number of segments ($G = 2$) has information criterion values less than the incorrect choice ($G = 3$). This pattern emerges consistently among all the sample sizes, both the classical ones and information-complexity based criteria.

In summary, the large-scale Monte Carlo simulation analysis highlights the performance of the *MIX-SPCR* model. As the sample size increases, the *MIX-SPCR* model improves its performance. As shown in Figure 3, the *MIX-SPCR* model can efficiently determine the structural change point and estimate the mixture proportions when the number of segments is unknown beforehand. Another key finding is that, by using the appropriate information criteria, the *MIX-SPCR* model can correctly identify the number of segments and the number of latent variables from the data. In other words, our approach can extract the main factors not only from the intercorrelated variables but also classify the data into several clearly defined segments on the time scale.

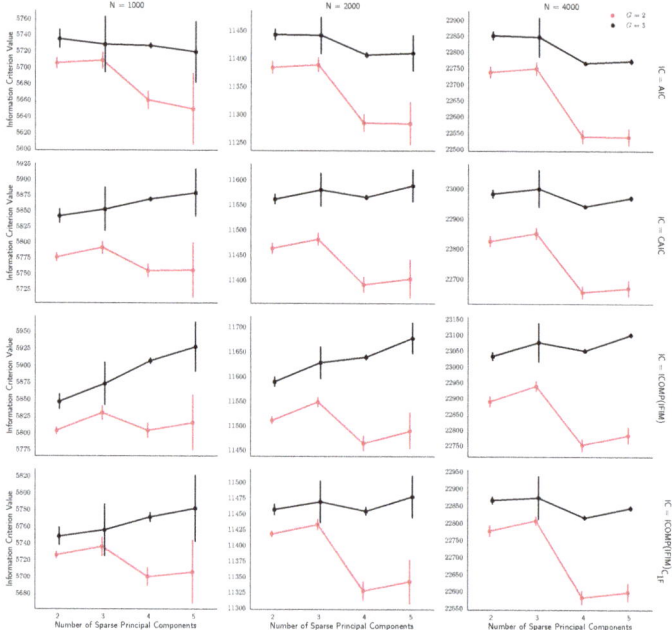

Figure 4. Plot of average and 1SD (standard deviation) of information criterion values over different sample sizes in all simulations with three Sparse Principal Components (SPCs) and $G = 2$ segments. The red line indicates the estimated *MIX-SPCR* model based on two groups ($G = 2$). Correspondingly, the black line indicates the estimated *MIX-SPCR* model for three groups ($G = 3$). Horizontal coordinates represent different numbers of SPCs.

6. Case Study: Segmentation of the S&P 500 Index

6.1. Description of Data

The financial market often generates a large amount of time-series data, and in most cases, the generated data is high-dimensional. In this paper, we use the S&P 500 index and its related hundreds of company stocks categorized into eleven sectors, which are high dimensional time-series data. The index value is the response variable mixed by plenty of companies' variations at each time point. These long time-series values often consist of different regimes and states. For example, the stock market experienced a boom period from 2017 to 2019, which is a dramatic change compared with the stock market during the 2008 financial crisis. If we analyze each sector or company, some industries perform more actively than others during a particular period.

In this section, we implement the *MIX-SPCR* model on the adjusted closing price of the S&P 500 (^GSPC) as a case study. We extract the daily adjusted closing prices from the Yahoo Finance database (https://finance.yahoo.com/) that spans the period from 1 January 1999 to 31 December 2019. By removing weekends and holidays, there are $n = 5292$ tradable days in total. The main focus of this section is to split the time-series into several self-contained segments. Besides, we expect the extracted sparse principal components to explain the variance and volatility in each segment.

6.2. Computational Results

To have a big picture of how the S&P 500 index values reflect the changes of 506 company stock prices, Figure 5 shows the plot of the normalized values of adjusted closing prices. We use the *MIX-SPCR* model with the information criteria to determine the number of segments and the number of sparse principal components. To achieve interpretable results, we limit our search space to a maximum of seven time-series and six sparse principal components. Table 2 shows the optimal combination of three self-contained segments and three sparse principal components for each of the segments by using the information complexity ICOMP(IFIM). The other three information criteria also choose this combination as the best-fitting model. Figure 6 illustrates the probability and time range of each segment. We can see that the first segment is from 1 January 1999, to 26 October 2007. The second time-series segment spans from 29 October 2007, to the end of 2016. The last segment extends from 30 December 2016 to 31 December 2019.

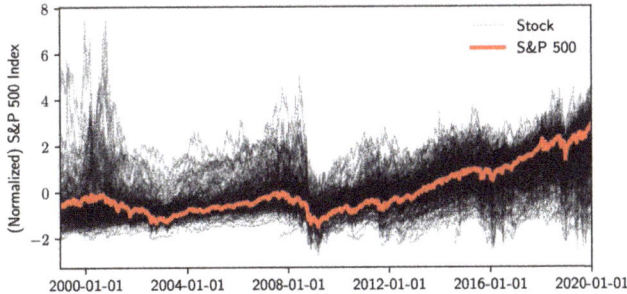

Figure 5. Normalized S&P 500 index and stock prices from January 1999 to December 2019.

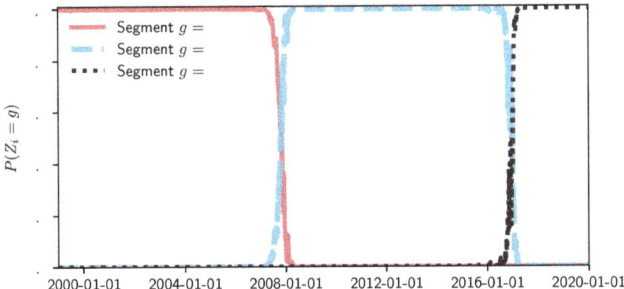

Figure 6. Segmented periods and probability. The plot's vertical coordinate indicates the probability that an individual time-series data point belongs to each segment.

Table 2. The ICOMP(IFIM) values of segmentation results for S&P 500 index data (Lower is better).

		\multicolumn{5}{c}{Number of Sparse Principal Components}
		1	2	3	4	5
Number of Segments	1	30,097.04	30,092.45	30,106.50	30,121.64	30,145.13
	2	29,975.01	30,058.40	30,293.55	30,234.65	30,347.94
	3	30,010.70	30,062.19	29,241.52	30,453.74	30,526.20
	4	29,877.27	29,825.73	29,811.53	30,571.39	30,628.61
	5	29,904.35	29,973.47	30,011.18	30,311.52	30,554.82
	6	30,111.35	30,361.39	30,388.47	30,665.26	30,581.29
	7	30,031.39	30,564.65	30,597.14	30,823.76	31,057.54

We emphasize that many factors may explain the stock market variation, and this is not a research on how the socioeconomic events influence the S&P 500 index. However, it does raise our interest in the distribution of two structural change points from the segmentation results. The first change point is October 2007, which is the early stage of the 2008 financial crisis. The second structural change point is December 2016, the transitional period of the USA presidential election. Identification of these two change points shows that our proposed method can detect the underlying physical and structural change from the available time-series data.

Table 3 lists the estimated coefficients (β_g) from sparse principal component regression. Because all the collected stock prices and S&P 500 index values are standardized before implementing the *MIX-SPCR* model, we make dimension reduction, remove the constant term, and perform regression analysis using the SPCR model. The R^2 values are above 0.8 across all three different time segments.

Table 3. SPCR coefficients (β_g) of three different segments.

	Segment 1 ($R^2 = 0.82$) 01-01-1999 ~ 26-10-2007	Segment 2 ($R^2 = 0.94$) 27-10-2007 ~ 29-12-2016	Segment 3 ($R^2 = 0.97$) 30-12-2016 ~ 31-12-2019
SPC1	0.0964	0.1240	0.1512
SPC2	0.0729	−0.0439	0.0359
SPC3	0.0079	0.0191	−0.0051

6.3. Interpretation of Computational Results

One may ask a question, "Can the *MIX-SPCR* model identify the key variables from the hundreds of companies?" If the constructed model is dense, the selected companies would include all the sectors whereby the dense model is limiting the interpretation of the data. Our analysis identifies all the companies with non-zero coefficient values and maps them back to each of the sectors in Tables A1–A3. Each calculated sparse principal component vector consists of around fifty companies, much less than the original data dimension ($p = 506$). We observe that these selected companies are grouped into a few sectors within different time segments. For example, energy companies load in the first sparse principal component vector from 1999 to 2007 (segment 1) and diminish after that.

To have a detailed analysis of how different sectors perform across three segments, we do the stem plot to show the sparse principal component coefficients P_g of four sectors, namely financials, real estate, energy, and information technology (IT). Figures 7 and 8 indicate a similar behavior that happened in financial and real estate companies. Both sectors play an essential role in the first two time-series segments but have no contribution in the third segment, which is the period after December 2016. Notice that in Figure 9, energy companies act as an essential player before 2016. However, during the recession in 2008, energy company loadings are negated from the first SPC to the second SPC. Compared with

Entropy 2020, 22, 1170

other industries, the variation in energy company stock prices does not contribute to the S&P 500 index after 2016.

Another question is "What sector/industry is the main contributing factor after the 2016 United States presidential election?" A possible answer is, as shown in Figure 10, the SPC coefficients of information technology companies. From 1999 to the recession in 2008, IT companies work mainly on the second SPC and the third SPC, which do not contribute much to the main variation. After the recession, the variations of IT companies do not contribute compared with other sectors. However, after December 2016, companies from the IT industry play an essential role in the primary stock price volatility.

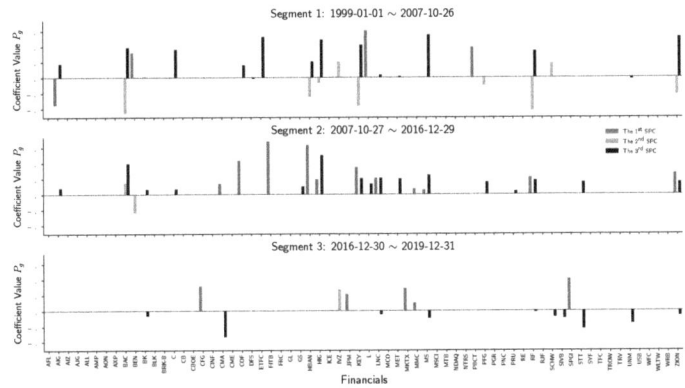

Figure 7. Stem plot of SPC coefficients P_g for financial companies within each time segment. From top to bottom, the three panels represent different segmented periods, respectively. The horizontal axis of each panel indicates the company in the industrial sector. The vertical axis shows the SPC coefficient values.

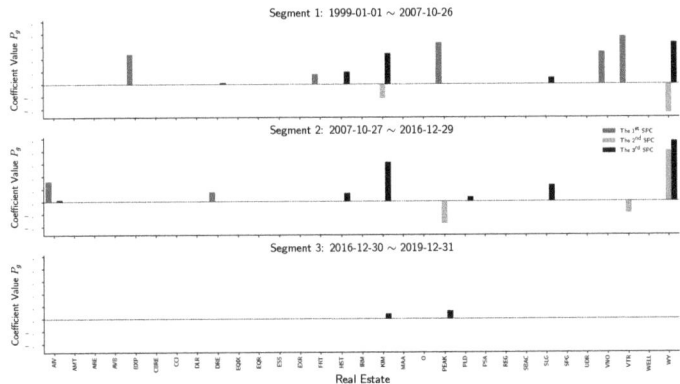

Figure 8. Stem plot of SPC coefficients P_g for real estate companies within each time segment. From top to bottom, the three panels represent different segmented periods, respectively. The horizontal axis of each panel indicates the company in the industrial sector. The vertical axis shows the SPC coefficient values.

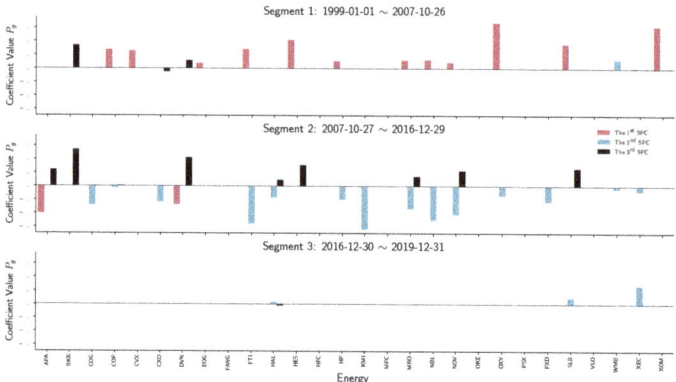

Figure 9. Stem plot of SPC coefficients $\mathbf{P}_g$ for energy companies within each time segment. From top to bottom, the three panels represent different segmented periods, respectively. The horizontal axis of each panel indicates the company in the industrial sector. The vertical axis shows the SPC coefficient values.

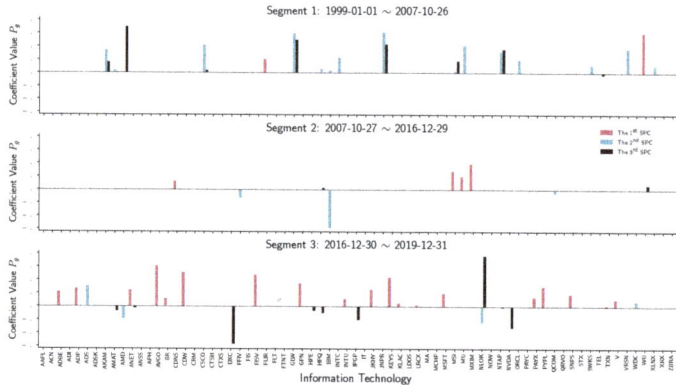

Figure 10. Stem plot of SPC coefficients $\mathbf{P}_g$ for information technology companies within each time segment. From top to bottom, the three panels represent different segmented periods, respectively. The horizontal axis of each panel indicates the company in the industrial sector. The vertical axis shows the SPC coefficient values.

As discussed above, Figures 7–10 provide a clear picture of how different sectors perform (via coefficient $\mathbf{P}_g$) without considering the effects on the S&P 500 index. It might raise the interest in how the SPCR coefficient $\mathbf{P}_g \beta_g$ changes before/after certain socioeconomic events. We follow the research implemented by Aït-Sahalia and Xiu [35] about how the Federal Reserve addressing heightened liquidity from March 10 to 14 March 2008, affects the stock market. The data analyzed by Aït-Sahalia and Xiu [35] are the S&P 100 index values using the traditional PCA, and the authors grouped stocks into financial and non-financial categories. Instead of PCA, we apply the SPCR model on the S&P 500 index and analyze how eleven sectors react before/after Federal Reserve operations. Figure 11 shows that financials, consumer discretionary, real estate, and industrials experienced more significant perturbations than other sectors in terms of SPCR coefficients $\mathbf{P}_g \beta_g$. This conclusion is consistent with the results from Aït-Sahalia and Xiu [35] that the average loadings of first and second principal components of financial companies are distinct from non-financial companies. However, considering that we have 506 companies in the raw data and make a sparse loading of companies for comparison, the excessive explanatory power is still maintained in this high-dimensional case using the SPCR model, which is more interpretable.

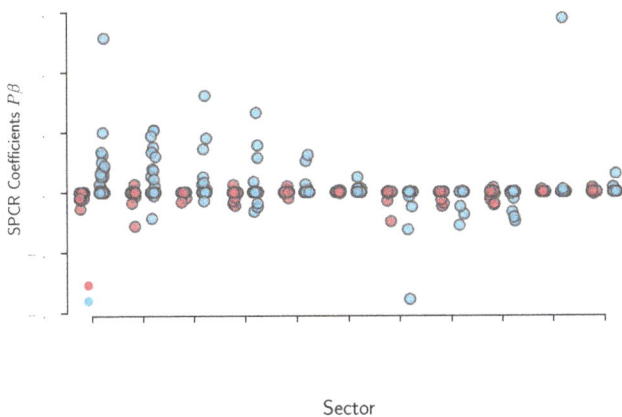

Figure 11. Overlay plot of the SPCR coefficients before/after 2008 financial crisis.

7. Conclusions and Discussions

In this paper, we presented a new and novel method to segment high-dimensional time-series data into different clusters or segments using the mixture model of the sparse principal components model (*MIX-SPCR*). The *MIX-SPCR* model considers both the relationships among the predictor variables and how various predictor variables contribute the explanatory power to the response variable through the sparsity settings. Information criteria have been introduced and derived for the *MIX-SPCR* model. These criteria are applied to study their performance under different sample sizes and to select the best-fitting model.

Our large-scale Monte Carlo simulation exercise showed that the *MIX-SPCR* model could successfully identify the real structure of the time-series data using the information criteria as the fitness function. In particular, based on our results, the information complexity-based criteria—i.e., ICOMP(IFIM) and ICOMP(IFIM)$_{C_{1F}}$—outperformed the conventional standard information criteria, such as the AIC-type criteria as the data dimension and the sample size increase.

Later, we empirically applied the *MIX-SPCR* model to uncover the S&P 500 index data (from 1999 to 2019) and identify two change points of this data set.

We observe that the first change point physically coincides with the early stages of the 2008 financial crisis. The second change point is immediately after the 2016 United States presidential election. This structural change point coincides with the election of President Trump and his transition.

Our findings showed how the S&P 500 index and company stock prices react within each time-series segment. The *MIX-SPCR* model presents excessive explanatory power by identifying how different sectors fluctuated before/after the Federal Reserve's addressing heightened liquidity from 10 March to 14 March 2008.

Although this is not a traditional event study paper, it is the first paper to use the sparse principal component regression model with mixture models in the time-series analysis. The proposed new and novel *MIX-SPCR* model enlightens us to explore more interpretable results on how macroeconomic factors/events influence the stock prices on the time scale. Later, in a separate paper, we will incorporate the event study in the *MIX-SPCR* model as our future research initiative.

This paper's time segmentation model builds on time-series data, constructs likelihood functions, and performs parameter estimation by introducing error information unique to each period. Researchers have recently realized that environmental background noise can positively affect the model building and analysis under certain circumstances ([36–42]). For example, in Azpeitia and Wagner [40], the authors highlighted that the introduction of noise is necessary to obtain information about the system. In our next study, we would like to explore this positive effect of environmental noise even further and use it to build better statistical models for analyzing high-dimensional time-series data.

Author Contributions: Conceptualization, H.B. and Y.S.; methodology, H.B. and Y.S.; software, Y.S.; validation, H.B. and Y.S.; formal analysis, H.B. and Y.S.; investigation, H.B. and Y.S.; resources, H.B. and Y.S.; data curation, Y.S.; writing–original draft preparation, H.B. and Y.S.; writing–review and editing, H.B. and Y.S.; visualization, H.B. and Y.S.; supervision, H.B.; project administration, H.B. and Y.S. All authors have read and agreed to the published version of the manuscript.

Funding: This research received no external funding.

Acknowledgments: The first author expresses his gratitude to Bozdogan in bringing this challenging problem to his attention as part of his doctoral thesis chapter and spending valuable time with him that resulted in this joint work. We also express our thanks to Ejaz Ahmed for inviting us to make a contribution to the Special Issue of Entropy. We extend our thanks and gratitude to anonymous reviewers. Their constructive comments further improved the paper.

Conflicts of Interest: The authors declare no conflict of interest.

Abbreviations

The following abbreviations are used in this manuscript:

MIX-SPCR	Mixture of the sparse principal component regression model
CV	Cross validation
TICC	Toeplitz inverse covariance-based clustering
GGS	Greedy Gaussian Segmentation
PCA	Principal Component Analysis
PC	Principal Component
SPCA	Sparse Principal Component Analysis
SPCR	Sparse Principal Component Regression
SPC	Sparse Principal Component
SPPCA	Sparse Probabilistic Principal Component Analysis
EM	Expectation–Maximization (Algorithm)
IC	Information Criterion
ICOMP	Information Complexity

Appendix A. Tables

Table A1. Sparse Principal Component (SPC) of Segment 1 (1 January 1999 ~ 26 October 2007).

	SPC1		SPC2		SPC3	
	Count	Percentage	Count	Percentage	Count	Percentage
Health Care	4	6.56	6	9.84	1	1.64
Industrials	6	8.57	3	4.29	6	8.57
Utilities	5	17.86	3	10.71	3	10.71
Materials	2	7.14	1	3.57	1	3.57
Consumer Discretionary	5	7.81	6	9.38	6	9.38
Energy	13	46.43	1	3.57	3	10.71
Financials	5	7.58	10	15.15	15	22.73
Real Estate	5	16.13	2	6.45	5	16.13
Consumer Staples	2	6.06	0	0.00	1	3.03
Communication Services	1	3.85	2	7.69	1	3.85
Information Technology	2	2.82	16	22.54	8	11.27

Table A2. Sparse Principal Component (SPC) of Segment 2 (27 October 2007 ~ 29 Decmeber 2016).

	SPC1		SPC2		SPC3	
	Count	Percentage	Count	Percentage	Count	Percentage
Health Care	7	11.48	2	3.28	1	1.64
Industrials	5	7.14	6	8.57	4	5.71
Utilities	0	0.00	0	0.00	5	17.86
Materials	6	21.43	2	7.14	3	10.71
Consumer Discretionary	7	10.94	14	21.88	3	4.69
Energy	2	7.14	14	50.00	9	32.14
Financials	12	18.18	3	4.55	16	24.24
Real Estate	2	6.45	3	9.68	6	19.35
Consumer Staples	0	0.00	0	0.00	0	0.00
Communication Services	5	19.23	3	11.54	1	3.85
Information Technology	4	5.63	3	4.23	2	2.82

Table A3. Sparse Principal Component (SPC) of Segment 3 (30 Decmeber 2016 ~ 31 Decmeber 2019).

	SPC1		SPC2		SPC3	
	Count	Percentage	Count	Percentage	Count	Percentage
Health Care	10	16.39	14	22.95	2	3.28
Industrials	9	12.86	4	5.71	4	5.71
Utilities	1	3.57	0	0.00	0	0.00
Materials	1	3.57	3	10.71	6	21.43
Consumer Discretionary	3	4.69	10	15.63	8	12.50
Energy	0	0.00	3	10.71	1	3.57
Financials	5	7.58	1	1.52	10	15.15
Real Estate	0	0.00	0	0.00	2	6.45
Consumer Staples	0	0.00	6	18.18	3	9.09
Communication Services	2	7.69	5	19.23	5	19.23
Information Technology	19	26.76	4	5.63	9	12.68

References

1. Barber, D.; Cemgil, A.T.; Chiappa, S. *Bayesian Time Series Models*; Cambridge University Press: Cambridge, UK, 2011.
2. Abonyi, J.; Feil, B. *Cluster Analysis for Data Mining and System Identification*; Springer Science & Business Media: New York, NY, USA, 2007.
3. Spagnolo, B.; Valenti, D. Volatility effects on the escape time in financial market models. *Int. J. Bifurc. Chaos* **2008**, *18*, 2775–2786.
4. Valenti, D.; Fazio, G.; Spagnolo, B. Stabilizing effect of volatility in financial markets. *Phys. Rev. E* **2018**, *97*, 062307.
5. S Lima, L. Nonlinear Stochastic Equation within an Itô Prescription for Modelling of Financial Market. *Entropy* **2019**, *21*, 530.
6. Ding, W.; Wang, B.; Xing, Y.; Li, J.C. Correlation noise and delay time enhanced stability of electricity futures market. *Mod. Phys. Lett. B* **2019**, *33*, 1950375.
7. Dillon, W.R.; Böckenholt, U.; De Borrero, M.S.; Bozdogan, H.; De Sarbo, W.; Gupta, S.; Kamakura, W.; Kumar, A.; Ramaswamy, B.; Zenor, M. Issues in the estimation and application of latent structure models of choice. *Mark. Lett.* **1994**, *5*, 323–334.
8. Quandt, R.E.; Ramsey, J. Estimating Mixtures of Normal Distributions and Switching Regressions. *J. Am. Stat. Assoc.* **1978**, *73*, 730–738.
9. Kiefer, N.M. Discrete parameter variation: Efficient estimation of a switching regression model. *Econometrica* **1978**, *46*, 427–434.
10. De Veaux, R.D. *Parameter Estimation for a Mixture of Linear Regressions*; Doctoral Dissertation and Tech. Rept. No. 247; Department of Statistics, Stanford University: Stanford, CA, USA, 1986.
11. DeSarbo, W.S.; Cron, W.L. A maximum likelihood methodology for clusterwise linear regression. *J. Classif.* **1988**, *5*, 249–282.
12. Wedel, M.; DeSarbo, W.S. A Review of Recent Developments in Latent Class Regression Models; In *Advanced Methods of Marketing Research*; Bagozzi, R., Ed.; Blackwell Pub.: Hoboken, NJ, USA, 1994; pp. 352–388.
13. Sclove, S.L. Time-series segmentation: A model and a method. *Inf. Sci.* **1983**, *29*, 7–25.
14. Zou, H.; Hastie, T.; Tibshirani, R. Sparse principal component analysis. *J. Comput. Graph. Stat.* **2006**, *15*, 265–286.
15. Yang, M.S.; Lai, C.Y.; Lin, C.Y. A robust EM clustering algorithm for Gaussian mixture models. *Pattern Recognit.* **2012**, *45*, 3950–3961.
16. Bozdogan, H. On the information-based measure of covariance complexity and its application to the evaluation of multivariate linear models. *Commun. Stat. Theory Methods* **1990**, *19*, 221–278.
17. Bozdogan, H. Choosing the number of component clusters in the mixture-model using a new informational complexity criterion of the inverse-Fisher information matrix. In *Information and Classification*; Springer: New York, NY, USA, 1993; pp. 40–54.
18. Bozdogan, H. Choosing the number of clusters, subset selection of variables, and outlier detection in the standard mixture-model cluster analysis. In *New approaches in Classification and Data Analysis*; Springer: New York, NY, USA, 1994; pp. 169–177.
19. Bozdogan, H. Mixture-model cluster analysis using model selection criteria and a new informational measure of complexity. In *Proceedings of the First US/Japan Conference on the Frontiers of Statistical Modeling: An Informational Approach*; Springer: New York, NY, USA, 1994; pp. 69–113.
20. Bozdogan, H. A new class of information complexity (ICOMP) criteria with an application to customer profiling and segmentation. *İstanbul Üniversitesi İşletme Fakültesi Derg.* **2010**, *39*, 370–398.
21. Shen, H.; Huang, J.Z. Sparse principal component analysis via regularized low rank matrix approximation. *J. Multivar. Anal.* **2008**, *99*, 1015–1034.
22. Mairal, J.; Bach, F.; Ponce, J.; Sapiro, G. Online dictionary learning for sparse coding. In Proceedings of the 26th Annual International Conference on Machine Learning, Montreal, QC, Canada, 14–18 June 2009; pp. 689–696.
23. Guan, Y.; Dy, J. Sparse probabilistic principal component analysis. In Proceedings of the Twelfth International Conference on Artificial Intelligence and Statistics, Clearwater, FL, USA, 16–19 April 2009; pp. 185–192.
24. Williams, P.M. Bayesian regularization and pruning using a Laplace prior. *Neural Comput.* **1995**, *7*, 117–143.

25. Guarcello, C.; Valenti, D.; Spagnolo, B.; Pierro, V.; Filatrella, G. Josephson-based threshold detector for Lévy-distributed current fluctuations. *Phys. Rev. Appl.* **2019**, *11*, 044078.
26. Gaffney, S.; Smyth, P. Trajectory clustering with mixtures of regression models. In Proceedings of the Fifth ACM SIGKDD International Conference on Knowledge Discovery and Data Mining, San Diego, CA, USA, 15–18 August 1999; pp. 63–72.
27. Esling, P.; Agon, C. Time-series data mining. *ACM Comput. Surv. (CSUR)* **2012**, *45*, 1–34.
28. Gaffney, S. Probabilistic Curve-Aligned Clustering and Prediction with Regression Mixture Models. Ph.D. Thesis, University of California, Irvine, CA, USA, 2004.
29. Kullback, A.; Leibler, R. On Information and Sufficiency. *Ann. Math. Stat.* **1951**, *22*, 79–86.
30. Akaike, H. Information Theory and an Extension of the Maximum Likelihood Principle. In *Second International Symposium on Information Theory*; Petrox, B., Csaki, F., Eds.; Academiai Kiado: Budapest, Hungary, 1973; pp. 267–281.
31. Bozdogan, H. Akaike's Information Criterion and Recent Developments in Information Complexity. *J. Math. Psychol.* **2000**, *44*, 62–91.
32. Van Emden, H.M. An analysis of complexity. In *Mathematical Centre Tracts*; Mathematisch Centrum: Amsterdam, The Netherlands, 1971.
33. Bozdogan, H. Model Selection and Akaike's Information Criteria (AIC): The General Theory and its Analytical Extensions. *Psychometrica* **1987**, *52*, 317–332.
34. van der Walt, S.; Colbert, S.C.; Varoquaux, G. The NumPy array: A structure for efficient numerical computation. *Comput. Sci. Eng.* **2011**, *13*, 22–30.
35. Aït-Sahalia, Y.; Xiu, D. Principal component analysis of high-frequency data. *J. Am. Stat. Assoc.* **2019**, *114*, 287–303.
36. Spagnolo, B.; Valenti, D.; Guarcello, C.; Carollo, A.; Adorno, D.P.; Spezia, S.; Pizzolato, N.; Di Paola, B. Noise-induced effects in nonlinear relaxation of condensed matter systems. *Chaos Solitons Fractals* **2015**, *81*, 412–424.
37. Valenti, D.; Magazzù, L.; Caldara, P.; Spagnolo, B. Stabilization of quantum metastable states by dissipation. *Phys. Rev. B* **2015**, *91*, 235412.
38. Spagnolo, B.; Guarcello, C.; Magazzù, L.; Carollo, A.; Persano Adorno, D.; Valenti, D. Nonlinear relaxation phenomena in metastable condensed matter systems. *Entropy* **2017**, *19*, 20.
39. Serdukova, L.; Zheng, Y.; Duan, J.; Kurths, J. Stochastic basins of attraction for metastable states. *Chaos Interdiscip. J. Nonlinear Sci.* **2016**, *26*, 073117.
40. Azpeitia, E.; Wagner, A. The positive role of noise for information acquisition in biological signaling pathways. *bioRxiv* **2019**, *2019*, 762989.
41. Addesso, P.; Filatrella, G.; Pierro, V. Characterization of escape times of Josephson junctions for signal detection. *Phys. Rev. E* **2012**, *85*, 016708.
42. Li, J.h.; Łuczka, J. Thermal-inertial ratchet effects: Negative mobility, resonant activation, noise-enhanced stability, and noise-weakened stability. *Phys. Rev. E* **2010**, *82*, 041104.

Publisher's Note: MDPI stays neutral with regard to jurisdictional claims in published maps and institutional affiliations.

© 2020 by the authors. Licensee MDPI, Basel, Switzerland. This article is an open access article distributed under the terms and conditions of the Creative Commons Attribution (CC BY) license (http://creativecommons.org/licenses/by/4.0/).

Article

Forecasting Financial Time Series through Causal and Dilated Convolutional Neural Networks

Lukas Börjesson and Martin Singull *

Department of Mathematics, Linköping University, 581 83 Linköping, Sweden; lukbo072@student.liu.se
* Correspondence: martin.singull@liu.se

Received: 31 August 2020; Accepted: 25 September 2020; Published: 29 September 2020

Abstract: In this paper, predictions of future price movements of a major American stock index were made by analyzing past movements of the same and other correlated indices. A model that has shown very good results in audio and speech generation was modified to suit the analysis of financial data and was then compared to a base model, restricted by assumptions made for an efficient market. The performance of any model, trained by looking at past observations, is heavily influenced by how the division of the data into train, validation and test sets is made. This is further exaggerated by the temporal structure of the financial data, which means that the causal relationship between the predictors and the response is dependent on time. The complexity of the financial system further increases the struggle to make accurate predictions, but the model suggested here was still able to outperform the naive base model by more than 20% and 37%, respectively, when predicting the next day's closing price and the next day's trend.

Keywords: deep learning; financial time series; causal and dilated convolutional neural networks

1. Introduction

Deep learning has brought a new paradigm into machine learning in the past decade and has shown remarkable results in areas such as computer vision, speech recognition and natural language processing. However, one of the areas where it is yet to become a mainstream tool is in forecasting financial time series. This despite the fact that time series does provide a suitable data representation for deep learning methods such as a convolutional neural network (CNN) [1]. Researchers and market participants (Market participants is a general expression for individuals or groups who are active in the market, such as banks, investors, investment funds, or traders (for their own account); often, we use the term *trader* as a synonym for *market participant* [2]) are still, to the most part, sticking to more historically well known and tested approaches, but there has been a slight shift of interest to deep learning methods in the past years [3]. The reason behind the shift, apart from the structure of the time series, is that the financial market is an increasingly complex system. This means that there is a need for more advanced models, such as deep neural networks, that do a better job in finding the nonlinear relations in the data.

Omar Berat Sezer et al. [3] gives a very informative review of the published literature on the subject between 2005 and 2019 and states that there has been a trend towards more usage of deep learning methods in the past five years. The review covers a wide range of deep learning methods, applied to various time series such as stock market indices, commodities and forex. From the review, it is clear that CNNs is not the topmost used method and that developers have focused more on recurrent neural networks (RNNs) and long short-term memory (LSTM) networks. The CNNs are, however, very good at building up high-level features from lower-level features that were originally found in the input data, which is not something a LSTM network is primarily designed to do.

However, the CNNs and LSTM networks do not need to be used as two separate models, but they could be used as two separate parts of the same network. An example is to use a CNN to preprocess the data, in order to extract suitable features, which could then be used as the input to the LSTM part of the network [4].

Furthermore, the WaveNet structure considered in [5] suggests that the model can catch long- and short-term dependencies, which is what the LSTM is designed to do as well. Although this is something that will not be further explored in this paper, it does provide further research questions, such as if the WaveNet can be used as a sort of preprocessing to a LSTM network. Another example would be to process the data through a CNN and a LSTM separately and then combine them, before the final output, in a suitable manner. This is something that is explored, with satisfactory results, in [6], and the CNN part of the network is in fact an implementation of the WaveNet here as well. However, only the LSTM part of the network handles the exogenous series; therefore, for future work, it would be interesting to see if the performance could be improved by making the WaveNet handle the exogenous series as well.

Papers where the WaveNet composes the whole network instead of just being a component in one exist as well. Two examples are [7,8], and these models take into consideration exogenous series as well. However, they used ReLU as the activation function instead of SeLU, but they implemented normalization of the network in a similar way as will be done in this paper. Furthermore, when considering the exogenous series, their approach regarding the output from each residual layer was different. Instead of extracting the residual from each exogenous series, which will be done in this paper, only the combined residual was used. See Section 3.2 for more details.

In contrast to the approach taken in this paper, and by all who try to fit statistical models on financial time series, there are those who state that complexity is not the issue, but instead advocate for the Efficient Market Hypothesis (EMH) [9]. A theory that essentially suggests that no model, no matter how complex, can outperform the market, since the price is based on all available information. The theory rests upon three key assumptions—(1) no transaction costs, (2) cost is not a barrier to obtain available information and (3) all market participants agree on the implications of the available information—which are stated to be sufficient, but not necessary (sufficient but not necessary means, in this context, that the assumptions do not need to be fulfilled at all times; for example, the second assumption might be loosened from including all to only a sufficient number of traders). These assumptions, even with modifications, are very bold, and there are many who have criticized the theory over the years. However, whether one agrees with the theory or not, one would probably agree with the statement that a model which satisfies the assumptions made in EMH would indeed be suitable as a base model. This means that such a model can be used as a benchmark in order to assess the accuracy of other models.

Traders and researchers alike would furthermore agree that the price of any asset is, apart from its inner utility, based on the expectation of its future value. For example, the price of a stock is partially determined by the company's current financials, but also by the expectation of future revenues or future dividends. This expectation is, by the neoclassical economics, seen as completely rational, giving rise to the area of rational expectation [10]. However, the emergence of behavioral economics has questioned this rationality and proposes that traders (or more generally, decision-makers who act under risk and uncertainty) are sometimes irrational and many times affected by biases [11].

A trader that sets out to exploit this irrationality and these biases can only do so by looking into the past and, thereby, also go against the hypothesis of the efficient market. Upon reading this, it should be fairly clear that making predictions in the financial markets is no trivial task, and it should be approached with humility. However, one should not be discouraged since the models proposed in [3] do provide promising or, oftentimes, positive results.

An important note about the expectation mentioned above is that the definition of a trader, provided by Paul and Baschnagel, is not limited to a human being; it might as well be an algorithm. This is important since the majority of transactions in the market are now made by algorithms.

These algorithms are used in order to decrease the impact of biases and irrationality in decision making. However, the algorithms are programmed by people and are still making predictions under uncertainty, based on historical data, which means that they are by no means free of biases. Algorithms are also more prone to get stuck in a feedback loop, which has been exploited by traders in the past. An interesting example is the two Norwegians, Svend Egil Larsen and Peder Veiby, who in 2010 were accused of manipulating algorithmic trading systems. They were, however, acquitted in 2012, since the court found no wrongdoing.

The aim of this paper is to expand the research in forecasting financial time series, specifically with a deep learning approach. To achieve this, two models, which greatly differ in the approach towards the effectiveness of the market, are compared. The first model is restricted by the assumptions made on the market by the EMH and is seen as the base model. The second model is a CNN, inspired by the WaveNet structure [5], and is influenced by a model developed for audio and speech generation by researchers at Google. The models set out to predict the next day's closing price as well as the trend (either up or down) of Standard and Poor's 500 (S&P 500), a well-known stock market index comprising 500 large companies in the US.

The outline of this paper is as follows. In Section 2 we give a brief background to the theory needed and in Section 3 formulate the considered models and discuss the methodology. The results of the study are given in Section 4 with a discussion in Section 5. We conclude the paper in Section 6.

2. Theoretical Background

2.1. Time Series

When using time series as a forecasting model, one makes the assumption that future events, such as the next day's closing price of a stock, can be determined by looking at past closing prices in the time series. Most models, however, include a random error as a factor, meaning that there is some noise in the data which cannot be explained by past values in the series.

Furthermore, the models can be categorized as parametric or non-parametric, where the parametric models are the ones most regularly used. In the parametric approach, each data point in the series is accompanied by a coefficient, which determines the impact the past value has on the forecast of future values in the series.

The linear autoregressive model of order p, written as $AR(p)$, is given by

$$X_t = c + \sum_{i=1}^{p} \varphi_i X_{t-i} + \varepsilon_t, \tag{1}$$

with unknown parameters φ_i and ε_t as white noise. This is one of the most well known time series models, and it is a model where the variable is regressed against itself (auto meaning "oneself", when used as a prefix). It is often used as a building block to more advanced time series models, such as the autoregressive moving average (ARMA) or the generalized autoregressive conditional heteroskedasticity (GARCH) models. However, the AR process will not be considered as a building block in the models proposed in this paper. Instead, the proposed CNN models in this paper can be represented as the nonlinear version of the AR model, or $NAR(p)$ for short, given as

$$X_t = c + f(X_{t-1}, \ldots, X_{t-p}) + \varepsilon_t, \tag{2}$$

with a nonlinear function $f(\cdot, \ldots, \cdot)$. In fact, a large number of machine learning models, when applied to time series, can be seen as AR or NAR models. This might seem obvious to some, but it is something that is seldom mentioned in the scientific literature. Furthermore, the models can be generalized to a nonlinear autoregressive exogenous (NARX) model. Given Z_k as an exogenous time series and ψ_k its accompanied coefficient, we have the $NARX(p,r)$ model

$$X_t = c + f(X_{t-1}, \ldots, X_{t-p}, Z_{t-1}, \ldots, Z_{t-r}) + \varepsilon_t. \tag{3}$$

When determining the coefficients in the autoregressive models, most models need the underlying stochastic processes $\{X_t : t \geq 0\}$ to be stationary or at least weak-sense stationary. This means that we are assuming that the mean and the variance of X_t are constant over time. However, when looking at historical prices in the stock market, one can clearly see that this is not the case, for either the variance or the mean. All of the above models can be generalized to handle this non-stationarity by applying suitable transformations to the series. These transformations, or integrations, are a necessity when determining the values for the coefficients, for most of the well-known methods, although this need not be the case when using a neural network [12].

2.2. Neural Networks

The neural network, when applied on a supervised problem, sets out to minimize a certain predefined loss function. The loss function used in this paper, when calculating the next day's closing price, was the mean absolute percentage error (MAPE)

$$\epsilon(\mathbf{w}) = \frac{100}{n} \sum_{i=1}^{n} \left| \frac{y_i - t_i}{t_i} \right|,$$

where t_i is the ith target value and y_i is the model's prediction. The reason for this is that the errors are now made proportional with respect to the target value. This is important, since the mean and variance of financial series cannot be assumed to be stationary, and this would otherwise skew the accuracy of the model, unproportionally, to times characterized by a low mean. Meanwhile, the loss function used for classifying the next day's trend was the binary crossentropy

$$\epsilon(\mathbf{w}) = \frac{1}{n} \sum_{i=1}^{n} t_i \log(y_i) + (1 - t_i) \log(1 - y_i)).$$

The loss function is with respect to the weight $\mathbf{w}$, and the loss is minimized when choosing the weights that solve the function

$$\frac{\partial \epsilon(\mathbf{w})}{\partial \mathbf{w}} = 0.$$

However, this algebraic solution is seldom achievable, and numerical solutions are more often used. These numerical methods set out to find points in close proximity to a local (hopefully global, but probably not) optima.

Moreover, instead of calculating the gradient with respect to each weight individually, backpropagation uses the chain rule, where each derivative can be computed layerwise backward. This leads to a decrease in complexity, which is very important, since it is not unusual that the number of weights might be counted in thousands or in tens of thousands.

The neural network is, unless stated otherwise, considered to be a fully connected network, which means that all weights, in two adjacent layers, are connected to each other. Although backpropagation did a remarkable job in decreasing the complexity, the fully connected models are not good at scaling to many hidden layers. This problem can be solved by having a sparse network, which means that not all weights are connected. The CNN model, further explained in the next section, is an example of a sparse network, where not all units are connected and where some units also share weights.

2.3. Convolutional Neural Networks

The input to the CNN, when modeling time series, is a three-dimensional tensor, i.e., (nr of observations)×(width of the input)×(nr of series). The number of series is here the main series, for which the predictions will be made over, plus optional exogenous series.

Furthermore, in the CNN model, there is an array of hyperparameters that defines the structure and complexity of the network. Below is a short explanation of the most important parameters to be acquainted with in order to understand the networks proposed in this paper.

2.3.1. Activation Function

In its simplest form, when it only takes on binary values, the activation function determines if the artificial neuron fires or not. More complex activation functions are often used, and the sigmoid and tanh functions

$$g(x) = \frac{e^x}{e^x + 1}, \qquad (4)$$

$$g(x) = tanh(x),$$

are two examples, which have been used to a large extent in the past. They are furthermore two good examples of activation functions that can cause the problem of vanishing gradients (studied by Sepp Hochreiter in 1991, but further analyzed in 1997 by Sepp Hochreiter and Jürgen Schmidhuber [13]), which of course is something that should be avoided. A function that does not exhibit this problem is the rectified linear unit (ReLU) function

$$g(x) = \begin{cases} 0 & \text{if } x \leq 0, \\ x & \text{otherwise,} \end{cases}$$

which has gained a lot of traction in recent years, and is today the most popular one for deep neural networks. One could easily understand why ReLU avoids the vanishing gradient problem, by looking at its derivative

$$g'(x) = \begin{cases} 0 & \text{if } x \leq 0, \\ 1 & \text{otherwise,} \end{cases}$$

and from it conclude that the gradient is either equal to 0 or 1. However, the derivative also shows a different problem that comes with the ReLU function, which is that the gradient might equal zero, and that the output from many of the nodes might in turn become zero. This problem is called the dead ReLU problem, and it might cause many of the nodes to have zero impact on the output. This can be solved by imposing minor modifications on the function, and it therefore now comes in an array of different flavors. One such flavor is the exponential linear unit (ELU)

$$g(x) = \begin{cases} \alpha(e^x - 1) & \text{if } x \leq 0, \\ x & \text{otherwise,} \end{cases}$$

where the value of alpha is often chosen to be between 0.1 and 0.3. The ELU solves the dead ReLU problem, but it comes with a greater computational cost. A variant of the ELU is the scaled exponential linear unit (SELU)

$$g(x) = \lambda \begin{cases} \alpha e^x & \text{if } x \leq 0, \\ x & \text{otherwise,} \end{cases} \qquad (5)$$

which is a relatively new activation function, first proposed in 2017 [14]. The values of α and λ have been predefined by the authors, and the activation also needs the weights in the network to be initialized in a certain way, called lecun_normal. lecun_normal initialization means that the start value for each weight is drawn from a standard normal distribution.

Normalization can be used as a preprocessing of the data, due to its often positive effect on the model's accuracy, and some networks also implement batch normalization at some point or points inside the network. This is what is called external normalization. However, the beauty of SeLU is that the output of each node is normalized, and this process is fittingly called internal normalization. Internal normalization proved to be more useful than external normalization for the models in this paper, which is why SeLU was used throughout the network in the final models.

2.3.2. Learning Rate

The learning rate, often denoted by η, plays a large role during the training phase of the models. After each iteration, the weights are updated by a predefined update rule such as gradient descent

$$\mathbf{w}_{i+1} = \mathbf{w}_i - \eta \nabla \epsilon(\mathbf{w}_i),$$

where $\nabla \epsilon(\mathbf{w}_t)$ is the gradient for the loss function at the ith iteration. The learning rate, η, can here be seen as determining the rate of change in every iteration. Gradient descent is but one of many update rules, or optimizers (as they are more often called), and it is by far one of the simplest. More advanced optimizers are often used, such as the adaptive moment estimation (Adam) [15], which has, as one if perks, individual learning rates for each weight. The discussion about optimizers will not continue further in this paper, but it should be clear that the value of the learning rate and the choice of optimizer have a great impact on the overall performance of the model.

2.3.3. Filters

The filter dimensions need to be determined before training the model, and the appropriate dimensions depend on the underlying data and the model of choice. When analyzing time series, the filter needs to be one dimensional, since the time series is just an ordered sequence. The developer needs to determine just two things: the width of the filters (Figure 1 shows a filter with the width equal to two) and then how many filters to use for each convolutional layer. The types of features that the convolutional layer "searches" for are highly influenced by the filter dimensions, and having multiple filters means that the network can search for more features in each layer.

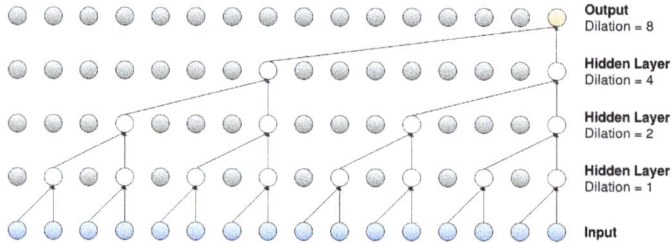

Figure 1. Dilated convolutional layers for an input series of length 16.

2.3.4. Dilation

A dilated convolutional filter is a filter that, not surprisingly, is widened but still uses the same number of parameters. The filter is widened by neglecting certain inputs, and an example of this can be observed in Figure 1. The bottom layer represents the input, in the form of a time series $x = (x_1, x_2, \ldots, x_n)$ (for some time n, onto which repeated dilated convolutions, with increasing dilation rates, are applied; the filter width is again set to equal two in the observed model). The first hidden layer applies dilated convolutions with the dilation rate equal to one, meaning that the layer applies the filter onto two adjacent elements, x_i and x_{i+1}, of the input series. The second layer applies dilated convolutions, with the rate now set to equal two, which means that the filter is applied onto elements x_i and x_{i+2} (notice here that the number of parameters remains the same, but the filter width has been "widened"). Lastly, the third and fourth layer have rates equal to four and eight, so the filter is applied onto elements x_i and x_{i+4}, and x_i and x_{i+8}, respectively.

2.3.5. Dropout

The dropout factor is a way to prevent the model from overfitting to the training data, and it does this by setting a fraction of the weights in a certain layer to zero. This leads to a decrease in complexity, but the developer does not have control over which nodes will be set to zero (i.e., the weights are chosen at random). Hence, dropout is not the same as changing the number of nodes in the network. For more details, see [16].

2.4. WaveNet

The CNN models proposed in this paper are inspired by the WaveNet structure, modeled by van den Oord et al. in 2016 [5]. The main part of the network in a WaveNet can be visualized in Figure 2, which incorporates a dilated (and causal) convolution and a 1×1 convolution (i.e., the width of the filter set to equal one). The input from the left side is the result of a casual convolution, with filter size equal to two, which has been applied to the original input series as a sort of preprocessing. The output on the right side of the layer is the residual, which can be used as the input to a new layer, with an identical set up. The number of residual connections must be predetermined by the developer, but the dilated convolution also sets an upper limit on how many connections can be used. Figure 1 displays repeated dilations on an input series with length equal to 16, and we can see that the number of layers has an upper limit of four.

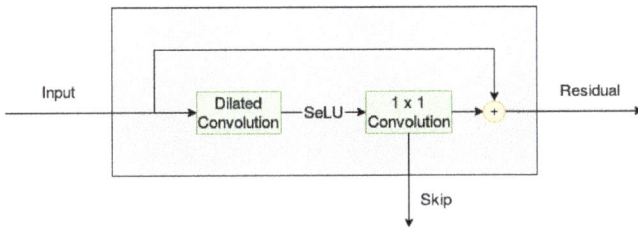

Figure 2. Overview of the residual layer, when only the main series is used as input.

Furthermore, the output from the bottom of each layer is the skip, which is the output that is passed on to the following layers in the network. If four layers are used, as in Figure 1, then the network would end up with four skip connections. These skip connections are then added (element-wise) together to form a single output series. This series is then passed through two 1×1 convolutions, and the result of this will be the output of the model.

The WaveNet has three important characteristics: it is dilated, causal and has residual connections. This means that the network is sparsely connected, that calculations can only include previous values in the input series (which can be observed in Figure 1) and that information is preserved across multiple layers. The sparsity is also further increased by having the width of the filters equal to only either one or two.

The WaveNet sets out to maximize the joint probability of the series $\mathbf{x} = (x_t, \ldots, x_{t-p})^T$, for any time t and length equal to p, which is factorized as a product of conditional probabilities

$$p(\mathbf{x}) = \prod_{i=1}^{t} p(x_i | x_1, \ldots, x_{i-1}),$$

where the conditional probability distributions are modeled by convolutions. Furthermore, the joint probability can be generalized to include exogenous series

$$p(\mathbf{x}|\mathbf{h}) = \prod_{i=1}^{t} p(x_i|x_1,\ldots,x_{i-1},h_1,\ldots,h_{i-1}),$$

where $\mathbf{h} = (h_t, h_{t-1}, \ldots, h_1)^T$ is the exogenous series.

The WaveNet, as proposed by the authors, uses a gated activation unit on the output from the dilated convolution layer in Figure 2

$$\mathbf{z} = tanh(\mathbf{w}_{t,k} * \mathbf{x}) \odot \sigma(\mathbf{w}_{s,k} * \mathbf{x}),$$

where $*$ is a convolution operator, $\odot$ is an element-wise multiplication error, σ is a sigmoid function, $\mathbf{w}_{*,k}$ is the weights for the filters and k denotes the layer index. However, the model proposed in this paper will be restricted to only use a single activation function

$$\mathbf{z} = SeLU(\mathbf{w}_k * \mathbf{x}) \tag{6}$$

and the reason behind this is, again, that the gated activation function did not generalize well to the analyzed time series data.

When using an exogenous series to help improve the predictions, the authors introduce two different ways to condition the main series by the exogenous series. The first way, termed global conditioning, uses a conditional latent vector $\mathbf{l}$ (not dependent on time), accompanied with a filter $\mathbf{v}_k$, and can be seen as a type of bias that influences the calculations across all timesteps

$$\mathbf{z} = SeLU(\mathbf{w}_k * \mathbf{x} + \mathbf{v}_k * \mathbf{l}).$$

The other way, termed local conditioning, uses one or more conditional time series $\mathbf{h} = (h_t, h_{t-1}, \ldots, h_1)^T$, that again influences the calculations across all timesteps

$$\mathbf{z} = SeLU(\mathbf{w}_k * \mathbf{x} + \mathbf{v}_k * \mathbf{h}) \tag{7}$$

and this is the approach that has been taken in this paper. This approach can further be observed in Figure 3.

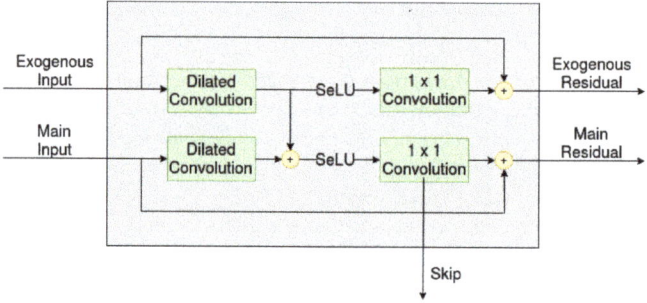

Figure 3. Overview of the residual layer, when the main series is conditioned by an exogenous series.

Lastly, the WaveNet originally used a softmax activation function on the last output of the network, with the target values (raw audio) quantized into 256 different values. However, the softmax did not generalize well to the predictions for the financial time series used here, where the use of no activation function performed better when predicting the next day's closing price of the S&P 500. In addition, when classifying the trend, the activation on the last output was seen as a hyperparameter, where the sigmoid and SeLU activations were compared against each other.

2.5. Walk-Forward Validation

Walk-forward validation, or walk-forward optimization, was suggested by Robert Pardo [17] and was brought forward since the ordinary cross-validation strategy is not well suited for time series data. The reason behind why cross-validation is not optimal for time series data is because temporal correlations exist in the data, and it should then be considered as "cheating" if one were to use future data points to predict past data points. This, most likely, leads to a lower training error, but should result in a higher validation/test error, i.e., it leads to poorer generalization due to overfitting. In order to avoid overfitting, the model should then, when making predictions at (or past) time t, only be trained on data points that were recorded before time t.

Depending on the data and the suggested model, one may choose between using all past observations (until the time of prediction) or using a fixed number of most recent observations as training data. The walk-forward scheme, using only a fixed number of observations, can be observed in Figure 4.

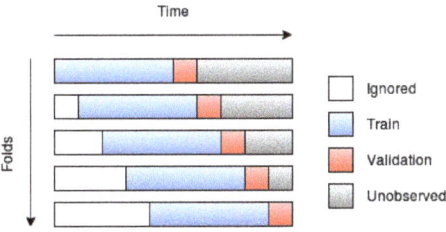

Figure 4. Walk-forward validation with five folds.

2.6. Efficient Market Hypothesis

Apart from the three sufficient assumptions, Eugen Fama (who can be seen as the father of modern EHM), lays out in [9] three different types of tests for the theory: weak form, where only past price movements are considered; semi-weak form, where other publicly available information is included, such as quarterly or annual reports; strong form, where some actors might have monopolistic access to relevant information. The tests done in this paper, outlined in the introduction, are clearly of the weak form.

Fama also brings to light three models that have historically been used to explain the movements of asset prices in an efficient market: the fair game model, the submartingale and the random walk. The fair game is by far the most general of the three, followed by the submartingale and then the random walk. However, this paper does not seek to explain the movements of the market, but merely to predict them, which means that any of the models can be used as the base model in the tests ahead.

Given the three assumptions on the market, the theory indicates that the best guess for any price in the future is the last known price (i.e., the best guess for tomorrow's price of an asset is the price of that asset today). This can be altered to include a drift term, which can be determined by calculating the mean increment for a certain number of past observations, and the best guess then changes to be the last known price added with that mean increment.

3. Model Formulations and Methodology

3.1. Base Model

The base model in this paper, when predicting the next day's closing price, was chosen to be that of a random walk, and this model can be modeled as an AR(1) process

$$X_t = c + \varphi_1 X_{t-i} + \varepsilon_t,$$

where φ_1 needs to equal one. ε_t is here again a white noise, which accounts for the random fluctuations of the asset price. The c parameter is the drift of the random walk, and it can be determined by taking the mean of k previous increments

$$c = \frac{1}{k}\sum_{j=1}^{k}(X_j - X_{j-1}).$$

The best guess of the next day's closing is obtained by taking the expectation of the random walk model (φ_1 equal to one)

$$E(X_t) = E(c + X_{t-1} + \varepsilon_t) = c + X_{t-1}, \tag{8}$$

which is the prediction that the base model used.

In the case of predicting the trend, the base model implemented a passive buy-and-hold strategy, which means that the model always predicts the trend to be up.

3.2. CNN Model

As stated in the introduction, a CNN model, inspired by the WaveNet, was compared to the base model, and two different approaches were needed in order to answer the research questions. The first approach was to structure the CNN as a univariate model, which only needed to be able to handle a single series (the series to make predictions over). This model can be expressed as a NAR model, which can be observed by studying Equation (2). Each element x_t in the sequence is determined by a non-linear function f (the CNN in this case), which takes the past p elements in the series as input. The second approach was to structure the CNN as a multivariate model, which needed to be able to handle multiple series (the series to make predictions over, together with exogenous series). This model, on the other hand, can be expressed as a NARX model, which can be observed by studying Equation (3). Again, each x_t is determined by a non-linear function f, which here takes the past p and r elements in the main and exogenous series as inputs.

The NAR and NARX models were here, for convenience, named the single- and multi-channel models. However, two different variants of the multi-channel model were tested in order to compare the different structures implemented in the original WaveNet paper and in [7,8].

As was mentioned in the introduction, the difference between the two variants is how the residuals from the exogenous series are handled. The implementation found in the WaveNet paper takes into account both the main series residuals as well as the exogenous series residuals, while the implementation in the second variant only takes into account the main series residuals. This can be visualized by observing Figure 3 and then ignoring the residual for each exogenous series. This, in turn, leads to each residual layer beyond the first layer having a similar structure to that of Figure 3. The two variants were named multi-channel-sparsely-connected model (multi-channel-sc) and multi-channel-fully-connected model (multi-channel-fc) in this paper. Furthermore, all models have adopted a dropout factor of 0.2 for all layers, since this leads to better performances for all three models.

The three models (the single-channel as well the two variants of the multi-channel) can be further observed in Figure 5, which is a side view of the dilated layer shown in Figure 1. (a) represents the single-channel model, where no exogenous series are considered; therefore, the model only has to handle a single residual. (b) represents the multi-channel-sc model, where exogenous series are considered, but again, only a single residual is taken into account, while (c) represents the multi-channel-fc, where all residuals are taken into account.

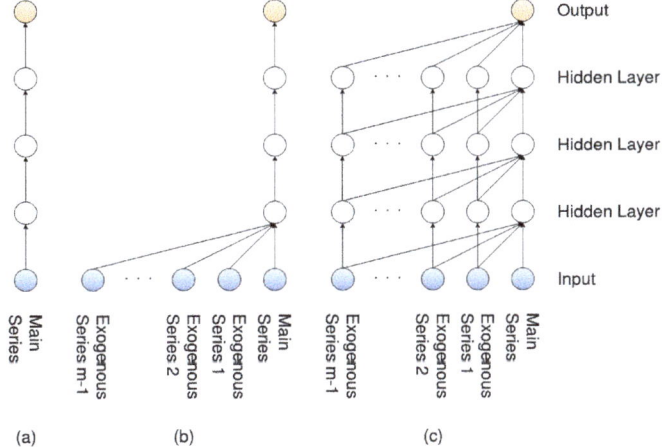

Figure 5. Side view of the dilated convolutional layers in Figure 1. (**a**) Only the main series is used; (**b**,**c**) when the main series is conditioned by an exogenous series, as in the model proposed in [7,8] and in the original WaveNet [5].

3.3. Data Sampling and Structuring

The financial data, for the single-channel model as well as the more complex multi-channel models, were collected from Yahoo! Finance. The time interval between the observations was chosen to equal a single day, since the objective was to predict, for any given time, the next day's closing price and the next day's trend (i.e., the time series $\mathbf{x} = (x_t, \ldots, x_{t-p})^T$, at any time t, was used to predict x_{t+1} in the first case and y_{t+1} in the second case, where $y_{t+1} = \{0, 1\}$).

For the single-channel model, the series under consideration, at any time t, was $\mathbf{x} = (x_t, \ldots, x_{t-p})^T$, which is composed of ordered closing prices from S&P 500. In the multi-channel models, different combinations of ordered OHLC (open, high, low and close) prices, of the S&P 500, VIX (implied volatility index of S&P 500), TY (10 year treasury note) and TYVIX (implied volatility index of TY) were considered. As mentioned before, the closing price of the S&P 500 was the main series, while the other series $\mathbf{Z} = (\mathbf{z}_1, \ldots, \mathbf{z}_m)$ were the exogenous series, where i, $\mathbf{z}_i = (z_{i,t}, \ldots, z_{i,t-r})^T$, for every i. The values of p and r determine the orders of the NAR and NARX, and different values will be tested during the validation phase. However, only combinations where p and r are equal will be tested, and p will therefore be used to denote the length for both the main and exogenous series in the continuation of this paper.

The time span of the observations was chosen to be between the first day in 2010 and the last day in 2019, which resulted in $2515 \times m$ observations, where again m denotes the number of exogenous input series. Furthermore, since the models require p preceding observations, $\mathbf{x} = (x_t, \ldots, x_{t-p})^T$, to make a prediction and then an additional observation, x_{t+1}, to evaluate this prediction, the number of time series that could be used for predictions were decreased to $(2515 - p - 1) \times m$. These observations were then structured into time series, resulting in a tensor with dimension $(2515 - p - 1) \times p \times m$.

The resulting tensor was then divided into folds of equal size, which were used in order to implement the walk-forward scheme. The complete horizontal bars in Figure 4 should here be seen as the whole tensor, while the subsets consisting of the blue and red sections are the folds. The blue and red sections (training and test set of that particular fold) should be seen as a sliding window that "sweeps" across the complete set of time series. Whenever the model is done evaluating a certain fold, the window sweeps a specific number of steps (determined by the size of the test set) in time in order to evaluate the next fold.

By further observing Figure 4, it should become clear that the number of folds is influenced by the size of the training and test sets. The size of each fold could (and most likely should) be seen as a hyperparameter. However, due to the interest of time, the size of each fold was set to 250 series, which means that each fold had a dimension of $250 \times p \times m$. Each fold was then further divided into a training set (first 240 series) and a test set (last 10 series), where the test set was used to evaluate the model for that particular fold.

The sizes chosen for the training and test sets gave as a result 226 folds. These folds were then split in half, where the first half was used in order to validate the models (i.e., determine the most appropriate hyperparameters), and the second half was used to test the generalization of the optimal model found during the validation phase.

One last note about the data sampling is that when predicting the closing price, the time series were the original prices collected from Yahoo!, while when predicting the trend, the time series were changed to model the increments each day.

3.4. Validation and Backtesting

During the validation phase, different values for the length of the input series (i.e., the value of p), the number of residual connections (i.e., number of layers stacked upon each other, see Figures 1 and 2) and the number of filters (explained in the theory section for CNNs) in each convolutional layer were considered. The values considered for p were 4, 6, 8 and 12, while the number of layers considered were 2 and 3, and the number of filters considered were 32, 64 and 96. When classifying the trend, the activation function applied to the last output was seen as a hyperparameter as well, and the sigmoid and SeLU activations were considered.

For the multi-channel models, all permutations of different combinations of the exogenous input series were considered. However, it was only for the mutli-channel-fc model that the exogenous series were seen as a hyperparameter. The optimal combination of exogenous series for the multi-channel-fc model was then chosen for the multi-channel-sc model as well. One final note regarding the hyperparameters is that the dilation rate was set to a fixed value equal to two, which is is the same rate as was proposed in the original WaveNet model, and the resulting dilated structure can be observed in Figure 1.

As was stated in the previous section, the validation was made on the first 113 folds. The overall mean for the error of these folds, for each combination of the hyperparameters above, was used in order to compare the different models, and the model with the lowest error was then used during the backtesting.

The batch size was set to equal one for all models, while the number of epochs was set to 300 in the single-channel model and 500 in the multi-channel models. The difference in epochs is here due to the added complexity that the exogenous series brings. An important note regarding the epochs and the evaluation of the models is that the model state, associated with the epoch with the lowest validation/test error, was ultimately chosen. This means that if a model made predictions over 300 epochs, but the lowest validation/test error was found during epoch 200, the model state (i.e., the value of the model's weights) associated with epoch 200 were chosen as the best performing model for that particular fold.

3.5. Developing the Convolutional Neural Network

The networks were implemented using the Keras API, from the well known open-source library TensorFlow. Keras provides a range of different models to work with, where the most intuitive might be the Sequential model, where developers can add/stack layers, and then compile them into a network. However, the Sequential model does not provide enough freedom to construct the complexity introduced in the residual and skip part of the WaveNet. Keras functional API (more information regarding Keras functional API can be found on Keras official documentation https://keras.io/models/model/) might be less intuitive at first, but it does provide more freedom, since the order of the layers in the network is defined by having the output of every layer explicitly defined as an input parameter to the next layer in the network.

Furthermore, Keras comes with TensorFlow's library of optimizers, which are used in order to estimate the parameters in the model and taken as an input parameter when compiling the model. The optimizer used here was the Adam optimizer, and the learning rate was set to equal 0.0001.

4. Results

4.1. Validation

Table 1 displays the validation error when predicting the next day's closing price, while Table 2 displays the validation accuracy when classifying the trend. The p is again the length of the input time series, while l is the number of layers in the residual part of the network (see Figure 1), and f is the number of filters used in each convolutional layer. In Table 2, a denotes the used activation function.

Table 1. Validation error for the different models, where p is the length of the time series, l is the number of residual layers and f is the number of filters.

Hyperparameters			MAPE		
p	l	f	Single-Channel	Multi-Channel-sc	Multi-Channel-fc
4	2	32	0.5787	0.5706	0.5599
4	2	64	0.5802	0.5527	0.5577
4	2	96	0.5783	0.5508	0.5587
6	2	32	0.5720	0.5491	0.5544
6	2	64	0.5752	0.5450	0.5426
6	2	96	0.5793	0.5462	0.5479
8	2	32	0.5764	0.5413	0.5498
8	2	64	0.5737	0.5445	0.5450
8	2	96	0.5796	0.5405	0.5495
8	3	32	0.5733	0.5518	**0.5251**
8	3	64	0.5692	**0.5259**	0.5377
8	3	96	0.5687	0.5468	0.5389
12	2	32	0.5714	0.5524	0.5526
12	2	64	0.5744	0.5478	0.5542
12	2	96	0.5744	0.5417	0.5422
12	3	32	0.5700	0.5298	0.5444
12	3	64	0.5672	0.5368	0.5290
12	3	96	**0.5584**	0.5312	0.5325

Table 2. Validation accuracy for the different models, where a is the activation function, p is the length of the time series, l is the number of residual layers and f is the number of filters.

Hyperparameters				Accuracy		
a	p	l	f	Single-Channel	Multi-Channel-sc	Multi-Channel-fc
Sigmoid	4	2	32	0.6035	0.6000	0.5779
Sigmoid	4	2	64	0.6150	0.5991	0.6283
Sigmoid	4	2	96	0.5956	0.6204	0.6637
Sigmoid	6	2	32	0.6283	0.6009	0.6283
Sigmoid	6	2	64	0.6115	0.6186	0.6319
Sigmoid	6	2	96	0.6097	0.6319	0.6389
Sigmoid	8	2	32	0.5938	0.6283	0.6146
Sigmoid	8	2	64	0.5956	0.6133	0.6248
Sigmoid	8	2	96	0.6248	0.5947	0.6363
Sigmoid	8	3	32	0.6150	0.6053	0.6000
Sigmoid	8	3	64	0.6265	0.6044	0.6451
Sigmoid	8	3	96	0.6487	0.6124	0.6327
Sigmoid	12	2	32	0.6009	0.6027	0.6115
Sigmoid	12	2	64	0.6035	0.6212	0.6327
Sigmoid	12	2	96	0.6168	0.5920	0.6398
Sigmoid	12	3	32	0.6168	0.5973	0.5973
Sigmoid	12	3	64	0.6195	0.6106	0.6248
Sigmoid	12	3	96	0.6442	0.6159	0.6292
SeLU	4	2	32	0.6752	0.6664	0.6894
SeLU	4	2	64	0.7053	0.7195	0.6920
SeLU	4	2	96	0.7018	0.7097	0.7469
SeLU	6	2	32	0.7062	0.6814	0.6938
SeLU	6	2	64	0.6947	0.7000	0.7133
SeLU	6	2	96	0.7204	0.7389	0.7230
SeLU	8	2	32	0.7018	0.6735	0.6717
SeLU	8	2	64	0.7071	0.6885	0.6841
SeLU	8	2	96	0.6947	0.7159	0.7257
SeLU	8	3	32	0.6549	0.6407	0.6619
SeLU	8	3	64	0.6973	0.7044	0.7133
SeLU	8	3	96	0.7336	**0.7345**	0.7425
SeLU	12	2	32	0.7097	0.6743	0.7018
SeLU	12	2	64	0.6752	0.7292	0.7027
SeLU	12	2	96	0.7434	0.7230	0.7142
SeLU	12	3	32	0.6699	0.6319	0.6637
SeLU	12	3	64	0.7257	0.7088	0.7434
SeLU	12	3	96	**0.7487**	0.7310	**0.7496**

The lowest validation error was achieved with p, l and f equal to 12, 3 and 96 for the single-channel model, while 8, 3, 64 and 8, 3, 32 were the optimal parameters for the multi-channel-sc model and the multi-channel-fc model, respectively. The highest validation accuracy was achieved with SeLU as the activation for all models and p, l and f equal to 12, 3 and 96 for the single-channel model, while 8, 3, 96 and 12, 3, 96 were the optimal parameters for the multi-channel-sc model and the multi-channel-fc model, respectively.

The validation error and validation accuracy for the multi-channel models are displayed only for the best combination of exogenous series found for the multi-fc-channel model, which proved to be just the highest daily value of the VIX in both cases.

4.2. Testing

Figure 6 shows the cumulative mean, of the test error, for all 113 test folds, while Figure 7 shows the cumulative mean for the last 50. These two figures paint two different pictures of the single-channel and multi-channel models. The means across all 113 folds were 0.5793, 0.4877, 0.4707 and 0.4621, for the base, the single-channel, multi-channel-sc and multi-channel-fc models, respectively, while the means across the last 50 were 0.6572, 0.5468, 0.5250 and 0.5416. By looking at these numbers, one can see that the performance of the multi-channel-fc model to the base model, when predicting the next day's closing price, was worse in the last 50 folds than for all 113 folds, while the reverse can be said about the single-channel and the multi-channel-sc models.

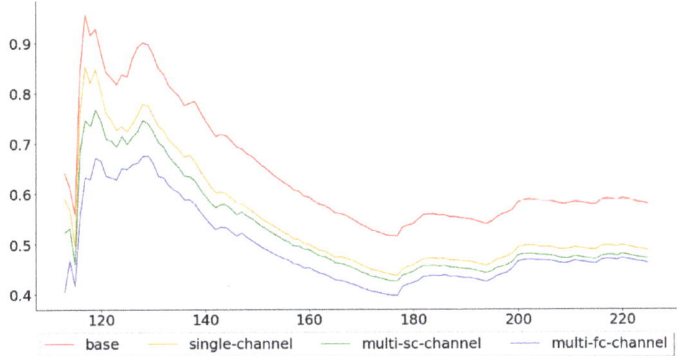

Figure 6. Cumulative mean of the MAPE for all 113 test folds.

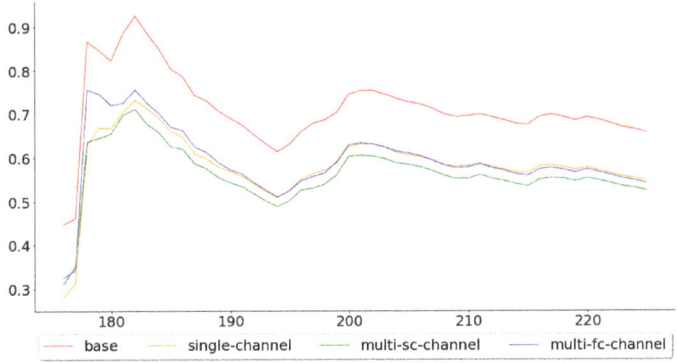

Figure 7. Cumulative mean of the MAPE for the last 50 test folds.

Figures 8 and 9, on the other hand, show the cumulative mean of the validation error for all 113 test folds and the last 50 test folds. Again, the multi-channel-fc model started out well, but the performance compared to the single-channel and multi-channel-sc model worsened over time. The means across all 113 test folds were 0.5442, 0.7504, 0.7451 and 0.7496 for the base, single-channel, multi-channel-sc and multi-channel-fc models, respectively, while the means for the last 50 test folds were 0.5600, 0.7580, 0.7600 and 0.7360.

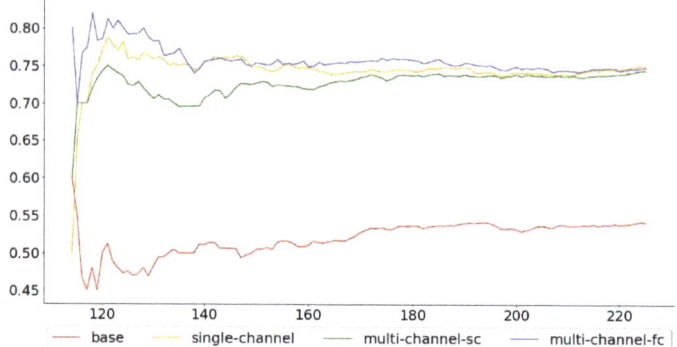

Figure 8. Cumulative mean for the accuracy for all 113 test folds.

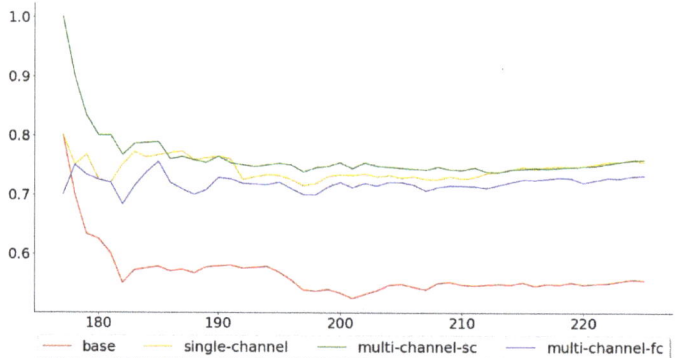

Figure 9. Cumulative mean for the accuracy for the last 50 test folds.

Both the single-channel and multi-channel models outperformed the base model over the test folds, which accounts for almost five years of observations. Furthermore, the multi-channel models clearly performed better than the single-channel model, when looking at the performance across all test folds. However, the positive effects of including the exogenous series seemed to wear off in the last folds for the complex multi-channel-fc model, while it actually increased for the simpler multi-channel-sc model. This suggests that the problem of generalization for the multi-channel-fc model probably lies in that the relationship between the main series and the exogenous series has been altered, which, interestingly enough, only affects the more complex model.

Figure 10 shows the gains for the models when applied to the test data. Both the single- and multi-channel models clearly outperformed the passive buy-and-hold strategy, which is to be expected, since the test accuracy for the base model was well below the other models. The gains were 1.5596, 26.3642, 25.0388 and 26.9360 for the base, single-channel, multi-channel-sc and multi-channel-fc, respectively, meaning that if one was to implement any of the WaveNet inspired models, during the specified period, he or she would have a profit of more than 24 times the original amount.

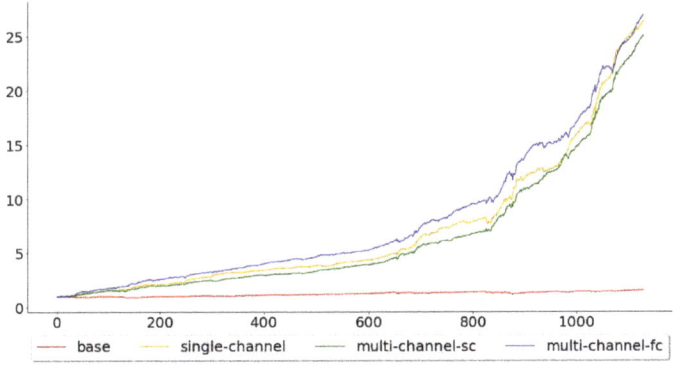

Figure 10. The calculated gain across all test folds, for all of the four models.

Lastly, while the multi-channel-fc model outperformed all other models across all folds, it is also of interest, for further work, to see in what settings the multivariate model performed the best and the worst. Figures 11 and 12 give an example of these settings, where it shows the folds for which the multi-channel-fc model outperformed (fold 139) and underperformed (fold 148) the most against the base model.

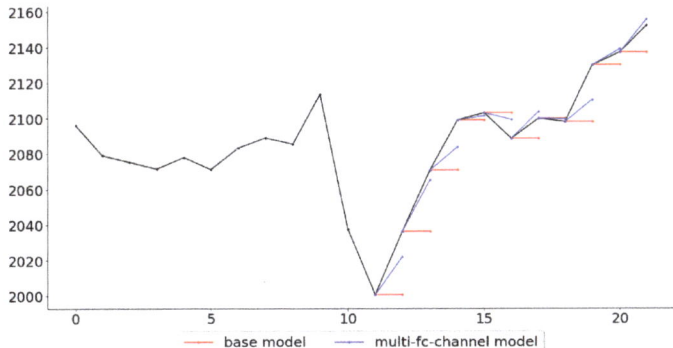

Figure 11. Predictions for the 10 test observations in test fold 25.

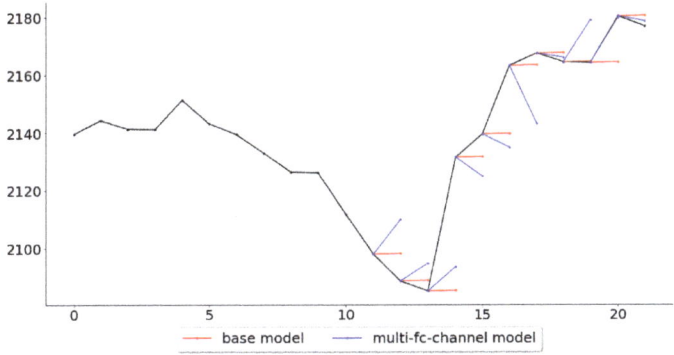

Figure 12. Predictions for the 10 test observations in test fold 34.

5. Discussion

There is no real scientific basis for having the training size equal to 240 and validation/test size equal to 10, although it did perform better than having the sizes equal 950 and 50 respectively. It might seem odd to someone, with little or no experience in analyzing financial data, that one would choose to have a limit on the training size and why the models, evidently, perform better using fewer observations, since having a larger set of training observations is generally seen as a good thing. However, the financial markets are ever-changing, and the predictors (the past values in the time series) usually change with it. New paradigms find their way into the markets, while old paradigms may lose their impact over time. These paradigms can be imposed by certain events, such as an increase in monetary spending, the infamous Brexit or the current Covid-19 pandemic (especially the response, by the governments and central banks, to the pandemic). Paradigms can also be recurrent, such as the ones that are imposed by where we are in the short- and long-term debt cycle. Because of these shifts, developers are restricted in how far back in time they can look and, therefore, need to put restrictions on the training size.

The paper brings forward two very positive results, which are that the predictions for the next day's closing price as well as the trend are made significantly better by the CNN models. However, the fact that the performance of the more complex multi-channel model decreases over time, against the other models, for both predicting the closing price and the trend, is indeed concerning. This became obvious when studying the change in performance between all 113 and the last 50 test folds. If both the multi-channel-sc and multi-channel-fc models performed worse in the last 50 folds, then it would have been easy to again "blame" the temporal structure of the financial data and, more specifically, the temporal dependencies between the main and exogenous series. However, only the more complex multi-channel model's performance degraded, which means that the complexity (i.e., the intermingling between the series in all residual layers) is the primary issue. A solution might then be to have the complexity as a hyperparameter as well and not differentiate between the two structures, as was done here. In other words, the two models might more appropriately be seen as two extreme cases of the same model, in a similar way as having the number of filters set to 32 and 96 (see Table 1, 32 and 96 are the extreme cases for the number of filters). By looking at Figure 5 (with p equal to 16 in this case), one can see that the hyperparameter for the complexity has two more values to chose from (having the exogenous series to directly influence the second and third hidden layers). It would also be appropriate to compare the models against different time frames and asset classes to see if the less complex model indeed generalizes better over time, or if the result here was just a special case. However, viewing the complexity as a hyperparameter could prove to be beneficial in both cases.

The tests made in this paper were not primarily intended to judge the suitability of the models as trading systems, but rather if a deep learning approach could perform better than a very naive base model. However, the multi-channel models outperformed the base model by more than 20%, and this difference is quite significant and begs the question of what changes could be done in order for the model to be used as a trading system. While most of the predictions in fold 25, Figure 11, are indeed very accurate, the predictions in fold 34, Figure 12, would be disheartening for any trader to see if it were to be used as a trading system. This suggests that one should try to look for market conditions, or patterns, similar to the ones that were associated with low error rates in the training data. This could be done by clustering the time series, in an unsupervised manner, and then assign a score for each class represented by the clusters, where the score can be seen as the probability of the model to make good predictions during the conditions specific to that class. A condition classed in a cluster with a high score, such as the pattern in Figure 11, would probably prompt the trader to trust the system and to take a position (either long or short, depending on the prediction), while a condition classed in a cluster with a lower score would prompt the trader to stay out of the market or to follow another trading system that particular day.

6. Conclusions

The deep learning approach, inspired by the WaveNet structure, proved successful in extracting information from past price movements of the financial data. The result was a model that outperformed a naive base model by more than 20%, when predicting the next day's closing price, and by more than 37% when predicting the next day's trend.

The performance of the deep learning approach is most likely due to its exceptional ability to extract non-linear dependencies from the raw input data. However, as the field of deep learning applied to the financial market progresses, the predictive patterns found in the data might become increasingly hard to find. This would suggest that the fluctuations in the market would come to more and more mirror a system, where the only predictive power lies in the estimation of the forces acting on the objects, which are heavily influenced by the current sentiment in the market. A way to extract the sentiment, at any current moment, might be to analyze unstructured data, extracted from, for example, multiple news sources or social media feeds. Further study in text mining, applied to financial news sources, might therefore be merited and might be an area that will become increasingly important to the financial sector in the future.

Author Contributions: Conceptualization, M.S. and L.B.; methodology, L.B.; software, L.B.; validation, L.B.; data curation, L.B.; writing—original draft preparation, L.B.; writing—review and editing, M.S.; project administration, M.S. and L.B.; funding acquisition, M.S. All authors have read and agreed to the published version of the manuscript.

Funding: This research received no external funding.

Acknowledgments: The authors would like to thank the anonymous reviewers for several valuable and helpful suggestions and comments to improve the presentation of the paper.

Conflicts of Interest: The authors declare no conflicts of interest.

References

1. Wang, Z.; Yan, W.; Oates, T. Time series classification from scratch with deep neural networks: A strong baseline. In Proceedings of the 2017 International Joint Conference on Neural Networks (IJCNN), Anchorage, AK, USA, 14–19 May 2017; pp. 1578–1585.
2. Paul, W.; Baschnagel, J. *Stochastic Processes from Physics to Finance*; Springer Science & Business Media: Heidelberg, Germany, 2013; Volume 2.
3. Sezer, O.B.; Gudelek, M.U; Ozbayoglu, A.M. Financial time series forecasting with deep learning: A systematic literature review: 2005–2019. *Appl. Soft Comput.* **2020**, *90*, 106181. [CrossRef]
4. Sainath, T.N.; Vinyals, O.; Senior, A.; Sak, H. Convolutional, long short-term memory, fully connected deep neural networks. In Proceedings of the 2015 IEEE International Conference on Acoustics, Speech and Signal Processing (ICASSP), Brisbane, Australia, 19–24 April 2015; pp. 4580–4584.
5. van den Oord, A.; Dieleman, S.; Zen, H.; Simonyan, K.; Vinyals, O.; Graves, A.; Kalchbrenner, N.; Senior, A.; Kavukcuoglu, K. Wavenet: A generative model for raw audio. *arXiv* **2016**, arXiv:1609.03499.
6. Shen, Z.; Zhang, Y.; Lu, J.; Xu, J.; Xiao, G. SeriesNet: A Generative Time Series Forecasting Model. In Proceedings of the 2018 International Joint Conference on Neural Networks (IJCNN), Rio de Janeiro, Brazil, 8–13 July 2018; pp. 1–8.
7. Borovykh, A.; Bohte, S.; Oosterlee, C.W. Conditional time series forecasting with convolutional neural networks. *arXiv* **2017**, arXiv:1703.04691.
8. Borovykh, A.; Bohte, S.; Oosterlee, C.W. Dilated convolutional neural networks for time series forecasting. *J. Comput. Financ.* **2018**. [CrossRef]
9. Malkiel, B.G.; Fama, E.F. Efficient capital markets: A review of theory and empirical work. *J. Financ.* **1970**, *25*, 383–417. [CrossRef]
10. Muth, J.F. Rational expectations and the theory of price movements. *Econom. J. Econom. Soc.* **1961**, *29*, 315–335. [CrossRef]
11. Thaler, R. Toward a positi theory of consumer choice. *J. Econ. Behav. Organ.* **1980**, *1*, 39–60. [CrossRef]

12. Kim, T.Y.; Oh, K.J.; Kim, C.; Do, J.D. Artificial neural networks for non-stationary time series. *Neurocomputing* **2004**, *61*, 439–447. [CrossRef]
13. Hochreiter, S.; Schmidhuber, J. Long short-term memory. *Neural Comput.* **1997**, *9*, 1735–1780. [CrossRef] [PubMed]
14. Klambauer, G.; Unterthiner, T.; Mayr, A.; Hochreiter, S. Self-normalizing neural networks. In Proceedings of the 31st International Conference on Neural Information Processing Systems, Long Beach, CA, USA, 4–9 December 2017; pp. 971–980.
15. Kingma, D.P.; Ba, J. Adam: A method for stochastic optimization. *arXiv* **2014**, arXiv:1412.6980.
16. Srivastava, N.; Hinton, G.; Krizhevsky, A.; Sutskever, I.; Salakhutdinov, R. Dropout: A simple way to prevent neural networks from overfitting. *J. Mach. Learn. Res.* **2014**, *15*, 1929–1958.
17. Pardo, R. *Design, Testing, and Optimization of Trading Systems*; John Wiley & Sons: New York, NY, USA, 1992; Volume 2.

© 2020 by the authors. Licensee MDPI, Basel, Switzerland. This article is an open access article distributed under the terms and conditions of the Creative Commons Attribution (CC BY) license (http://creativecommons.org/licenses/by/4.0/).

MDPI
St. Alban-Anlage 66
4052 Basel
Switzerland
Tel. +41 61 683 77 34
Fax +41 61 302 89 18
www.mdpi.com

Entropy Editorial Office
E-mail: entropy@mdpi.com
www.mdpi.com/journal/entropy

www.ingramcontent.com/pod-product-compliance
Lightning Source LLC
LaVergne TN
LVHW070434100526
838202LV00014B/1595